Editor: Liu Zizhu, Zhang Hua

# The Vegetable

Groups and Cultivars of Guangzhou

# 广州蔬菜品种志

主编 刘自珠 张 华

SPM 南方出版传媒
广东科技出版社 | 全国优秀出版社
·广州·

### 图书在版编目（CIP）数据

广州蔬菜品种志 / 刘自珠，张华主编. —广州：广东科技出版社，2016.9

ISBN 978-7-5359-6512-7

Ⅰ. ①广… Ⅱ. ①刘…②张… Ⅲ. ①蔬菜—品种—广州市 Ⅳ. ① S630.292

中国版本图书馆 CIP 数据核字（2016）第 084922 号

---

**广州蔬菜品种志**

责任编辑：罗孝政
封面设计：柳国雄
责任校对：蒋鸣亚　梁小帆　盘婉薇
责任印制：彭海波
出版发行：广东科技出版社
　　　　　（广州市环市东路水荫路 11 号　邮政编码：510075）
http：//www.gdstp.com.cn
E-mail：gdkjyxb@gdstp.com.cn（营销中心）
E-mail：gdkjzbb@gdstp.com.cn（总编办）
经　　销：广东新华发行集团股份有限公司
印　　刷：广州市岭美彩印有限公司
　　　　　（广州市荔湾区花地大道南海南工商贸易区 A 幢　邮政编码：510385）
规　　格：889mm×1 194mm　1/16　印张 19.75　字数 475 千
版　　次：2016 年 9 月第 1 版
　　　　　2016 年 9 月第 1 次印刷
定　　价：280.00 元

---

**如发现因印装质量问题影响阅读，请与承印厂联系调换。**

## 《广州蔬菜品种志》编辑委员会

编委会主任：姚玉凡

副 主 任：谭增华　杨华杰

编　　委：姚玉凡　谭增华　杨华杰　陈顺鑫　黎翠梨　刘自珠
　　　　　张　华　谭耀文　罗少波　郑康炎　林明宝　王子明
　　　　　邓彩联　肖旭林　钟广建　叶少球　尹国强　黄锦明
　　　　　刘景业　曹学文　刘成枝　洪启金

主　　编：刘自珠　张　华

编写组成员：刘自珠　张　华　秦晓霜　林春华　林鉴荣　田耀加

摄　　影：黄绍力　张文胜　陈胜文　张　华

英文翻译：刘厚诚　乔燕春　郭　爽　田耀加　刘　峰　吴　蓓

英文校订：刘厚诚

参编人员（按姓氏拼音排列）：
　　　　　艾　卉　常绍东　陈胜文　郭　爽　黄　贞　黄红弟
　　　　　黄绍力　李伯寿　李光光　李莲芳　李向阳　林春华
　　　　　林鉴荣　林锦英　刘自珠　刘振翔　乔燕春　秦晓霜
　　　　　丘漫宇　谭　雪　田耀加　吴宇军　夏秀娴　谢秀菊
　　　　　徐勋志　叶伟忠　张　华　张　晶　张素平　张文胜
　　　　　赵守光　郑岩松　钟国君

## 内容简介

　　本志编入广州主要蔬菜 15 类 127 种 375 个品种，包括广州的名特产蔬菜品种，早熟、高产、优质、抗逆、抗病以及适合出口品种，反映了《广州蔬菜品种志》(1993) 出版后 20 年间广州蔬菜品种变化的新面貌。全部品种均有彩色照片。前言、编写说明、目录、各种蔬菜简介和品种名称均有英文对照。还以附录的形式收录了广州蔬菜产销概况、主要名特产蔬菜营养成分等。

　　本志可供国内外蔬菜生产、科研、教学、推广及蔬菜种子企业等部门和个人参考。

## Introduction

　　More than 15 categories, 127 species and 375 varieties in Guangzhou are selected in the book. The selected cultivars include the famous and special vegetable cultivars, early-maturity, high-yield, good-quality, disease resistant and exported cultivars. This book reflects the new features of vegetable cultivars variation in Guangzhou during nearly 20 years since *Record of Vegetable Cultivars of Guangzhou* was published in 1993. All cultivars have beautiful pictures. Preface, explanation of compilation, the table of contents, vegetable introduction and species name are expressed in both Chinese and English. The appendices include the general survey of vegetable production and market consumption in Guangzhou, the nutritional values of some vegetables, and so on.

　　This book can be used as reference by departments and persons engaged in vegetable production, scientific research, teaching and vegetable seed business.

# 前　言

《广州蔬菜品种志》(1993) 于 1994 年出版后，深受国内外蔬菜生产、科研、教学、推广及种业等单位和个人的欢迎。近 20 年来，广州的蔬菜产销体制不断深化改革，蔬菜的产供销面貌焕然一新，蔬菜品种结构与栽培技术也发生了很大的变化，一些不适应市场需要的品种逐渐被淘汰，本市新育成及从国内外引进的许多优良品种迅速应用于生产，原有的品种志已不能反映广州市蔬菜生产的新面貌。为了适应蔬菜产业和市场经济的需求，推动蔬菜生产向高产、优质、高效益方向发展，向国内外介绍广州的优良蔬菜品种，有必要编写主要反映广州蔬菜的新品种志。为与 1993 版相衔接，新编品种志 2013 版名为《广州蔬菜品种志》。

本志由广州市农业局于 2014 年立项资助，由广州市农业局蔬菜办公室组织编写工作，广州市农业科学研究院具体承担。邀请广东省、广州市的蔬菜专家教授及区熟悉蔬菜科研、生产的领导或专家组成编辑委员会。编委会单位有华南农业大学园艺学院、广东省农业科学院蔬菜研究所以及广州市白云区、花都、南沙区、番禺区、从化区、增城区和黄埔区等农业（蔬菜）管理部门。编写过程首先对全市的蔬菜品种进行调查登记，由编委会提出初步入编品种名录；然后分三个层面邀请省、市、区蔬菜专家教授，各区熟悉蔬菜业务的领导或专家，种业界的代表分别对编写方案和入编品种进行论证，确定入编品种；由编写人员在产地观察记录，撰写品种说明初稿，编写组修改；完成样稿后再邀请有关人员座谈；最后定稿。品种彩照由广州市农业科学研究院专职摄影人员到现场拍摄。

本志入编蔬菜按农业生物学特性分为 15 类 127 种 375 个品种，包括了广州的主要蔬菜和著名的特产蔬菜。保留原品种志中 101 个品种，占总数的 27%。新编入的品种 274 个，占总数的 73%，其中新育成品种和杂种一代 191 个，引进品种 51 个，新发展的地方品种 32 个。品种结构的变化，主要是早熟品种，特色品种，高产、优质和抗逆、抗病品种，以及适合出口品种增多，围绕周年供应的配套品种有所发展，展现出高产、优质和高效益的发展趋势。同时，还编入一些稀少蔬菜和野生半野生蔬菜，并附录了广州蔬菜产销概况、广州气候概况、广州主要名特产蔬菜营养成分及广州蔬菜主要病虫害与防治简介。全部品种有彩照，全书共 570 余幅。前言、编写说明、目录、各种蔬菜简介和品种名称均有英文说明。实际上，广州地区的蔬菜品种不止这些，因时间、人力和篇幅所限，有少数农家品种未能编入，有些正在试种的新选育品种和杂种一代也未编入。

本志于 2013 年 3 月筹备，7 月正式开展工作，经过两年多的努力，顺利完成编写工作。除参加人员充分合作外，还得到广州市气候与农业气象中心、广州各区蔬菜研究和技术推广等单位的热情支持，华南农业大学刘厚诚教授对蔬菜学名及英文进行了校订。对他们的合作和支持，在此表示衷心的感谢！

由于我们的学识所限，调查和编撰时间较仓促，内容可能有误，有的照片也不够理想，恳请读者批评指正。

编　者

2015 年 7 月

# Preface

The original *Record of Vegetable Cultivars of Guangzhou* published in 1993 has been widely greeted by vegetable production, scientific research, teaching, extension, seed industry, and other departments and individuals. The past 20 years, vegetable production and marketing system has continued to deepen reform in Guangzhou. Vegetable production and supply emerge a new look, and the structure of vegetable varieties and cultivation techniques have undergone great changes. Some species that didn't meet the market needs were gradually being phased out. Many new cultivars bred in the city and introduced from abroad quickly used in production. Original *Record of Vegetable Cultivars of Guangzhou* published in 1993 did not reflect the new face of Guangzhou vegetable production. In order to adapt to the market economy and vegetable industry demand, promote vegetable production to high-yield, high-quality, cost-effective direction, to introduce fine cultivars of vegetables, it is necessary to rewrite the book. To link up with the 1993 edition, new book (2013 edition) entitled *Record of Vegetable Cultivars of Guangzhou*.

This book is financially supported by the Guangzhou Agricultural Bureau in 2014, and organized by the Vegetable Office of Guangzhou Agricultural Bureau. Guangzhou Academy of Agricultural Sciences is responsible for compiling. Professors and leaders who are familiar with vegetable research and production compose editorial committee. Those units that invite to take part in compiling include College of Horticulture in South China Agricultural University, Vegetable Research Institute in Guangdong Academy of Agricultural Sciences, vegetables administration from Baiyun District, Huadu District, Nansha District, Panyu District, Conghua District, Zengcheng District, Huangpu District, and so on. First procedure of compiling this book is to investigate and register the vegetable cultivars. The editorial committee propose preliminary list of selected cultivars. Professors and experts who were invited to determine the cultivars into the book. The investigators begin to write the data of cultivars (primary manuscript) after their actual observation and taking notes in the local vegetable producing area. Then the revision of primary manuscript is done by the editorial group. The final manuscript is to be determined by the editorial board after discussion. The color pictures were taken on the spot by the professional photographers from Guangzhou Academy of Agricultural Sciences.

According to agricultural biological characteristics, the selected vegetables are 15 categories, 127 species and 375 cultivars. The main and famous vegetables in Guangzhou are included in it. Of which, 101 cultivars belonged to the original "*Record*", 27% the total number of selected cultivars, while 274 cultivars newly added, 73% the total number. These newly added cultivars include: 191 new cultivars, 51 exotic cultivars and 32 newly developed farmer cultivars. The character of vegetable cultivars change is: Early cultivars, special cultivars, high-yield, better quality and resistant cultivars, exported cultivars are increased gradually. The cultivar series for all seasons growing are developed, the trend of vegetable growing is appeared to obtain high yield, better quality and more

significant benefit. Some rare vegetable cultivars and wild, semi-wild vegetables also are selected. The appendices include vegetable production profile in Guangzhou, climate survey in Guangzhou, the nutritional values of some vegetables, vegetable diseases and control profile. There are about 570 color pictures in the book. Preface, explanation of compilation, the table of contents, vegetable introduction and species name are expressed in both Chinese and English. As a matter of fact, this book can not include all the vegetable cultivars in Guangzhou due to time, manpower and space constraints.

The preparation of this book was began in March, 2013 and our actual editorial work started in July, 2013. After more than two years, all the work successfully complete. In compiling the book, we obtain timely cooperation and supports from Climate and Agricultural Meteorology Center in Guangzhou and some departments about vegetable research and demonstration. The scientific names of vegetables and English were checked by professor Liu Houcheng from South China Agricultural University. We thank them for their cooperation and support.

Because of our limited knowledge, research and compilation time, there might be fault in content, some photos are not ideal. Any comments and suggestions are sincerely welcomed.

The editors<br>
July, 2015

# 编写说明

1. 入编本志的蔬菜品种，是截至2013年12月在本市辖区内收集到的有一定栽培面积的地方品种、新育成品种和杂种一代，以及从国内外引进推广的品种。有发展前途的外来品种，则有选择地编入。

2. 入编蔬菜依农业生物学特性分类，并按其在广州市蔬菜生产的重要位置，兼顾植物学科属关系编排。每种蔬菜的品种，一般以早、中、晚熟次序编排，同一熟性的品种按地方品种、新育成品种和引进品种次序编排，引进品种中先国内品种，后国外品种。

3. 每种蔬菜用中英文介绍科属、种名（学名）、别名、染色体数、历史和分布、主要植物学性状、类型、栽培环境与方法、收获、病虫害、营养及功效等。每个品种有彩照、名称、来源、分布地区、特征特性和栽培要点。

4. 各种蔬菜的名称以国内外通用名称为正名，其他为别名。品种名称则以本地惯用的为正名。正名排前，别名排后，并以字体大小相区别。其英文名称，属特征性为意译，其他为音译。

5. 品种来源：栽培久远的统称地方品种，新育成品种和引进品种说明其来源和年代。

6. 分布地区：按各区调查结果列出。

7. 品种特征：描述品种的植株高度、开展度，根、茎、叶、花、果实和种子的形状、大小、重量与颜色，其中着重商品部分的特征。数据为代表性10株的平均值或是一般的变幅。

8. 品种特性：介绍品种的熟性、播种至初收的生长期、适应性、品质、产量（以中等水平计算）、加工、贮藏等。

9. 栽培要点：介绍本地的播种期、株行距、田间管理、收获期等。

10. 年份用公历，面积用公顷（$hm^2$）、667米$^2$（亩），长度用厘米（cm）、米（m），重量用克（g）、千克（kg）、吨（t）。彩照内标尺为10厘米。

# Explanation of compilation

1. Vegetable cultivars compiled in the book are the farmer's cultivars, new cultivars, hybrids, domestic cultivars from other places and exotic cultivars which can be propagated locally. Some cultivars with potential utilization value are selectively recorded.

2. Vegetables are arranged in such order according to 3 basis:

(1) Classification based on agricultural biological characters.

(2) The important position in vegetable production of Guangzhou.

(3) Plant classification into families and genera.

Vegetable cultivars are arranged in the order of early maturity, mid-maturity and late maturity. When the cultivars have the same maturity, they are arranged in the order of:

(1) The farmer's cultivars.

(2) New cultivars.

(3) Introduced domestic cultivars.

(4) Exotic cultivars.

3. Expressed in both Chinese and English are the following items of each vegetable: plant family, genus and species, synonyms, chromosome number, history, distribution, prominent botanical character, species, cultural environment and methods, major pests, nutritional uses, and so on. Each cultivar is shown by picture, name, origin, distribution region, characteristics and cultivation tips.

4. The generally used name is the common name of each vegetable, and the other names of the same vegetable are synonyms. The English common name of each vegetable is literally translated (in case of characteristics) or translated by sound.

5. Origin of cultivar: The era and the origin of farmer's cultivars, new cultivars, introduced cultivars are to be noted and explained.

6. Distribution areas are listed based on investigation results of each district.

7. Item of cultivar special features: Plant height, growth divergence, root, stem, leaf, flower, fruit and seed are described, especially edible (commercial) parts. The number represents the average of 10 typical plants or general variable.

8. Items of cultivar characters: Maturity of cultivar, growth period from sowing to first harvest, adaptability, quality, edibility processing, handling, storage, etc. are described.

9. Cultivation tips: Local sowing dates, row and plant spacing, field management, harvesting dates, etc. are introduced.

10. Gregorian calendar is adopted, hectare and 667 $m^2$ for area, centimeter and meter for length measurement, gram, kilogram, ton for weight. The length of scale in the color pictures is 10 cm.

# 目 录

## 一、白菜类

### 菜心 …… 002
四九-19号 …… 004
油绿501 …… 004
碧绿粗薹 …… 005
碧绿四九 …… 005
农苑45天油青 …… 006
菜场4号 …… 006
中南尖叶油青甜 …… 006
新西兰杷洲甜4560 …… 007
强盛尖叶 …… 007
绿宝70天 …… 007
油绿701 …… 008
油绿702 …… 008
油绿80天 …… 009
油绿802 …… 009
迟心2号 …… 010
特青迟心4号 …… 010
穗美89号迟花 …… 011
玉田2号 …… 011
增城迟菜心 …… 012
鹤哥脷菜心 …… 012

### 小白菜 …… 013
矮脚黑叶 …… 014
鹤斗奶白 …… 015
农普奶白 …… 015
潮汕黄叶 …… 015
夏盛 …… 016
白玫瑰 …… 016
华冠青江 …… 017
冠华青江 …… 017

### 大白菜 …… 018
早熟5号 …… 019
夏阳白菜早50 …… 020
春福白菜 …… 020

## 二、甘蓝类

### 芥蓝 …… 022
夏翠芥蓝 …… 024
秋盛芥蓝 …… 024
金绿芥蓝 …… 025
银绿芥蓝 …… 025
绿宝芥蓝 …… 026
顺宝芥蓝 …… 026
中花芥蓝 …… 026
四季粗条芥蓝 …… 027
迟花芥蓝 …… 027

### 花椰菜 …… 028
广泰45 …… 029
白狮王65 …… 030
GL-65松花 …… 030
五山65天松花菜 …… 030
喜雪 …… 031
华艺菜花王80 …… 031
利卡 …… 031

### 青花菜 …… 032
珠绿 …… 034
曼陀绿 …… 034
炎秀 …… 035
格福 …… 035

### 西兰薹 …… 036
芊秀 …… 037
秀丽 …… 038
翠钰2号 …… 038

### 结球甘蓝 …… 039
京丰1号 …… 040
KK甘蓝 …… 041
元绿甘蓝 …… 041
中甘11号 …… 042
紫甘1号 …… 042

### 抱子甘蓝 …… 043

### 羽衣甘蓝 …… 044

## 三、芥菜类

### 叶用芥菜 …… 046
南风芥 …… 047
特选客家芥 …… 047
竹芥 …… 048
水东芥菜 …… 048

### 结球芥菜 …… 049
大坪埔11号包心芥菜 …… 049

### 根用芥菜 …… 050
光头芥菜 …… 050

## 四、绿叶菜类

### 蕹菜 …… 052
白梗柳叶 …… 053
白梗大叶 …… 054
青梗柳叶 …… 054
青梗大叶 …… 054
青白柳叶 …… 055
丝蕹 …… 055

### 莴苣 …… 056
皱叶莴苣 …… 058
玻璃生菜 …… 058
意大利生菜 …… 058
奶油生菜 …… 059
红叶生菜 …… 059
长叶莴苣 …… 059
罗马生菜 …… 059
尖叶无斑油荬菜 …… 060
香水油荬菜 …… 060
结球莴苣 …… 061
万利包心生菜 …… 061
茎用莴苣（莴笋） …… 061

001

白皮尖叶莴笋 ……061
苦荬菜 ……062
　　尖叶苦荬 ……063
　　花叶苦荬 ……063
苦苣 ……064
　　美国大叶苦苣 ……065
　　荷兰细叶苦苣 ……065
菠菜 ……066
　　新时代A级菠菜 ……067
　　亨达利菠菜 ……067
茼蒿 ……068
　　大叶茼蒿 ……069
　　小叶茼蒿 ……069
　　广良803中叶茼蒿 ……069
芹菜 ……070
　　黄叶香芹 ……071
　　青梗香芹 ……072
　　荷兰西芹 ……072
苋菜 ……073
　　尖叶绿苋 ……074
　　圆叶绿苋 ……074
　　红苋 ……075
　　花红苋 ……075
落葵 ……076
　　青梗藤菜 ……078
　　红梗藤菜 ……078
叶菾菜 ……079
　　青梗莙达菜 ……079
　　白梗莙达菜 ……080
　　红梗莙达菜 ……080
枸杞 ……081
　　大叶枸杞 ……082
　　细叶枸杞 ……082
千宝菜 ……083
银丝菜 ……083
芝麻菜 ……084
番杏 ……084
冬寒菜 ……085
香麻叶 ……085
益母草 ……086
藤三七 ……086

菊花脑 ……087
一点红 ……087
马齿苋 ……088
鱼腥草 ……088
紫背菜 ……089
珍珠菜 ……089
人参菜 ……090
无瓣蓳菜 ……090
刺芫荽 ……091
车前草 ……091
野苋 ……092
番薯叶 ……092
观音菜 ……092

## 五、瓜类

丝瓜 ……094
　　有棱丝瓜 ……094
　　绿胜1号 ……096
　　绿胜3号 ……096
　　夏绿3号 ……097
　　雅绿二号 ……097
　　雅绿六号 ……098
　　雅绿八号 ……098
　　翠丰 ……098
　　满绿 ……099
　　夏晖3号 ……099
　　华绿宝丰 ……099
　　新秀五号 ……100
　　夏胜2号 ……100
　　翠雅 ……100
　　粤优2号 ……101
　　美味高朋 ……101
　　步步高 ……102
　　夏优4号 ……102
　　普通丝瓜 ……103
　　中度水瓜 ……104
　　短度水瓜 ……104
苦瓜 ……105
　　绿宝石 ……107
　　丰绿 ……107

　　长绿 ……108
　　碧绿三号 ……108
　　碧丰3号 ……108
　　金秀3号 ……109
　　玉船2号 ……109
　　GL 924 ……109
　　万绿早 ……110
　　越秀粗瘤苦瓜 ……110
　　新秀128 ……110
　　德宝1号 ……111
　　大顶苦瓜 ……111
　　翠绿3号 ……112
　　碧珍1号 ……112
　　珠江苦瓜 ……112
节瓜 ……113
　　冠星2号 ……115
　　冠华4号 ……115
　　粤农 ……116
　　夏冠1号 ……116
　　丰乐 ……117
　　玲珑 ……117
　　粤秀 ……117
　　七星仔 ……118
　　碧绿翡翠 ……118
　　冠玉1号 ……119
　　宝玉 ……119
　　杂优连环节 ……120
　　丰冠 ……120
冬瓜 ……121
　　黑优2号 ……122
　　铁柱 ……123
　　灰皮冬瓜 ……123
　　黑皮冬 ……124
　　迷你小冬瓜 ……124
瓠瓜 ……125
　　葫芦瓜 ……126
　　青秀蒲瓜 ……127
　　油青早1号蒲瓜 ……127
　　绿富短蒲瓜 ……127
黄瓜 ……128
　　粤秀8号 ……130

中农 8 号 ……………… 130
沃林 9 号 ……………… 131
早青 4 号 ……………… 131
金山黄瓜 ……………… 132
夏丰 606 大吊瓜 ……… 132
先锋大吊瓜 …………… 133
夏美伦小黄瓜 ………… 133
南瓜 ……………………… 134
蜜本南瓜 ……………… 135
金铃南瓜 ……………… 135
盒瓜 …………………… 135
一串铃 ………………… 136
东升红皮南瓜 ………… 136
越瓜 ……………………… 137
长丰白瓜 ……………… 138
青筋白瓜 ……………… 138
西葫芦 …………………… 139
欧冠皇 ………………… 140
华丽 …………………… 141
百盛 …………………… 141
佛手瓜 …………………… 142

## 六、豆类

长豇豆 …………………… 144
丰产 2 号 ……………… 145
丰产 6 号 ……………… 146
宝丰油豆角 …………… 146
穗丰 8 号 ……………… 146
穗丰 5 号 ……………… 147
穗丰 9 号 ……………… 147
珠豇 1 号 ……………… 148
珠豇 3 号 ……………… 148
碧玉油白 ……………… 148
中度珠仔豆 …………… 149
全能王油白 …………… 149
长丰黑仁油青 ………… 150
穗青豆角 ……………… 150
八月角 ………………… 150
菜豆 ……………………… 151
12 号菜豆 ……………… 153

穗丰 4 号 ……………… 153
35 号玉豆 ……………… 154
丰田 39 号 ……………… 154
新丰玉豆 2202 ………… 154
豌豆 ……………………… 155
改良 11 号 ……………… 157
大荚荷兰豆 …………… 157
红花中花 ……………… 158
奇珍 76 号甜豌豆 ……… 158
农普甜豌豆 …………… 159
美国手牌豆苗 ………… 159
红花麦豆 ……………… 159
四棱豆 …………………… 160
菜用大豆 ………………… 162
黄豆 …………………… 162

## 七、茄果类

茄子 ……………………… 164
长丰 2 号 ……………… 166
长丰 3 号 ……………… 166
紫荣 8 号 ……………… 167
农丰 …………………… 167
园丰 …………………… 168
公牛 …………………… 168
润丰 3 号 ……………… 169
紫丰 2 号 ……………… 169
绿霸新秀 ……………… 170
玫瑰紫花茄 …………… 170
观音手指 ……………… 171
翡翠绿 ………………… 171
象牙白 ………………… 172
白玉 …………………… 172
辣椒 ……………………… 173
牛角椒 ………………… 175
辣优 4 号 ……………… 175
辣优 15 号 ……………… 175
汇丰 2 号 ……………… 176
湘研 158 ……………… 176
东方神剑 ……………… 177
茂椒 4 号 ……………… 177

线椒 …………………… 178
香妃 …………………… 178
金田 3 号 ……………… 178
永利 …………………… 179
朝天椒 ………………… 179
广良 5 号 ……………… 179
艳红 …………………… 180
甜椒 …………………… 180
卡尔顿 ………………… 180
黄欧宝 ………………… 181
辣椒叶 ………………… 181
农普辣椒叶 …………… 181
番茄 ……………………… 182
大番茄 ………………… 184
金丰 1 号 ……………… 184
年丰 …………………… 184
益丰 …………………… 185
金石 …………………… 185
新星 101 ……………… 186
托美多 ………………… 186
拉菲 …………………… 187
先丰 …………………… 187
樱桃番茄 ……………… 188
红艳 …………………… 188
金艳 …………………… 188
黄樱 1 号 ……………… 189
绿樱 1 号 ……………… 189
华喜珍珠 ……………… 190
红箭 …………………… 190

## 八、根茎类

萝卜 ……………………… 192
短叶 13 号 ……………… 194
耙齿萝卜 ……………… 194
南畔洲 ………………… 195
白玉春 ………………… 195
胡萝卜 …………………… 196
新黑田五寸人参 ……… 196

## 九、薯芋类

- 芋 ... 198
  - 红芽芋 ... 199
  - 槟榔芋 ... 199
- 马铃薯 ... 200
  - 粤引85-38 ... 201
- 沙葛 ... 202
  - 牧马山 ... 202
- 葛 ... 203
  - 细叶粉葛 ... 204
- 姜 ... 205
  - 疏轮大肉姜 ... 206
  - 密轮细肉姜 ... 206
- 山药 ... 207
  - 桂淮2号 ... 208
  - 紫玉淮山 ... 208
- 大薯 ... 209
  - 早白薯 ... 210
  - 糯米薯 ... 210
  - 紫薯 ... 210

## 十、葱蒜类

- 分葱 ... 212
  - 疏轮香葱 ... 213
  - 密轮香葱 ... 214
  - 四季葱 ... 214
- 大葱 ... 215
  - 软尾水葱 ... 216
  - 硬尾水葱 ... 216
- 大蒜 ... 217
  - 硬尾大蒜 ... 218
  - 软尾大蒜 ... 218
- 薤 ... 219
  - 丝荞 ... 220
  - 头荞 ... 220
- 韭 ... 221
  - 细叶韭菜 ... 222
  - 大叶韭菜 ... 222
  - 年花韭菜 ... 222

## 十一、香辛类

- 罗勒 ... 224
- 香茅 ... 224
- 薄荷 ... 225
- 香花菜 ... 225
- 紫苏 ... 226
  - 大鸡冠紫苏 ... 227
  - 青紫苏 ... 227
- 芫荽 ... 228
  - 沙滘芫荽 ... 229
  - 大叶芫荽 ... 229
  - 留香芫荽 ... 229
- 山柰 ... 230
- 迷迭香 ... 230
- 球茎茴香 ... 231
- 莳萝 ... 231
- 欧芹 ... 232
- 香蜂草 ... 232
- 艾叶 ... 232

## 十二、水生蔬菜类

- 莲藕 ... 234
  - 海南洲 ... 236
  - 京塘丝藕 ... 236
  - 鄂莲5号 ... 237
  - 鄂莲6号 ... 237
- 茭白 ... 238
  - 软尾茭笋 ... 238
- 慈姑 ... 239
  - 沙姑 ... 240
  - 白肉慈姑 ... 240
- 菱 ... 241
  - 红菱 ... 241
- 荸荠 ... 242
  - 水马蹄 ... 243
  - 桂林马蹄 ... 243
- 豆瓣菜 ... 244
  - 大叶西洋菜 ... 245
  - 小叶西洋菜 ... 245
- 水芹 ... 246

## 十三、多年生和杂类蔬菜

- 黄花菜 ... 248
- 竹笋 ... 249
  - 毛竹 ... 250
  - 麻竹 ... 251
  - 吊丝单 ... 251
- 黄秋葵 ... 252
  - 粤海 ... 253
  - 五福 ... 254
  - 红箭 ... 254
- 夜香花 ... 255
- 霸王花 ... 256
- 鸡肉花 ... 258

## 十四、食用菌类

- 草菇 ... 260
- 香菇 ... 261
- 金针菇 ... 262
- 蘑菇 ... 262
- 平菇 ... 263
- 茶树菇 ... 263
- 鸡腿菇 ... 264
- 杏鲍菇 ... 264
- 木耳 ... 265
- 猴头菇 ... 265
- 鸡枞菌 ... 266
- 香魏蘑 ... 266

## 十五、菜用玉米

- 甜玉米 ... 270
  - 粤甜9号 ... 270
  - 粤甜13号 ... 270
  - 粤甜16号 ... 271
  - 正甜68 ... 271

| | | | |
|---|---|---|---|
| 农甜 88 | 272 | 仲鲜甜 3 号 | 277 |
| 华宝甜 8 号 | 272 | 珍珠 | 277 |
| 新美夏珍 | 273 | 糯玉米 | 278 |
| 穗甜 1 号 | 273 | 广糯 1 号 | 278 |
| 广甜 2 号 | 274 | 广糯 2 号 | 278 |
| 广甜 3 号 | 274 | 仲糯 1 号 | 279 |
| 华美甜 8 号 | 275 | 粤彩糯 2 号 | 279 |
| 华美甜 168 | 275 | 粤白糯 6 号 | 280 |
| 金银粟 2 号 | 276 | 美玉糯 7 号 | 280 |
| 先甜 5 号 | 276 | | |

## 附录

附录 1　广州蔬菜产销概况 ······281

附录 2　广州气候概况 ······286

附录 3　广州主要名特产蔬菜营养成分 ······287

附录 4　广州蔬菜主要病虫害与防治简介 ······288

## 参考文献 ······292

# Contents

## CHINESE CABBAGE GROUP

**Flowering Chinese cabbage** ········· 003
   Sijiu No.19 flowering Chinese cabbage ········004
   Youlü 501 flowering Chinese cabbage ········004
   Bilücutai flowering Chinese cabbage ········005
   Bilüsijiu flowering Chinese cabbage ········005
   Nongyuan 45 tian youqing flowering Chinese cabbage ········006
   Caichang No.4 flowering Chinese cabbage ········006
   Zhongnan jianye youqing sweet flowering Chinese cabbage ········006
   Xinxilan pazhou sweet flowering Chinese cabbage 4560 ········007
   Qiangsheng jianye flowering Chinese cabbage ······007
   Lübao 70 tian flowering Chinese cabbage ········007
   Youlü 701 flowering Chinese cabbage ········008
   Youlü 702 flowering Chinese cabbage ········008
   Youlü 80 tian flowering Chinese cabbage ········009
   Youlü 802 flowering Chinese cabbage ········009
   Late No.2 flowering Chinese cabbage ········010
   Teqing late No.4 flowering Chinese cabbage ········010
   Suimei late No.89 flowering Chinese cabbage ·····011
   Yutian No.2 flowering Chinese cabbage ········011
   Zengcheng late flowering Chinese cabbage ········012
   Liaogeli flowering Chinese cabbage ········012

**Pak-choi** ········· 014
   Short-petiole dark leaf pak-choi ········014
   Hedou naibai pak-choi ········015
   Nongpu naibai pak-choi ········015
   Chaoshan yellow pak-choi ········015
   Xiasheng pak-choi ········016
   Baimeigui pak-choi ········016
   Huaguan qingjiang pak-choi ········017
   Guanhua qingjiang pak-choi ········017

**Chinese cabbage** ········· 019
   Early maturity No.5 Chinese cabbage ········019
   Xiayang early 50 Chinese cabbage ········020
   Chunfu Chinese cabbage ········020

## COLE CROPS

**Chinese kale** ········· 023
   Xiacui Chinese kale ········024
   Qiusheng Chinese kale ········024
   Jinlü Chinese kale ········025
   Yinlü Chinese kale ········025
   Lübao Chinese kale ········026
   Shunbao Chinese kale ········026
   Mid-maturity Chinese kale ········026
   Sijicutiao Chinese kale ········027
   Late maturity Chinese kale ········027

**Cauliflower** ········· 029
   Guangtai 45 cauliflower ········029
   Baishiwang 65 cauliflower ········030
   GL-65 songhua cauliflower ········030
   Wushan 65 tian songhua cauliflower ········030
   Xixue cauliflower ········031
   Huayi caihuawang 80 cauliflower ········031
   Lika cauliflower ········031

**Broccoli** ········· 033
   Zhulü broccoli ········034
   Mantuolü broccoli ········034
   Yanxiu broccoli ········035
   Gefu broccoli ········035

**Broccolini** ········· 037
   Qianxiu broccolini ········037
   Xiuli broccolini ········038
   Cuiyu No.2 broccolini ········038

**Cabbage** ········· 040

| | |
|---|---|
| Jinfeng No.1 cabbage ·················040 | Stem lettuce ·················061 |
| KK cabbage ·················041 | Baipi jianye lettuce ·················061 |
| Yuanlü cabbage ·················041 | Common sowthistle ·················062 |
| Zhonggan No.11 cabbage ·················042 | Acute leaf commom sowthistle ·················063 |
| Zigan No.1 cabbage ·················042 | Parted leaf common sowthistle ·················063 |
| Brussels sprout ·················043 | Endive ·················064 |
| Kale ·················044 | Broad leaf endive ·················065 |
| | Narrow leaf endive ·················065 |

## MUSTARD VEGETABLES

| | |
|---|---|
| Leaf mustard ·················046 | Spinach ·················066 |
| Nanfeng mustard ·················047 | Xinshidai A spinach ·················067 |
| Texuan hakka mustard ·················047 | Hengdali spinach ·················067 |
| Zhu mustard ·················048 | Garland chrysanthemum ·················068 |
| Shuidong mustard ·················048 | Large leaf garland chrysanthemum ·················069 |
| Head mustard ·················049 | Small leaf garland chrysanthemum ·················069 |
| Dapinpu No.11 head mustard ·················049 | Guangliang 803 mid-leaf garland chrysanthemum ·················069 |
| Root mustard ·················050 | Celery ·················071 |
| Guangtou root mustard ·················050 | White petiole celery ·················071 |

## GREEN VEGETABLES

| | |
|---|---|
| | Green petiole celery ·················072 |
| | Helan xiqin celery ·················072 |
| Water spinach ·················053 | Edible amaranth ·················073 |
| White petiole willow leaf water spinach ·················053 | Green acute leaf edible amaranth ·················074 |
| White petiole large leaf water spinach ·················054 | Green round leaf edible amaranth ·················074 |
| Green petiole willow leaf water spinach ·················054 | Red edible amaranth ·················075 |
| Green petiole large leaf water spinach ·················054 | Central red edible amaranth ·················075 |
| Green white willow leaf water spinach ·················055 | Malabar spinach ·················077 |
| Filiform leaf water spinach ·················055 | Green petiole malabar spinach ·················078 |
| Lettuce ·················057 | Red petiole malabar spinach ·················078 |
| Wrinkled leaf lettuce ·················058 | Swiss chard ·················079 |
| Boli lettuce ·················058 | Green petiole swiss chard ·················079 |
| Yidali lettuce ·················058 | White petiole swiss chard ·················080 |
| Naiyou leaf lettuce ·················059 | Red petiole swiss chard ·················080 |
| Red leaf lettuce ·················059 | Chinese wolfberry ·················081 |
| Long leaf lettuce ·················059 | Large leaf Chinese wolfberry ·················082 |
| Luoma lettuce ·················059 | Small leaf Chinese wolfberry ·················082 |
| Jianye wuban lettuce ·················060 | Senposai ·················083 |
| Xiangshui lettuce ·················060 | Jade silk vegetable ·················083 |
| Leaf head lettuce ·················061 | *Eruca sativa* ·················084 |
| Wanli head lettuce ·················061 | New zealand spinach ·················084 |
| | *Malva verticillata* ·················085 |
| | Sesame leaves ·················085 |

*Leonurus artemsisia* ·········· 086
Madeira-vine ·········· 086
*Chrysanthemum nankingense* ·········· 087
*Emilia sonchifolia* ·········· 087
*Portulaca oleracea* ·········· 088
*Houttuynia cordata* ·········· 088
*Gynura* ·········· 089
Ghostplant wormwood ·········· 089
Panicled fameflower ·········· 090
Indian rorippa ·········· 090
Foecid eryngo ·········· 091
Plantain herb ·········· 091
Wild amaranth ·········· 092
Sweet potato leaves ·········· 092
Wild chives ·········· 092

## GOURDS

Loofah ·········· 095
  Angular sponge gourd ·········· 095
    Lüsheng No.1 angular sponge gourd ·········· 096
    Lüsheng No.3 angular sponge gourd ·········· 096
    Xialü No.3 angular sponge gourd ·········· 097
    Yalü No.2 angular sponge gourd ·········· 097
    Yalü No.6 angular sponge gourd ·········· 098
    Yalü No.8 angular sponge gourd ·········· 098
    Cuifeng angular sponge gourd ·········· 098
    Manlü angular sponge gourd ·········· 099
    Xiahui No.3 angular sponge gourd ·········· 099
    Hualübaofeng angular sponge gourd ·········· 099
    Xinxiu No.5 angular sponge gourd ·········· 100
    Xiasheng No.2 angular sponge gourd ·········· 100
    Cuiya angular sponge gourd ·········· 100
    Yueyou No.2 angular sponge gourd ·········· 101
    Meiweigaopeng angular sponge gourd ·········· 101
    Bubugao angular sponge gourd ·········· 102
    Xiayou No.4 angular sponge gourd ·········· 102
  Sponge gourd ·········· 103
    Medium body sponge gourd ·········· 104
    Short body sponge gourd ·········· 104
Bitter gourd ·········· 106
  Lübaoshi bitter gourd ·········· 107
  Fenglü bitter gourd ·········· 107
  Changlü bitter gourd ·········· 108
  Bilü No.3 bitter gourd ·········· 108
  Bifeng No.3 bitter gourd ·········· 108
  Jinxiu No.3 bitter gourd ·········· 109
  Yuchuan No.2 bitter gourd ·········· 109
  GL 924 bitter gourd ·········· 109
  Wanlüzao bitter gourd ·········· 110
  Yuexiuculiu bitter gourd ·········· 110
  Xinxiu 128 bitter gourd ·········· 110
  Debao No.1 bitter gourd ·········· 111
  Dading bitter gourd ·········· 111
  Cuilü No.3 bitter gourd ·········· 112
  Bizhen No.1 bitter gourd ·········· 112
  Zhujiang bitter gourd ·········· 112
Chieh-qua ·········· 114
  Guanxing No.2 chieh-qua ·········· 115
  Guanhua No.4 chieh-qua ·········· 115
  Yuenong chieh-qua ·········· 116
  Xiaguan No.1 chieh-qua ·········· 116
  Fengle chieh-qua ·········· 117
  Linglong chieh-qua ·········· 117
  Yuexiu chieh-qua ·········· 117
  Qixingzai chieh-qua ·········· 118
  Bilüfeicui chieh-qua ·········· 118
  Guanyu No.1 chieh-qua ·········· 119
  Baoyu chieh-qua ·········· 119
  Zayoulianhuanjie chieh-qua ·········· 120
  Fengguan chieh-qua ·········· 120
Wax gourd ·········· 122
  Heiyou No.2 wax gourd ·········· 122
  Tiezhu wax gourd ·········· 123
  Grey skin wax gourd ·········· 123
  Dark skin wax gourd ·········· 124
  Mini wax gourd ·········· 124
Bottle gourd ·········· 126
  Gourd calabash ·········· 126
  Qingxiu calabash ·········· 127
  Youqingzao No.1 calabash ·········· 127
  Lüfuduan calabash ·········· 127

Cucumber ·········· 129
    Yuexiu No.8 cucumber ·········· 130
    Zhongnong No.8 cucumber ·········· 130
    Wolin No.9 cucumber ·········· 131
    Zaoqing No.4 cucumber ·········· 131
    Jinshan cucumber ·········· 132
    Xiafeng 606 cucumber ·········· 132
    Xianfeng cucumber ·········· 133
    Xiameilun mini cucumber ·········· 133
Pumpkin and squashes ·········· 134
    Miben pumpkin ·········· 135
    Jinling pumpkin ·········· 135
    He pumpkin ·········· 135
    Yichuanling pumpkin ·········· 136
    Dongsheng red skin pumkin ·········· 136
Oriental pickling melon ·········· 137
    Changfeng oriental pickling melon ·········· 138
    Green rid oriental pickling melon ·········· 138
Summer squash ·········· 140
    Ouguanhuang summer squash ·········· 140
    Huali summer squash ·········· 141
    Baisheng summer squash ·········· 141
Chayote ·········· 142

## VEGETABLE LEGUMES

Asparagus bean ·········· 145
    Fengchan No.2 asparagus bean ·········· 145
    Fengchan No.6 asparagus bean ·········· 146
    Baofeng asparagus bean ·········· 146
    Suifeng No.8 asparagus bean ·········· 146
    Suifeng No.5 asparagus bean ·········· 147
    Suifeng No.9 asparagus bean ·········· 147
    Zhujiang No.1 asparagus bean ·········· 148
    Zhujiang No.3 asparagus bean ·········· 148
    Biyuyoubai asparagus bean ·········· 148
    Zhongduzhuzai asparagus bean ·········· 149
    Quannengwangyoubai asparagus bean ·········· 149
    Changfeng black seed asparagus bean ·········· 150
    Suiqing asparagus bean ·········· 150
    Bayue asparagus bean ·········· 150

Kidney bean ·········· 152
    No.12 kidney bean ·········· 153
    Suifeng No.4 kidney bean ·········· 153
    No.35 kidney bean ·········· 154
    Fengtian No.39 kidney bean ·········· 154
    Xinfeng 2202 kidney bean ·········· 154
Vegetable pea ·········· 156
    Improved No.11 pea ·········· 157
    Large pod pea ·········· 157
    Red flower mid-pod pea ·········· 158
    Qizhen No.76 sweet pea ·········· 158
    Nongpu sweet pea ·········· 159
    Meiguo shoupai pea shoot ·········· 159
    Red flower pea ·········· 159
Winged bean ·········· 161
Vegetable soybean ·········· 162
    Soybean ·········· 162

## SOLANACEOUS FRUITS

Eggplant ·········· 165
    Changfeng No.2 eggplant ·········· 166
    Changfeng No.3 eggplant ·········· 166
    Zirong No.8 eggplant ·········· 167
    Nongfeng eggplant ·········· 167
    Yuanfeng eggplant ·········· 168
    Gongniu eggplant ·········· 168
    Runfeng No.3 eggplant ·········· 169
    Zifeng No.2 eggplant ·········· 169
    Lübaxinxiu eggplant ·········· 170
    Meiguizi eggplant ·········· 170
    Guanyinshouzhi eggplant ·········· 171
    Feicuilü eggplant ·········· 171
    Xiangyabai eggplant ·········· 172
    Baiyu eggplant ·········· 172
Pepper ·········· 174
    Ox horn pepper ·········· 175
    Layou No.4 pepper ·········· 175
    Layou No.15 pepper ·········· 175
    Huifeng No.2 pepper ·········· 176
    Xiangyan 158 pepper ·········· 176

Dongfangshenjian pepper ·············· 177
Maojiao No.4 pepper ················· 177
Thin cayenne pepper ················· 178
　　Xiangfei thin cayenne pepper ······· 178
　　Jintian No.3 thin cayenne pepper ···· 178
　　Yongli thin cayenne pepper ········· 179
Pod pepper ·························· 179
　　Guangliang No.5 pod pepper ········ 179
　　Yanhong pod pepper ··············· 180
Sweet pepper ······················· 180
　　Kaerdun sweet pepper ············· 180
　　Huangoubao sweet pepper ·········· 181
Leaf-edible pepper ·················· 181
　　Nongpu leaf-edible pepper ·········· 181
Tomato ····························· 183
　　Large fruit tomato ················· 184
　　　　Jinfeng No.1 tomato ············ 184
　　　　Nianfeng tomato ··············· 184
　　　　Yifeng tomato ················· 185
　　　　Jinshi tomato ·················· 185
　　　　Xinxing 101 tomato ············ 186
　　　　Tuomeiduo tomato ············· 186
　　　　Lafei tomato ·················· 187
　　　　Xianfeng tomato ··············· 187
　　Cherry tomato ··················· 188
　　　　Hongyan cherry tomato ········· 188
　　　　Jinyan cherry tomato ··········· 188
　　　　Huangying No.1 cherry tomato ·· 189
　　　　Lüying No.1 cherry tomato ····· 189
　　　　Huaxizhenzhu cherry tomato ···· 190
　　　　Hongjian cherry tomato ········ 190

## ROOT VEGETABLES

Radish ····························· 193
　　Short leaf No.13 radish ············· 194
　　Pachi radish ····················· 194
　　Nanpanzhou radish ················ 195
　　Baiyuchun radish ················· 195
Carrot ····························· 196
　　Xinheitian wucun carrot ············ 196

## TUBER CROPS

Taro ······························ 198
　　Red bud taro ····················· 199
　　Binglang taro ···················· 199
Potato ····························· 201
　　Yueyin 85-38 potato ·············· 201
Yam bean ··························· 202
　　Mumashan yam bean ·············· 202
Kudzu ····························· 204
　　Thin leaf kudzu ·················· 204
Ginger ····························· 205
　　Fat flesh ginger ·················· 206
　　Thin flesh ginger ················· 206
Chinese yam ························ 207
　　Guihuai No.2 Chinese yam ········· 208
　　Ziyu Chinese yam ················ 208
Yam ······························· 209
　　Zaobaishu yam ··················· 210
　　Nuomishu yam ··················· 210
　　Purple yam ······················ 210

## BULB CROPS

Bunching onion ····················· 213
　　Thin tiller bunching onion ·········· 213
　　Dense tiller bunching onion ········· 214
　　Four seasons bunching onion ······· 214
Welsh onion ························ 215
　　Soft leaf welsh onion ·············· 216
　　Hard leaf welsh onion ············· 216
Garlic ····························· 217
　　Hard leaf garlic ·················· 218
　　Soft leaf garlic ··················· 218
Scallion ····························· 219
　　Vegetable scallion ················ 220
　　Head scallion ···················· 220
Chinese chive ······················· 221
　　Small leaf Chinese chive ··········· 222
　　Large leaf Chinese chive ··········· 222
　　Nianhua Chinese chive ············ 222

## AROMATIC VEGETABLES

- Basil ·············· 224
- Citronella ·············· 224
- Mint ·············· 225
- Spearmint ·············· 225
- Perilla ·············· 226
  - Cockscomb perilla ·············· 227
  - Green perilla ·············· 227
- Coriander ·············· 228
  - Shajiao coriander ·············· 229
  - Large leaf coriander ·············· 229
  - Liuxiang coriander ·············· 229
- Rhizoma kaempferiae ·············· 230
- Rosemary ·············· 230
- Fennel ·············· 231
- Dill ·············· 231
- Parsley ·············· 232
- Lemon balm ·············· 232
- Mugwort ·············· 232

## AQUATIC VEGETABLES

- Lotus root ·············· 235
  - Hainanzhou lotus root ·············· 236
  - Jingtang lotus root ·············· 236
  - Elian No.5 lotus root ·············· 237
  - Elian No.6 lotus root ·············· 237
- Water bamboo ·············· 238
  - Soft leaf water bamboo ·············· 238
- Chinese arrowhead ·············· 239
  - Shagu Chinese arrowhead ·············· 240
  - White flesh Chinese arrowhead ·············· 240
- Water caltrop ·············· 241
  - Red water caltrop ·············· 241
- Chinese water chestnut ·············· 242
  - Chinese water chestnut ·············· 243
  - Guilin Chinese water chestnut ·············· 243
- Water cress ·············· 244
  - Large leaf water cress ·············· 245
  - Small leaf water cress ·············· 245
- Water dropwort ·············· 246

## PERENNIAL AND MIXED VEGETABLES

- Day lily ·············· 248
- Bamboo shoot ·············· 250
  - Moso bamboo ·············· 250
  - Broad flower dendrocalamus ·············· 251
  - Vario-striata dendrocalamopsis ·············· 251
- Okra ·············· 252
  - Yuehai okra ·············· 253
  - Wufu okra ·············· 254
  - Hongjian okra ·············· 254
- Chinese violet ·············· 255
- Nightblooming cereus ·············· 257
- Rose of sharon ·············· 258

## EDIBLE FUNGI

- Straw mushroom ·············· 260
- Shiitake ·············· 261
- Enokitake ·············· 262
- White mushroom ·············· 262
- Oyster mushroom ·············· 263
- Poplar mushroom ·············· 263
- Shaggy mane ·············· 264
- King oyster mushroom ·············· 264
- Jew's ear ·············· 265
- Lion's mane mushroom ·············· 265
- Termite mushroom ·············· 266
- *Lentinula edodes* × *Pleurotus nebrodensi* ·············· 266

## VEGETABLE CORN

- Sweet corn ·············· 270
  - Yuetian No.9 sweet corn ·············· 270
  - Yuetian No.13 sweet corn ·············· 270
  - Yuetian No.16 sweet corn ·············· 271
  - Zhengtian 68 sweet corn ·············· 271
  - Nongtian 88 sweet corn ·············· 272
  - Huabaotian No.8 sweet corn ·············· 272

Xinmeixiazhen sweet corn ······273
Suitian No.1 sweet corn ······273
Guangtian No.2 sweet corn ······274
Guangtian No.3 sweet corn ······274
Huameitian No.8 sweet corn ······275
Huameitian 168 sweet corn ······275
Jinyinsu No.2 sweet corn ······276
Xiantian No.5 sweet corn ······276
Zhongxiantian No.3 sweet corn ······277
Zhenzhu sweet corn ······277
**Waxy Corn** ······ 278
Guangnuo No.1 waxy corn ······278
Guangnuo No.2 waxy corn ······278
Zhongnuo No.1 waxy corn ······279
Yuecainuo No.2 waxy corn ······279
Yuebainuo No.6 waxy corn ······280
Meiyunuo No.7 sweet waxy corn ······280

**Appendices** ······ 281
  Appendix 1  The General Survey of Vegetable Production and Market Consumption in Guangzhou ······281
  Appendix 2  The Climatic Conditions of Guangzhou ······ 286
  Appendix 3  The Nutrient Composition of Guangzhou Main Vegetables ······ 287
  Appendix 4  Synopsis of Main Vegetable Diseases and Insects of Guangzhou and Their Control Methods ······288

**References** ······ 292

# 一、白菜类
# Chinese Cabbage Group

- 菜心
- 小白菜
- 大白菜

# 菜心

十字花科芸薹属一、二年生草本植物，学名 *Brassica campestris* L. ssp. *chinensis* (L.) var. *utilis* Tsen et Lee，别名菜薹、绿菜薹、广东菜、菜尖等，古称"薹心菜"，染色体数 $2n=2x=20$。

菜心原产中国，为广东特产蔬菜之一，南宋时培育成功，栽培历史悠久，类型与品种丰富。菜心的称谓，最早文字记载见于清代道光二十一年（1841 年）广东《新会县志》。菜心食用器官为花茎和薹叶，薹质柔嫩多汁，风味独特，深受群众喜爱，被誉为"蔬品之冠"。广州可周年生产供应，年栽培面积约 16 800 公顷。

**植物学性状**：根系浅。株型直立或半直立。抽薹前茎短缩。基叶多为卵形、阔卵形或圆形。抽生的花茎（即菜薹）横切面圆形，白绿色、淡绿色或青绿色，具光泽。薹叶狭卵形或披针形，短柄或无柄，黄绿色、绿色或深绿色。总状花序，花冠多为黄色，个别品种为白色。虫媒花，异花授粉。长角果。种子细小，近圆形，棕褐色或黑褐色，千粒重 1.3~1.7 克。

**类型**：按品种的生长期长短和对栽培季节的适应性，可分为早熟、中熟和迟熟 3 种类型。

1. **早熟类型** 耐热、耐湿能力较强，适播期 5—10 月，生长期 28~50 天。植株和菜薹较小，生长期短，短缩茎不明显，腋芽萌发力弱，以收主薹为主。对低温敏感，遇低温容易提早抽薹。目前，主栽品种有四九 -19 号、油绿 501、碧绿粗薹、农苑 45 天油青、菜场 4 号、碧绿四九、中南尖叶油青甜、新西兰杷洲甜 4560、强盛尖叶等。

2. **中熟类型** 耐热性中等，适播期 3—4 月及 9—11 月，生长期 60~80 天。植株中等，有短缩茎，菜薹较大，腋芽有一定的萌发力，主侧薹兼收，以收主薹为主，品质好。遇低温易抽薹。目前，主栽品种有油绿 701、油绿 702、绿宝 70 天等。

3. **迟熟类型** 较耐寒，不耐热，适播期 11 月至翌年 3 月，生长期 70~90 天。植株较大，薹粗壮，短缩茎明显，腋芽萌发力强，主侧薹兼收。低温下抽薹慢，冬性强，采收期较长，菜薹产量较高。目前，主栽品种有油绿 80 天、油绿 802、特青迟心 4 号、迟心 2 号、穗美 89 号迟花、玉田 2 号、增城迟菜心、鹅哥脷菜心等。

**栽培环境与方法**：菜心喜温，耐热，喜湿，怕涝。菜薹形成适温为 15~25℃。大多数品种对光周期要求不严格，早、中熟品种遇低温通过春化易早抽薹，晚熟品种对春化低温要求稍严。直播和育苗移栽均可。每 667 米$^2$ 用种量：直播 0.35~0.6 千克，育苗移栽 0.1~0.2 千克。

**收获**：菜薹与植株叶片等高并有初花（俗称"齐口花"）或接近时采收。近年菜场采收的标准为现蕾后即采收。

**病虫害**：病害主要有炭疽病、软腐病、霜霉病、病毒病等，虫害主要有黄曲条跳甲、小菜蛾、菜青虫、斜纹夜蛾、甜菜夜蛾、蚜虫等。

**营养及功效**：富含纤维素、维生素和矿物质。菜薹味甘，性平，具除烦解渴、利尿通便、杀菌等功效。

# Flowering Chinese cabbage

*Brassica campestris* L. ssp. *chinensis* (L.) var. *utilis* Tsen et Lee., an anual or biennial herb. Family: Cruciferae. Synoym: Tsai Tai, Choisum. The chromosome number: $2n=2x=20$.

Flowering Chinese cabbage is originated from China, and a special localized vegetable in Guangdong, has a long be cultivated since the Southern Song Dynasty, was first recorded in *Xinhui Topography* of Guangdong in 1841. The edible part is flower stalk and leaf, with a special flavor, fleshy and mentioned as the "Crown of vegetable". It can be grown year round and the total acreage in Guangzhou per annum is about 16800 $hm^2$.

Botanical characters: Erect or half erect. Root system, shallow. The stem dwarf and green in color before the formation of the flower stalk. Round in cross-section and yellow-green or green in color. Leaf, broad oval or oblong, yellow-green to dark-green in color, the leaves on the flower stalk, lanceolate or narrow oval, short or without stalk, yellow-green, green or dark green in color. The inflorescence: a raceme, yellow corolla, white corolla for specific cultivars, pollination by insect. The fruit is siliques. The seed small, roundish in shape, red brown to brown in color. The weight per 1000 seeds is 1.3~1.7 g.

Types: According to the growth period and the adaptability to season, flowering Chinese cabbage is divided into 3 types:

1. Early maturity type: Plant has high resistance to heat and wet. Sowing from May to October, 28~50 days from sowing to harvest. These cultivars are small in size, mainly harvest main flower stalk and without lateral flower stalk. They are sensitive to low temperature, easy to bolting under low temperature. The cultivars include Sijiu No.19, Youlü 501, Bilücutai, Nongyuan 45 tian youqing, Caichang No.4, Bilü 49, Zhongnan jianye youqing sweet, Xinxilan pazhou sweet 4560, Qiangsheng jianye, etc.

2. Medium maturity type: Plant has medium heat-resistance. Sowing from March to April and Septemper to October, 60~80 days from sowing to harvest. These cultivars are medium in size, with larger and good texture main flower stalk, with few lateral flower stalk in some case. These cultivars include strong adaptablity to temperature, while they are easy to bolting under low temperature. The cultivars are Youlü 701, Youlü 702, Lübao 70 tian.

3. Late maturity type: Plant has strong cold resistance while weak heat resistance. Sowing from November to March the following year, 70~90 days from sowing to harvest. These cultivars are large in size and easy to develop lateral flower stalk, and they are grown for both the main and lateral flower stalk. They bolt slowly under low temperature. The cultivars include Youlü 80 tian, Youlü 802, Teqing late No. 4, Late No. 2, Suimei late No. 89, Yutian No. 2, Zengcheng late, Liaogeli, etc.

Cultivation environment and methods: High temperature and humidity are favorable to growth of flowering Chinese cabbage. The suitable temperature for flower stalk formation is 15~25℃. Most cultivars are not strict in photoperiod requirement. Under low temperature the early and mid mature type cultivars are bolting easily. However the late mature type cultivars have strict low temperature requirements for bolting. Seeding and transplanting. The seed dosage per ha 5.25~9.0 kg for seeding and 1.5~3.0 kg for transplanting.

Harvest: Flowering Chinese cabbage is harvested when plant has few flower. In recent years, it can harvested when plant has flower bud.

Nutrition and efficacy: Flowering Chinese cabbage is rich in fibers, vitamins and minerals.

## 四九-19号

Sijiu No.19 flowering Chinese cabbage

> 品种来源

广州市农业科学研究院育成，1982年通过广东省农作物品种审定。

分布地区｜全市各区。

特　　征｜株高38厘米，开展度23厘米。基叶5~6片，叶倒卵形，长23厘米，宽13厘米，淡绿色；叶柄长6厘米，宽0.6厘米，青绿色。主薹高约20厘米，横径1.5~2.0厘米，薹叶4~6片，长卵形，淡绿色。主薹重约40克。

特　　性｜早熟，播种至初收33天，延续采收约10天。根群发达。侧芽萌发力弱，以收主薹为主。耐热、耐湿，抗逆性强，适应性广，田间表现较耐软腐病。品质中等。每公顷产量15~22吨。

栽培要点｜播种期5—10月，宜直播。直播苗期可用纱网覆盖防雨降温。株行距15厘米×18厘米。宜施足基肥，生长期及时追肥，注意病虫害防治。

## 油绿501

Youlü 501 flowering Chinese cabbage

> 品种来源

广州市农业科学研究院育成，2011年通过广东省农作物品种审定。

分布地区｜全市各区。

特　　征｜株高24.4厘米，开展度21.3厘米。基叶4~5片，叶卵圆形，长18.3厘米，宽10.2厘米，油绿有光泽；叶柄长5.2厘米，宽1.6厘米，油绿色。主薹高约18.4厘米，横径1.7~2.0厘米，薹叶4~6片，短卵形，油绿色。主薹重约40克。

特　　性｜早熟，播种至初收32~35天，延续采收6~8天。根群发达。侧芽萌发力较强，以收主薹为主。适应性较广，耐热、耐湿，抗逆性强，田间表现较耐霜霉病和炭疽病。品质优。每公顷产量15~22吨。

栽培要点｜适播期5—10月，可直播或育苗移植，育苗移植苗龄13~16天，苗具3~4片真叶。

## 碧绿粗薹

Bilücutai flowering Chinese cabbage

**品种来源**

广东省农业科学院蔬菜研究所育成，2010年通过广东省农作物品种审定。

📍 分布地区｜全市各区。

特　　征｜株高28.7厘米，开展度23.1厘米。基叶5~7片，叶椭圆形，长17.6厘米，宽10.0厘米，油绿色；叶柄长10厘米，宽1.3厘米。主薹高约19.9厘米，横径1.6~2.0厘米，薹叶6~7片，短卵形，油绿色。主薹重约30.3克。

特　　性｜早熟，播种至初收30~35天，延续采收5~7天。主根不发达，须根多。侧芽萌发力一般，以收主薹为主。适应性广，田间表现耐热性、耐涝性强，抗炭疽病中等。品质较优。每公顷产量15~20吨。

栽培要点｜适播期5月至11月上旬。其余参照油绿501。

## 碧绿四九

Bilüsijiu flowering Chinese cabbage

**品种来源**

广州市丰顺种子经营部。

📍 分布地区｜全市各区。

特　　征｜株高20.3厘米，开展度18.8厘米。基叶6~7片，叶卵形，长15.3厘米，宽10.5厘米，油绿色；叶柄长5.5厘米，宽1.3厘米，油绿色。主薹高约17.6厘米，横径1.3~1.5厘米，薹叶5~6片，卵形，油绿色。主薹重约40克。

特　　性｜早熟，播种至初收33天左右。田间表现耐热、耐雨水能力较强。质脆爽甜，食味佳，品质中等。每公顷产量18~20吨。

栽培要点｜适播期5月中旬至10月。其余参照油绿501。

## 农苑 45 天油青

Nongyuan 45 tian youqing flowering Chinese cabbage

**品种来源** ｜ 佛山市南海区大沥农苑园艺种子种苗行。

**分布地区** ｜ 全市各区。

**特　　征** ｜ 株高 20.2 厘米，开展度 19.5 厘米。基叶 5~7 片，叶卵圆形，长 14.3 厘米，宽 12.5 厘米，油绿色；叶柄长 4 厘米，宽 1.4 厘米，油绿色。主薹高约 15 厘米，横径 1.4~1.8 厘米，薹叶 5~6 片，卵形，油绿色。主薹重约 45 克。

**特　　性** ｜ 早熟，播种至初收 35 天左右。侧芽少。抗性强，田间表现较抗霜霉病。质脆爽甜，食味佳，品质较优。每公顷产量 16~18 吨。

**栽培要点** ｜ 适播期 4—6 月、10—11 月。直播或育苗移植均可。其余参照油绿 501。

## 菜场 4 号（45 天）

Caichang No.4 flowering Chinese cabbage

**品种来源** ｜ 广东湛江市新苗种子有限公司。

**分布地区** ｜ 全市各区。

**特　　征** ｜ 株高 20.3 厘米，开展度 19.8 厘米。基叶 6~8 片，叶卵圆形，长 14.7 厘米，宽 12.5 厘米，油绿色；叶柄长 4.5 厘米，宽 1.6 厘米，油绿色。主薹高约 15.6 厘米，横径 1.5~1.6 厘米，有棱沟，薹叶 5~6 片，卵圆形，油绿色。主薹重约 48 克。

**特　　性** ｜ 早熟，播种至初收 37 天左右。田间表现耐热、耐雨水能力较强。质脆爽甜，食味佳，品质较优。每公顷产量 18~20 吨。

**栽培要点** ｜ 适播期 4 月中旬至 10 月。其余参照油绿 501。

## 中南尖叶油青甜

Zhongnan jianye youqing sweet flowering Chinese cabbage

**品种来源** ｜ 广州广南农业科技有限公司。

**分布地区** ｜ 全市各区。

**特　　征** ｜ 株高 25.2 厘米，开展度 18.5 厘米。基叶 5~6 片，叶卵形，长 15.6 厘米，宽 8.5 厘米，油绿色；叶柄长 4.5 厘米，宽 1.1 厘米，油绿色。主薹高约 21 厘米，横径 1.3~1.5 厘米，薹叶 5~6 片，稍柳叶形，油绿色。主薹重约 38 克。

**特　　性** ｜ 早熟，播种至初收 28~35 天。侧芽少。质脆爽甜，食味佳，品质优。每公顷产量 15~17 吨。

**栽培要点** ｜ 适播期 4 月中旬至 10 月。直播或育苗移植均可。其余参照油绿 501。

## 新西兰杷洲甜 4560

Xinxilan pazhou sweet flowering Chinese cabbage 4560

**品种来源** 广州华绿种子有限公司。

**分布地区** 全市各区。

**特　征** | 株高 21.8 厘米，开展度 19.5 厘米。基叶 5~7 片，叶卵圆形，长 13.6 厘米，宽 11.5 厘米，油绿色；叶柄长 3.8 厘米，宽 1.5 厘米，油绿色。主薹高约 17 厘米，横径 1.5~1.7 厘米，有棱沟，薹叶 5~6 片，卵圆形，油绿色。主薹重约 50 克。

**特　性** | 早熟，播种至初收 37~40 天。田间表现耐热、耐雨水能力较强。质脆爽甜，食味佳，品质较优。每公顷产量 18~20 吨。

**栽培要点** | 适播期 4 月中旬至 10 月。其余参照油绿 501。

## 强盛尖叶

Qiangsheng jianye flowering Chinese cabbage

**品种来源** 广州恒丰种子行。

**分布地区** 全市各区。

**特　征** | 株高 21.5 厘米，开展度 19.6 厘米。基叶 6~7 片，叶卵形，长 16.7 厘米，宽 11.3 厘米，油绿色；叶柄长 5.8 厘米，宽 1.3 厘米，油绿色。主薹高约 18.5 厘米，横径 1.3~1.5 厘米，薹叶 5~6 片，稍柳叶形，油绿色。主薹重约 40 克。

**特　性** | 早熟，播种至初收 28~35 天。质脆爽甜，食味佳，品质优。每公顷产量 16~18 吨。

**栽培要点** | 适播期 4 月中旬至 10 月。其余参照油绿 501。

## 绿宝 70 天

Lübao 70 tian flowering Chinese cabbage

**品种来源** 广州市农业科学研究院育成，2003 年通过广东省农作物品种审定。

**分布地区** 全市各区。

**特　征** | 株高 33 厘米，开展度 32 厘米。基叶 7~9 片，叶长卵形，长 18 厘米，宽 10 厘米，深绿色；叶柄长 10 厘米，宽 1.3 厘米。主薹高 22~26 厘米，横径 1.5~1.8 厘米，青绿色，薹叶 6~7 片，柳叶形，主薹重约 45 克。

**特　性** | 中迟熟，播种至初收 39~45 天，延续采收 10 天。根群发达。侧芽萌发力中等，以收主薹为主，可兼收侧薹。耐病毒病能力强，抗逆性强，适应性较广。品质优。每公顷产量 15~22 吨。

**栽培要点** | 适播期 9—11 月及 2 月下旬至 3 月。其余参照油绿 501。

## 油绿 701

Youlü 701 flowering Chinese cabbage

**品种来源**

广州市农业科学研究院育成,2005 年通过广东省农作物品种审定。

📍 **分布地区** | 全市各区。

特　　征 | 株高 30.4 厘米,开展度 26.7 厘米。基叶 5~6 片,稍柳叶形,最大叶片长 23.7 厘米,宽 9.2 厘米,油绿有光泽;叶柄长 8.7 厘米,宽 1.72 厘米,油绿色。主薹高 23~25 厘米,横径 1.5~2.0 厘米,薹叶 4~6 片,柳叶形,油绿色。主薹重 45~50 克。

特　　性 | 中熟,播种至初收 37~40 天,延续采收 7~10 天。侧芽萌发力中等,以收主薹为主,可兼收侧薹。田间表现耐病毒病、霜霉病。适应性广,抗逆性强。品质优。每公顷产量 15~18 吨。

栽培要点 | 适播期 9—11 月上旬及 3 月中旬。其余参照油绿 501。

## 油绿 702

Youlü 702 flowering Chinese cabbage

**品种来源**

广州市农业科学研究院育成,2012 年通过广东省农作物品种审定。

📍 **分布地区** | 全市各区。

特　　征 | 株高 24.9 厘米,开展度 15.9 厘米。基叶 5~6 片,叶短卵形,长 18.6 厘米,宽 8.2 厘米,深油绿色;叶柄长 7.6 厘米,宽 1.9 厘米,碧绿有光泽。主薹高约 22.4 厘米,横径 1.7 厘米,薹叶 4~6 片,油绿色。主薹重约 40 克。

特　　性 | 中迟熟,播种至初收 36~40 天,延续采收 7~10 天。根群发达。侧芽萌发力中等,以收主薹为主,可兼收侧薹。抽薹整齐,花球大,齐口花。田间表现耐软腐病、霜霉病,适应性广,抗逆性强。味甜,爽脆,红维少,品质优。每公顷产量 18~22 吨。

栽培要点 | 适播期 9 月至 12 月上旬及 3 月上旬至 4 月中旬。其余参照油绿 501。

## 油绿 80 天

Youlü 80 tian flowering Chinese cabbage

**品种来源**

广州市农业科学研究院育成，2005 年通过广东省农作物品种审定。

📍 **分布地区** | 全市各区。

特　　征 | 株高 35.5 厘米，开展度 29.2 厘米。基叶 5~6 片，长椭圆形，浅绿色，最大叶片长 28.2 厘米，宽 10.5 厘米，油绿有光泽；叶柄长 11.5 厘米，宽 1.7 厘米，油绿色。主薹高 25~27 厘米，横径 1.5~2.0 厘米，薹叶 5~6 片，柳叶形，油绿色。主薹重 55~65 克。

特　　性 | 迟熟，播种至初收 43~48 天，延续采收 7~10 天。根群发达。侧芽萌发力中等，以收主薹为主，可兼收侧薹。耐寒性较强，耐热性中，耐涝性较强，田间表现耐病毒病、霜霉病。适应性广，品质优。每公顷产量 18~22.5 吨。

栽培要点 | 适播期 11 月至翌年 1 月上旬及 2 月下旬至 3 月。直播或育苗移植，移植苗龄 25 天左右，播种后用遮阳网覆盖保湿，3 叶期前后及时间苗，株行距 15 厘米 ×20 厘米。施足基肥，生长期每隔 7~10 天追施一次速效肥；保持土壤湿润；注意防治黄曲条跳甲、菜青虫、小菜蛾及蚜虫。

## 油绿 802

Youlü 802 flowering Chinese cabbage

**品种来源**

广州市农业科学研究院育成，2012 年通过广东省农作物品种审定。

📍 **分布地区** | 全市各区。

特　　征 | 株型紧凑。株高 22.7 厘米，开展度 19.3 厘米。基叶 5~6 片，圆形，叶片主脉较明显，最大叶片长 22.0 厘米，宽 10.1 厘米，深油绿色；叶柄长 8.0 厘米，宽 1.5 厘米，碧绿有光泽。主薹高约 22.4 厘米，横径 1.7 厘米，薹叶 4~6 片，油绿色。主薹重约 40 克。

特　　性 | 迟熟，播种至初收 38~45 天，延续采收 7~10 天。根群发达。侧芽萌发力较强，以收主薹为主，可兼收侧薹。抽薹整齐，花球大，齐口花。耐寒性强、抗霜霉病能力强，适应性广。肉质紧实，味甜，爽脆，纤维少，品质优。每公顷产量 18~22 吨。

栽培要点 | 适播期 10 月下旬至 12 月及 2 月下旬至 3 月。其余参照油绿 80 天。

## 迟心 2 号

Late No.2 flowering Chinese cabbage

**品种来源**

广州市农业科学研究院育成，1991 年通过广东省农作物品种审定。

**分布地区** | 全市各区。

**特　　征** | 株型紧凑。株高 37.5 厘米，开展度 26.2 厘米。基叶 12 片，长椭圆形，深绿色，最大叶片长 28.5 厘米，宽 10.8 厘米，油绿有光泽；叶柄长 7.5 厘米，宽 1.7 厘米，油绿色。主薹高约 25 厘米，横径 2.0 厘米，薹叶 8 片，柳叶形，深油绿色。主薹重 55~65 克。

**特　　性** | 迟熟，播种至初收 60 天左右，延续采收 10~12 天。根群发达。侧芽萌发力中等，以收主薹为主，可兼收侧薹。耐寒性较强，耐热性中等，耐涝性较强，耐病毒病、霜霉病，适应性广。不易空心，味甜，品质较优。每公顷产量 18~22.5 吨。

**栽培要点** | 适播期 11 月至翌年 2 月。其余参照油绿 80 天。

## 特青迟心 4 号

Teqing late No.4 flowering Chinese cabbage

**品种来源**

广州市农业科学研究院育成，2002 年通过广东省农作物品种审定。

**分布地区** | 全市各区。

**特　　征** | 株高 28.5 厘米，开展度 29.2 厘米。基叶 8~10 片，长椭圆形，深绿色，最大叶片长 20 厘米，宽 9.5 厘米，深油绿色；叶柄长 8.5 厘米，宽 1.6 厘米，油绿色。主薹高约 22.5 厘米，横径 2 厘米，薹叶 6~8 片，卵叶形，深油绿色。主薹重 45.2~62.5 克。

**特　　性** | 迟熟，播种至初收 53~56 天，延续采收 7~10 天。根群发达。侧芽萌发力中等，以收主薹为主，可兼收侧薹。耐寒性较强，耐热性中等，耐涝性较强，田间表现耐病毒病、霜霉病，适应性广。质脆嫩，风味好，纤维少，品质较优，商品性状佳。每公顷产量 18~22.5 吨。

**栽培要点** | 播种期 11 月至翌年 2 月，最适播期 12 月至翌年 1 月。其余参照油绿 80 天。

# 穗美 89 号迟花

Suimei late No.89 flowering Chinese cabbage

**品种来源**

佛山农苑种子种苗商行。

📍 分布地区｜全市各区。

特　　征｜株高33.5厘米，株型半开展，开展度27.2厘米。基叶9片，椭圆形，油绿色，最大叶片长25.5厘米，宽12.8厘米，叶柄长7.5厘米，宽1.8厘米，油绿色。主薹高约25厘米，横径1.9厘米，薹叶8片，卵圆形。主薹重50~65克。

特　　性｜迟熟，播种至初收50天左右，延续采收10~12天。根群发达。侧芽萌发力中等，以收主薹为主，可兼收侧薹。耐寒性中等，抗逆性中等。品质中等。每公顷产量18~20吨。

栽培要点｜适播期11月至12月中旬及2月至4月中旬。其余参照油绿80天。

# 玉田 2 号

Yutian No.2 flowering Chinese cabbage

**品种来源**

广州市农业科学研究院2013年育成的杂种一代。

📍 分布地区｜全市各区。

特　　征｜株高27.5厘米，开展度25.2厘米。基叶6~7片，长椭圆形，碧绿色，最大叶片长24.1厘米，宽11.1厘米，油绿有光泽；叶柄长5.8厘米，宽2.0厘米，油绿色。主薹高24~25厘米，横径1.8~2.0厘米，薹叶6~7片，卵叶形，油绿色。主薹重60~65克。

特　　性｜中迟熟，播种至初收43~46天，延续采收7~10天。根群发达。侧芽萌发力较强，以收主薹为主，可兼收侧薹。耐寒性较强，耐热性中等，耐涝性较强，耐病毒病、霜霉病，适应性广。品质优。每公顷产量18~22.5吨。

栽培要点｜适播期10月至12月上旬及3月中旬至4月下旬。其余参照油绿501。

## 增城迟菜心

Zengcheng late flowering Chinese cabbage

> 品种来源

地方品种，又名增城菜心、高脚菜心、山白菜、山婆菜。

📍 分布地区｜增城、从化、白云等区均有栽培，但以增城为主，该品种在增城于2010年11月被批准为国家地理标志产地保护产品。

特　　征｜植株高大，分枝力强。株高60厘米，开展度55厘米。基叶12~15片，叶长卵形，淡绿色，叶面皱，背部叶脉明显，叶缘波状缺刻，长58厘米，宽26厘米；叶柄匙形，粉白色或青白色，长32厘米，宽2.5厘米。菜薹粗壮，有沟纹，被白粉，主薹高约50厘米，横径3~4厘米，薹叶10~12片。主薹重约500克。

特　　性｜迟熟，播种至初收100天左右。冬性强，晚抽薹，抗逆性较强。菜质鲜嫩、香脆、爽甜，风味独特，品质优。每公顷产量38~42吨。

栽培要点｜播种期为10月至翌年1月，每667米$^2$用种量200克。苗龄18~30天。4~5片真叶定植，株距25~30厘米，行距30~40厘米，每667米$^2$种植4 000株左右。施足基肥，植后6~7天开始追肥，以后每隔6~8天追肥一次，每667米$^2$用复合肥10~15千克进行追肥。主薹采收后应施重肥促侧薹生长。注意防治病虫害。

## 鹩哥脷菜心

Liaogeli flowering Chinese cabbage

> 品种来源

地方品种，又名了哥利菜心。

📍 分布地区｜在增城、黄埔均有栽培，主要分布于增城朱村街、石滩镇及黄埔镇龙镇等。

特　　征｜植株高大，生长势旺，分枝力强。株高55厘米，开展度58厘米。基叶14~17片。柳叶形，淡绿色，叶缘稍缺刻，长62厘米，宽22厘米；叶柄匙形，青白色，叶柄长30厘米，宽2.3厘米。菜薹粗壮，有沟纹，被白粉，主薹高约46厘米，横径3~4厘米，薹叶10~12片。主薹重约400克。

特　　性｜迟熟，播种至初收约80天。冬性较强，晚抽薹，抗逆性好，纤维少，品质优，风味佳。可采收主薹和侧薹，采收期长。每公顷产量35~38吨。

栽培要点｜播种期7—9月，苗期18~25天，适宜移植栽培，4~5片真叶定植，株距25~30厘米，行距30~35厘米，每667米$^2$种植4 500株左右。其余参照增城迟菜心。

# 小白菜

十字花科芸薹属芸薹种白菜亚种的一个变种，一、二年生草本植物，学名 *Brassica campestris* L. ssp. *chinensis* (L.) Makino var. *communis* Tsen et Lee，别名普通白菜、白菜、青菜、油菜，古称"菘"，染色体数 $2n=2x=20$。

小白菜起源于中国。自唐代起，广州就有栽培，目前年栽培面积约 6 200 公顷。

**植物学性状**：植株较矮小。浅根系。短缩茎，花茎高达 1.5~1.6 米。叶片柔嫩多汁，叶色浅绿、油绿到深绿，甚至墨绿；叶形有圆形、卵圆形、倒卵形或椭圆形。叶缘全缘或有锯齿、波状皱褶，少数基部有缺刻或叶耳。叶柄肥厚，扁平或匙形，白色、绿白色、浅绿色或绿色。内轮叶片舒展或呈束腰状。总状花序，花冠淡黄色至深黄色。虫媒花，异花授粉。长角果。种子近圆形，红色、黑色或黄褐色，千粒重 1.7~2.2 克。

**类型**：按栽培季节可分为 4 种类型。

1. 春白菜 播种期 1—3 月，收获期 3—5 月。宜选用冬性强、抽薹迟的品种。

2. 夏白菜 播种期 4—8 月，收获期 6—10 月。宜选用耐热、抗风雨力强的品种。

3. 秋白菜 播种期 9—10 月，收获期 11—12 月。秋季气候凉爽，适宜各品种生长。

4. 冬白菜 播种期 11—12 月，收获期翌年 1—2 月。宜选用冬性强的品种。

目前，主栽品种有矮脚黑叶、农普奶白、潮汕黄叶、夏盛、白玫瑰、鹤斗奶白、华冠青江、冠华青江等。

**栽培环境与方法**：种子发芽适温为 20~25℃，可耐 4~8℃低温、40℃高温。长日照作物，春化后，在 12~14 小时长日照和 18~30℃下可迅速抽薹开花。直播或育苗移栽均可。

**收获**：在抽薹前任何时期均可采收，也可采摘抽薹后的白菜薹。

**病虫害**：病害主要有软腐病、菌核病、根肿病及干烧心病等，虫害主要有黄曲条跳甲、小菜蛾、菜青虫、斜纹夜蛾、蚜虫等。

**营养及功效**：富含维生素、矿物质和粗纤维。小白菜味甘，性温，具行气祛瘀、除烦宽胸、消肿散结、通利胃肠等功效。

# Pak-choi

*Brassica campestris* L. ssp. *chinensis* (L.) Makino var. *communis* Tsen et Lee, an annual or binnial herb. Family: Cruciferae. Synonym: Spoon Cabbage. The chromosome number: $2n=2x=20$.

Pak-choi is naitve to China and cultivated in Guangzhou since the Tang dynasy. The cultivated acreage per annum is about 6200 hm$^2$.

**Botanical characters:** Short plant, fibrous and shallow root. Leaves are soft and juicy, light green, green to dark green, even to black green. Leaves are spoon shaped, round, oval, obovate or oval, with entire undulate or serriate margin. Sometime there is few notch or ear leaf at the leaf base. Petiole swollen, flat, or spoon-shaped. White, white green and green in color. Inner leaves stretch or embrace to waist-shape. The inflorescence is raceme with light to dark yellow corolla, pollination by insect. Siliques fruit. The seeds are roundish, red, black or brown in color. The weight per 1000 seeds is 1.7~2.2 g.

**Types:** Pak-choi in Guangzhou can be classified into four types according to growing seasons.

1. Spring pak-choi: Sowing dates: January to March. Harvesting dates: March to May. The cultivars used should be high cold-tolerance and late bolting.

2. Summer pak-choi: Sowing dates: April to August. Harvesting dates: June to October. The cultivars should have high tolerance to heat, rain and wind.

3. Autumn pak-choi: Sowing dates: Septmber to October. Harvesting dates: November to December. Due to the cool weather, most cultivars can be cultivated.

4. Winter pak-choi: Sowing dates: November to December. Harvesting dates: January to February the following year. The cultivars should be high cold-tolerance.

At present, the main pak-choi cultivars are Short-petiole dark leaf pak-choi, Nongpu naibai pak-choi, Chaoshan yellow pak-choi, Xiasheng pak-choi, Baimeigui pak-choi, Hedou naibai pak-choi, Huaguan qingjiang pak-choi, Guanhua qingjiang pak-choi.

**Cultivation environment and methods:** Suitable temperature for seed germination: 20~25℃. Pak-choi can tolerate 4~8℃ and over 40℃. Long day plant, it would rapid bolting under 12~14 hours sunlight and 18~30℃ after vernalization. Seeding or seedling transplanting.

**Harvest:** Pak-choi plant usually be harvested before bolting, and the flower stalk is harvested after bolting.

**Nutrition and efficacy:** Pak-choi is rich in vitamins, minerals and crude fibers.

## 矮脚黑叶

Short-petiole dark leaf pak-choi

**品种来源**

广州市农业科学研究院提纯选育而成。

**分布地区** | 全市各区。

**特　征** | 株高20~22厘米，开展度32~34厘米。叶片墨绿色，叶面皱缩，全缘，长31.4厘米，宽22.5厘米；叶柄短、肥厚、匙形，长10.5厘米，宽3.8厘米，厚1厘米。单株重250~300克。

**特　性** | 中熟，播种至初收45~55天。耐热，抗性强。味甜，纤维少，品质优。每公顷产量22.5~37.5吨。

**栽培要点** | 3—11月播种，最适播期8—10月，宜直播，每667米$^2$用种量400克。株行距16厘米×20厘米。适施基肥，每667米$^2$施腐熟农家肥0.8吨，定植5天后或直播苗龄15天后追肥，每周追肥1次，共3~4次，前两次可用较稀薄的肥水，后几次每667米$^2$用复合肥5千克淋施或撒施。保持土壤湿润，雨季注意排水，注意防治病虫害。

## 鹤斗奶白

Hedou naibai pak-choi

**品种来源** 地方品种。

**分布地区** | 全市各区。

**特　　征** | 株型矮小，株高12厘米，开展度27厘米。叶色深绿，叶面皱缩，叶柄短，洁白宽厚；叶柄宽4.8厘米，厚1.2厘米。单株重190克。

**特　　性** | 早熟，播种至初收30~40天。抗病性及耐热性较好。纤维少，口感甜脆，品质优，为奶白类型的上品。每公顷产量20~30吨。

**栽培要点** | 可周年播种，最适播期为4—10月。宜直播，每667米²用种量400~500克。及时间苗，株行距约12厘米×12厘米或根据采收上市的大小而定。冬季播种期需加强水肥管理，防止早抽薹。

## 农普奶白

Nongpu naibai pak-choi

**品种来源** 广州市农业科学研究院提纯选育而成。

**分布地区** | 全市各区。

**特　　征** | 株型矮，株高12~14厘米，开展度20~26厘米。叶近圆形，长18.4厘米，宽13厘米，深绿色，有光泽，叶面皱，全缘；叶柄肥短，匙形，长5厘米，宽3.7厘米，肉厚1厘米，雪白。单株重约210克。

**特　　性** | 中熟，播种至初收45~50天。不耐热，抗病能力较差。纤维少，味清甜，品质优。每公顷产量20~30吨。

**栽培要点** | 播种期3—11月，最适播期8—10月，宜直播，每667米²用种量400克。苗期用纱网覆盖防雨降温。施足基肥，生长期间及时追肥，注意防治病虫害。

## 潮汕黄叶

Chaoshan yellow pak-choi

**品种来源** 引自潮汕地区。

**分布地区** | 全市各区。

**特　　征** | 株高21.7厘米，开展度34.4厘米。叶片卵圆形，长29.6厘米，宽18.5厘米，浅黄色，全缘；叶柄长7.4厘米，宽2.5厘米，浅白色。单株重约80克。

**特　　性** | 播种至采收36~45天。耐热性、耐涝性较强，较耐软腐病。品质优。每公顷产量15~20吨。

**栽培要点** | 周年均可播种，最适播期5—10月。其余参照矮脚黑叶。

## 夏盛

Xiasheng pak-choi

**品种来源**

广东省农业科学院蔬菜研究所育成,2012 年通过广东省农作物品种审定的杂种一代。

📍 **分布地区** | 全市各区。

特　　征 | 株型直立,株高 26.9 厘米,开展度 33.6 厘米。叶绿色,椭圆形,叶长 26.0 厘米,宽 11.7 厘米,叶面平滑,全缘;叶柄绿白色,长 8.8 厘米,宽 3.6 厘米,厚 0.7 厘米。单株重约 136 克。

特　　性 | 早熟,播种至采收 35~45 天。耐热性、耐涝性较强,较耐软腐病。品质优。每公顷产量 26~40 吨。

栽培要点 | 适播期 5—10 月。育苗移栽,苗龄 20~25 天,株行距 17 厘米 ×25 厘米。施足基肥,注意防治黄曲条跳甲和菜青虫。

## 白玫瑰

Baimeigui pak-choi

**品种来源**

广东省良种引进服务公司引进的杂种一代。

📍 **分布地区** | 全市各区。

特　　征 | 株高 21.6 厘米,开展度 37.5 厘米。叶倒卵形,长 30.4 厘米,宽 17.7 厘米,墨绿色,有光泽,叶面微皱,叶脉疏,叶肉厚,全缘;叶柄洁白,匙形,长 10.7 厘米,宽 3.8 厘米,厚 1 厘米。单株重 325 克。

特　　性 | 早熟,播种至初收 35~55 天。主根发达,生长势强,抗逆性好,耐涝、耐热和耐寒性好。商品率高,纤维少,品质优。除鲜食外,还适合加工成菜干。每公顷产量 60~70 吨。

栽培要点 | 2—11 月播种,最适播期 9—11 月。宜直播,每 667 米$^2$ 用种量 200~250 克。其余参照矮脚黑叶。

## 华冠青江

Huaguan qingjiang pak-choi

**品种来源** 广东省良种引进服务公司 2008 年从日本引进的杂种一代。

📍 **分布地区** | 全市各区。

**特　　征** | 株型直立，株高 15~19 厘米，开展度 22~24 厘米，束腰。叶片椭圆，长 22 厘米，宽 14 厘米，浅绿色；叶缘平滑，无齿。叶柄长 5~6 厘米，宽 4~6 厘米，厚 1.0 厘米，青绿色。单株重约 200 克。

**特　　性** | 早熟，播种至初收约 43 天。叶片多，晚抽薹，生长整齐，耐寒、耐热、耐湿性强，抗病性强。品质中等。每公顷产量 45~60 吨。

**栽培要点** | 周年均可播种，直播、育苗移栽均可，直播每 667 米² 用种量 250~400 克。株行距按季节而定，一般以 15 厘米 × 20 厘米为宜。施足基肥，注意防治黄曲条跳甲和菜青虫。

## 冠华青江

Guanhua qingjiang pak-choi

**品种来源** 广州市农业科学研究院 2006 年引进的杂种一代。

📍 **分布地区** | 全市各区。

**特　　征** | 株型直立，株高 19~20 厘米，开展度 23~25 厘米，束腰。叶片椭圆形，长 18 厘米，宽 16 厘米，浅绿色；叶缘平滑，无齿；叶柄长 6~7 厘米，宽 4~5 厘米，厚 0.5 厘米，匙形，浅绿色。单株重约 180 克。

**特　　性** | 早熟，播种至初收约 45 天。株型整齐，耐寒、耐热、耐湿性强，抗病性强。品质较好。每公顷产量 45~55 吨。

**栽培要点** | 参照华冠青江白菜。

# 大白菜

十字花科芸薹属芸薹种中能形成叶球的亚种，一、二年生草本植物，学名 Brassica campestris L. ssp. pekinensis (Lour.) Olsson，别名结球白菜、黄芽白、绍菜、包心白菜等，染色体数 $2n=2x=20$。

大白菜原产中国，最早可追溯到 2 500 年前的西周时期。经过漫长的进化，在公元 14—16 世纪结球白菜诞生，公元 17 世纪由北方引至肇庆。百余年前广州已普遍种植，目前年栽培面积约 1 000 公顷。

**植物学性状**：直根系，主根发达。茎短缩。叶片表现为多种形态，包括子叶、初生叶、莲座叶、球叶和茎生叶。莲座叶互生，叶片肥大，倒披针形至阔倒卵圆形，无明显叶柄，边缘锯齿状。总状花序，花冠黄色或淡黄色。异花授粉。长角果。种子呈微扁的圆球形，红褐色至褐色，或黄色，千粒重 2.0~2.4 克。

**类型**：根据熟性可以分为早熟和晚熟两种类型，早熟类型生长期 70~80 天，晚熟类型生长期 90~110 天。20 世纪 70 年代以前有高脚大青及屈尾黄两个类型。20 世纪 80 年代以后，引入早皇杂、小杂 56、中熟白麻叶及夏阳白等优良杂种一代代替了原有品种。目前春福和夏阳白种植面积较大。

**栽培环境与方法**：半耐寒性蔬菜，生长适温 12~22℃，高于 30℃ 则出现逆境伤害，并难以形成叶球。整个生长期的需水量较大，在进入结球后期需减少用水，避免叶片提前衰老，降低叶球耐贮存性和减少病害发生。广州地区秋冬季播种可直播或育苗移栽。

**收获**：在叶球生长较为紧实时即可收获。

**病虫害**：病害主要有软腐病、菌核病、病毒病、霜霉病、根肿病及干烧心病等，虫害主要有黄曲条跳甲、小菜蛾、菜青虫、斜纹夜蛾、蚜虫等。

**营养及功效**：富含维生素、蛋白质、矿物质及纤维素等多种营养物质。大白菜味甘，性温，具养胃生津、除烦解渴、利尿通便、清热解毒等功效。

# Chinese cabbage

## 早熟 5 号
### Early maturity No.5 Chinese cabbage

*Brassica campestris* L. ssp. *pekinensis* (Lour.) Olsson, an annual or bienial herb. Family: Cruciferae. Synonyms: Pe-tsai, Celery cabbage. The chromosome number: $2n=2x=20$.

Chinese cabbage is native to China, was introduced to Zhaoqing, Guangdong province, from northern China during the 17th century, it was widely cultivated in Guangzhou about 100 years ago. The cultivated area is about 1000 $hm^2$ per annum.

Botanical characters: The leaves born on the dwarfed stem, and are diversification, include cotyledon, primary leaf, rosette leaf, leaf head and cauline leaf. The rosette leaves are alternate, fleshy, oblanceolate to broad obovate, no obvious petiole, serrate leaf margin. The infloresoence: a raceme with yellow or light yellow corolla. Cross-pollination. Fruit is siliques. Seed is slightly flat spherical, reddish brown, brown or yellow in color. The weight per 1000 seeds is 2.0~2.4 g.

Types: There are two types cultivars of Chinese cabbage: early-maturing and late-maturing cultivar, the early-maturing cultivar is 70~80 days from sowing to harvest, and the late-maturing cultivar is 90~110 days from sowing to harvest. There were Long-petiole large-green type and the Curved tailed yellow type grown in Guangzhou before the 1970s. After the 1980s, the domestic hybrid cultivars of Early-king hybrid, Xiaoza 56 and Mid-maturity white hemp-leaf and the exotic hybrid cultivar of Summer white were introduced, which replaced the former cultivars. The popular cultivars in Guangzhou are Chunfu and Xiayangbai.

Cultivation environment and methods: The mild temperature is favorable for the growth of Chinese cabbage, the suitable temperature for growth is 12~22℃. Under the temperature above 30℃, Chinese cabbage is damaged, and the leaf head is difficult to form. The growth of Chinese cabbage is need sufficient water supply, the reduced water supply during later leaf head formating is benefit to leaf head storage and disease-control. Chinese cabbage is sowing in autumn and winter, seeding or seedling transplant.

Harvest: It can be harvested when the leaf head is tight.

Nutrition and efficacy: Chinese is rich in vitamins, proteins, minerals and crude fibers.

**品种来源**

浙江省农业科学院园艺研究所育成并于 1989 年通过该省审定的杂种一代。

**分布地区** | 白云、花都、增城、从化等区。

**特　　征** | 株高 31 厘米，开展度 40~45 厘米。叶长 36.7 厘米，宽 25.5 厘米，深绿色，无毛。叶柄白色，中肋长 20 厘米，宽 6 厘米。叶球高 25 厘米，横径 15.5 厘米。单球重约 1 300 克。

**特　　性** | 早熟，播种至收获叶球 50~55 天。也可作小白菜栽培，一般 35 天以后就能采收上市，并能连续多茬播种。抗病毒病、炭疽病等能力较强，耐热、耐涝性较强，适应性强。商品性好，品质优。每公顷产量 40~50 吨。

**栽培要点** | 可周年播种。选前茬为非十字花科蔬菜土地种植，每 667 米² 施用 1 吨左右有机肥作基肥，根据长势进行追施肥水，注意防治病虫害。4—9 月作小白菜播种，直播或撒播，每 667 米² 用种量 450~500 克，分批采收。8 月上旬至翌年 1 月上旬播种作结球白菜，采用育苗移栽，每 667 米² 用种量约 100 克，苗期 25 天左右，每 667 米² 种植 3 600~4 000 株。

## 夏阳白菜早 50

Xiayang early 50 Chinese cabbage

品种来源 | 广西灵山县强坤蔬菜种子有限公司引进。

分布地区 | 白云、花都、增城、从化等区。

特　征 | 株型直立，株高 25~30 厘米，开展度 35~50 厘米。外叶少，卵圆形，长 32 厘米，宽 23 厘米。叶色浓，稍有皱褶。叶柄白色。叶球卵圆形，高 25~30 厘米，横径 12~14 厘米，白色。单球重约 1 千克。

特　性 | 早熟，播种至收获 50~60 天。耐热，耐湿，抗软腐病、黑斑病能力较强。生长势强，可密植。结球紧实，耐贮运。播后 50 天可长至 1 千克，延后采收可达 1.5 千克。不耐低温，春季种植要提防低温抽薹。质嫩，商品性好，品质优。每公顷产量 42~48 吨。

栽培要点 | 最适播期 9—11 月。育苗移栽或直播。其余参照早熟 5 号。

## 春福白菜

Chunfu Chinese cabbage

品种来源 | 广州市兴田种子公司引进的杂种一代。

分布地区 | 白云、花都、增城、从化等区。

特　征 | 株型直立，高 33 厘米，开展度 63.4 厘米。叶片卵圆形，长 27 厘米，宽 16.5 厘米，叶片边缘有微小针毛，叶面稍平滑，下部叶缘呈锯齿状；包球叶片浅黄色，长卵形，尖头，上部小，下部大。单球重 2.5~3 千克。

特　性 | 中熟，播种至收获 65~75 天。适应性广，耐热，耐寒，对软腐病抵抗能力较差。味淡，品质优。每公顷产量 35~40 吨。

栽培要点 | 播种期 8 月至翌年 1 月，最适播期 9—11 月。直播或育苗移栽，苗期 30~35 天，株行距 30 厘米×35 厘米。8 月播种，生长前期用遮阳网覆盖栽培。

# 二、甘蓝类
# COLE CROPS

- 芥蓝
- 花椰菜
- 青花菜
- 西兰薹
- 结球甘蓝
- 抱子甘蓝
- 羽衣甘蓝

# 芥蓝

十字花科芸薹属一、二年生草本植物，学名 Brassica alboglabra L. H. Bailey，别名芥兰、白花芥蓝，染色体数 $2n=2x=18$。

芥蓝原产我国南方，是广州著名的特产蔬菜之一，栽培历史悠久，类型与品种丰富，相传公元 8 世纪广州已有栽培。目前年栽培面积约 2 800 公顷。

**植物学性状**：须根系。抽薹前茎短缩，绿色。基叶互生，多为卵形至近圆形。叶面平滑或稍皱，绿色或深绿色，被蜡粉；叶柄长，浅绿色。抽生的花茎（即菜薹）肉质，横切面圆形，节间较疏，绿色。薹叶狭卵形或披针形，短柄或无柄，绿色或深绿色。总状花序，花冠多为白色，个别品种为黄色。虫媒花。长角果。种子近圆形，褐色或黑褐色。千粒重 4.5~5 克。食用器官为花茎和薹叶。

**类型**：按品种生长期的长短和对栽培季节的适应性，可分为早熟、中熟和迟熟 3 种类型。

1. 早熟类型　耐热、耐湿能力较强，适播期 7—9 月，播种至初收 60~65 天，采收期 9—11 月。植株和菜薹较小，生长期较短，短缩茎不明显，腋芽萌发力弱，以收主薹为主。目前品种主要有夏翠、绿宝、银绿等。

2. 中熟类型　耐热性中等，适播期 9—11 月及翌年 2 月下旬至 3 月上旬，播种至初收 65~75 天，采收期 11 月至翌年 2 月。植株中等，有短缩茎，菜薹较大，腋芽有一定的萌发力，主侧薹兼收，菜薹品质好。不适宜冬季播种。目前品种主要有秋盛、金绿、顺宝、中花、四季粗条芥蓝等。

3. 迟熟类型　耐寒，适播期 11 月至翌年 2 月，播种至初收 75~85 天，采收期翌年 1—4 月。植株较大，薹粗壮，短缩茎明显，腋芽萌发力强，主侧薹兼收。低温下抽薹性好，冬性强。目前品种主要有迟花芥蓝等。

**栽培环境与方法**：芥蓝喜冷凉，耐寒，耐湿，耐涝。菜薹形成适温 15~25℃。大多数品种对光周期要求不严格，晚熟品种对春化低温要求稍严。直播和育苗移栽均可。

**收获**：薹与植株叶片等高并有初花（俗称"齐口花"）或接近时采收，也可根据市场需要提前采收。

**病虫害**：病害主要有软腐病、黑腐病、立枯病、霜霉病、炭疽病、花叶病等，虫害主要有黄曲条跳甲、小菜蛾、菜青虫、斜纹夜蛾、甜菜夜蛾、蚜虫等。

**营养及功效**：富含维生素 C、粗纤维、蛋白质、可溶性糖和萝卜硫素。芥蓝味甘，性温，具除烦解渴、利尿通便、散血消肿、杀菌、降血压、降血脂等功效。

# Chinese kale

*Brassica alboglabre* L. H. Bailey, an annual or biennial herb. Family: Cruciferae. Synonym: Kale borecole, Collard, Curly kale. The chromosome number: $2n=2x=18$.

Chinese kale, originated from the southern China, is one of famous special local vegetable in Guangzhou. It was grown in Guangzhou during the 8th century and rich in cultivars. At present the cultivated area is about 2800 $hm^2$ per annum.

Botanical characters: Fibrous root system. Stem dwarfed before bolting and green in color. Leaves alternate, oval to approximate round. Leaf surface smooth or slight wrinkled, green or dark green in color and wax-covered. Petiole long and green. Flower stalk fleshy, with sparse nodes, round and green. Stalk leaf. Long oval or lanceolate, green or dark green in color, with short leaf petole or not. Raceme, the entomophilous flower, with white corollas for most of cutivars and yellow for the others. Silique, with approximate round seed, brown or dark brown. The weight per 1000 seeds is 4.5~5 g. The edible part is flower stalk and stalk leaf.

Types: There are three types of Chinese kale according to maturity and seasonal adaptability.

1. Early maturing type: It is strong resistant to heat and waterlogging. Sowing dates: July to September, 60~65 days from sowing to harvest. Harvesting dates: September to November. Both plant and flower stalk are small. Dwarfed stem is not obvious, with few axillary buds. Main flower stalk is havested. The cultivars include Xiacui, Lübao, Yinlü, etc.

2. Medium maturing type: It is mid-resistant to heat. Sowing dates: September to November and late February to early March the following year, 65~75 days from sowing to harvest. Harvesting dates: November to February the following year. The plant is middle size with dwarf stem. Flower stalk is large, with some lateral flower stalk. The main flower stalk and lateral flower stalk are harvested. The cultivars include Qiusheng, Jinlü Shunbao, Zhonghua, Sijicutiao.

3. Late maturing type: It is resistant to cold. Sowing dates: November to February the following year, 75~85 days from sowing to harvest. Harvesting dates: January to April the following year. Both plant and flower stalk are large, with obviously dwarf stem. The main flower stalk and lateral flower stalk are harvested. The plant grow and flower stalk format well under low temperature. The cultivars include Late Chinese kale.

Cultivation environment and methods: Chinese kale is tolerant to cold and waterlogging. The suitable temperature for formation of flower stalk is 15~25℃. Most of cultivars are non-sensitive to photoperiod. Vernalization of late maturing cultivars requried rigid lower temperature. Sowing or seedling transplanting.

Harvest: Chinese kale is harvested when plant has few flower.

Nutrition and efficacy: Chinese kale is rich in vitamin C, crude fibers, proteins, soluble sugars, sulforaphane, etc.

## 夏翠芥蓝

Xiacui Chinese kale

**品种来源**

广东省农业科学院蔬菜研究所育成的杂种一代，2010年通过广东省农作物品种审定。

**分布地区** | 全市各区。

**特　　征** | 株型直立，株高36厘米，开展度34厘米。叶片椭圆形，肥厚微皱，浅绿色，蜡粉少。菜薹圆粗壮均匀，节间中长，主薹高17~20厘米，粗2.0~2.5厘米，重110~150克，薹色绿。花白色。

**特　　性** | 早熟，播种至初收45~51天。生长势较强，在适播期内表现较好的适应性。耐热，耐涝，抗逆性强，耐炭疽病、软腐病、霜霉病。菜薹鲜嫩，爽甜可口，品质优。每公顷产量18~24吨。

**栽培要点** | 广州地区适播期8月至10月上旬，宜选择前作不是十字花科作物的壤土栽植。一般采用穴盘基质育苗，1穴1株。苗龄20~25天时移栽。株行距17厘米×20厘米。苗期应注意控制水分，防止幼苗徒长。定植成活后追肥2~3次，前期主要追施复合肥或少量氮肥。整个生育期注意防治黄曲条跳甲、小菜蛾和菜粉蝶等。

## 秋盛芥蓝

Qiusheng Chinese kale

**品种来源**

广东省农业科学院蔬菜研究所育成的杂交一代，2010年通过广东省农作物品种审定。

**分布地区** | 全市各区。

**特　　征** | 株型半披，株高33厘米，开展度33厘米。叶片圆形，肥厚，深绿色，叶面皱缩，叶长20厘米，宽21厘米；叶柄长5.5厘米。主薹高15~20厘米，粗2.2~2.8厘米，重120~180克。花白色。

**特　　性** | 早中熟，播种至初收62~65天，主薹延续采收8天。生长势强，抽薹整齐。商品综合性状好，品质优，产量稳定。耐热、耐涝、耐寒，抗逆性强，耐病毒病、黑腐病、软腐病、霜霉病。菜薹质爽脆味甜，每公顷产量18~26吨。

**栽培要点** | 播种期2月中旬至3月中旬、8—11月。一般采用穴盘基质育苗，1穴1株。苗期可用纱网覆盖抗热防雨。幼苗4~5片真叶或苗龄20~25天时及时定植。株行距（18~20）厘米×（20~25）厘米。生长期间及时追肥，一般全期追肥2~3次，前期追施薄肥，中后期每次每667米$^2$施复合肥10~16千克。出现齐口花时及时采收。收主薹后施速效肥，以促进侧薹生长。注意防治病虫害。

## 金绿芥蓝

Jinlü Chinese kale

### 品种来源

广州市农业科学研究院育成的杂交一代。

📍 分布地区 | 全市各区。

特　　征 | 株型直立，健壮，较紧凑。株高 32.0 厘米，开展度 35.0 厘米。基叶数 11 片；叶片卵圆形，叶长 22.5 厘米，宽 16.5 厘米，深绿色，叶面微皱，全缘，具叶耳；叶柄长 5.0 厘米，浅绿色。菜薹紧实匀条，主薹高 23.5 厘米，粗 2.3 厘米，重 165 克，薹叶较小，披针形，薹色绿。花白色。

特　　性 | 早中熟，播种至初收秋植 62 天、春植 61 天，主薹延续采收 8 天。生长势强，抽薹整齐。耐热，耐涝，耐寒，抗逆性强，耐病毒病、黑腐病、软腐病、霜霉病。菜薹质爽脆味甜，商品综合性状好，品质优。每公顷产量 18~26 吨。

栽培要点 | 参照秋盛芥蓝。

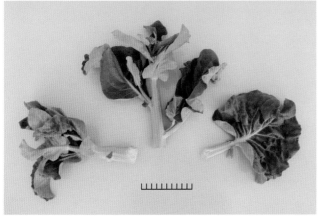

## 银绿芥蓝

Yinlü Chinese kale

### 品种来源

广州市农业科学研究院 2011 年育成。

📍 分布地区 | 全市各区。

特　　征 | 株型直立紧凑，较矮壮。株高 28.0 厘米，开展度 32.0 厘米。基叶数 11 片；叶片卵圆形，深绿色，叶面微皱，全缘，具叶耳；叶柄较短，浅绿色。菜薹均匀，节间中长，主薹高 18~22 厘米，粗 1.7~2.2 厘米，重 70~120 克。薹叶较小，薹色绿。花白色。

特　　性 | 早熟，播种至初收 55~58 天，主薹延续采收 10 天。生长势较强。耐热、耐涝、较耐寒，耐病毒病、黑腐病、软腐病、霜霉病。菜薹质爽脆味甜，商品综合性状好，品质优。每公顷产量 15~20 吨。

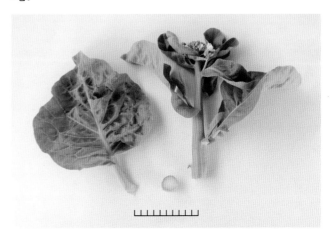

栽培要点 | 宜直播或育苗移栽，直播每公顷用种量 3~5 千克。幼苗 2 叶 1 心期时及时疏苗。其余参照夏翠芥蓝。

## 绿宝芥蓝 Lübao Chinese kale

**品种来源** | 广东省良种引进服务公司从日本引进推广的杂种一代，2009年通过广东省农作物品种审定。

**分布地区** | 全市各区。

**特　征** | 株型直立紧凑，较矮壮。株高29厘米，开展度34厘米。叶片近圆形，微皱，绿色，蜡粉少；叶长20.4厘米，宽16.3厘米；叶柄长5.7厘米。菜薹均匀，节间中长，主薹高18~21厘米，粗1.7~2.1厘米，重65~115克。薹叶较小，薹色绿。花白色。

**特　性** | 早熟，播种至初收45~55天。生长势强，抗病、抗逆性强。菜薹鲜嫩，爽甜可口，品质优。每公顷产量15~20吨。

**栽培要点** | 参照夏翠芥蓝。

## 顺宝芥蓝 Shunbao Chinese kale

**品种来源** | 广东省良种引进服务公司从日本引进推广的杂种一代，2010年通过广东省农作物品种审定。

**分布地区** | 全市各区。

**特　征** | 株型直立，健壮，较紧凑。株高33厘米，开展度33厘米。叶片圆形，肥厚，深绿色，叶面皱缩，叶长20厘米，宽19.6厘米；叶柄长5.5厘米。主薹高22~28厘米，粗2.2~2.7厘米，重110~130克。花白色。

**特　性** | 中熟，播种至初收60~65天，主薹延续采收8天。生长势强，抽薹整齐，抗病抗逆性强。菜薹质爽脆味甜，商品综合性状好，品质优。每公顷产量18~26吨。

**栽培要点** | 参照秋盛芥蓝。

## 中花芥蓝 Mid-maturity Chinese kale

**品种来源** | 从香港引进。

**分布地区** | 全市各区。

**特　征** | 株型直立，健壮，较紧凑。株高35.0厘米，开展度35.0厘米。叶片卵圆形，深绿色，叶面微皱，全缘，具叶耳；叶柄长7.0厘米，浅绿色。菜薹紧实，主薹高23.5厘米，粗2.3厘米，重120克，薹叶较小，披针形，薹色绿。花白色。

**特　性** | 中熟，播种至初收65天，主薹延续采收8天。生长势强，抽薹整齐，抗病性较强。菜薹质爽脆味甜，商品综合性状好，品质优。每公顷产量20~24吨。

**栽培要点** | 参照秋盛芥蓝。

## 四季粗条芥蓝

Sijicutiao Chinese kale

**品种来源**

地方品种。

📍 分布地区 | 全市各区。

特　征 | 株型直立，健壮，较紧凑。株高 32.0 厘米，开展度 35.0 厘米。基叶数 12 片；叶片卵圆形，叶长 22.5 厘米，宽 16.5 厘米，绿色，叶面微皱，全缘，具叶耳；叶柄长 9.0 厘米，浅绿色。菜薹粗紧实，主薹高 25~30 厘米，粗 3.5~4.2 厘米，重 150~250 克。薹叶较小，披针形，薹色绿。花白色。

特　性 | 中熟，播种至初收 65 天，主薹延续采收 8 天。生长势强，较耐热、耐涝、耐寒，耐病毒病、黑腐病、软腐病、霜霉病。菜薹质爽脆味甜，商品性好，品质优。每公顷产量 22~26 吨。

栽培要点 | 适播期 8—12 月。一般采用穴盘基质育苗，1 穴 1 株。苗期可用纱网覆盖抗热防雨。幼苗 4~5 片真叶或苗龄 20~25 天时及时定植。株行距 20 厘米×20 厘米或 20 厘米×25 厘米。生长期间及时追肥，一般全期追肥 2~3 次，前期主要结合浇水追施薄肥，中后期每次每 667 米$^2$施复合肥 10~16 千克。

## 迟花芥蓝

Late maturity Chinese kale

**品种来源**

地方品种。

📍 分布地区 | 全市各区。

特　征 | 株型较高大。株高 45.0 厘米，开展度 43.0 厘米。基叶数 13 片；叶片卵圆形，叶长 25.0 厘米，宽 23.5 厘米，绿色，叶面微皱，全缘，具叶耳；叶柄长 8.0 厘米，浅绿色。菜薹较粗紧实，主薹高 25 厘米，粗 3.5 厘米，重 250 克，薹色绿。花白色。

特　性 | 迟熟，播种至初收 80 天。生长势强，耐寒，抗逆性强，耐病毒病、黑腐病、软腐病、霜霉病。菜薹质爽脆味甜，商品综合性状好，品质优。耐涝，每公顷产量 25~30 吨。

栽培要点 | 适播期 11 月至翌年 2 月。其余参照秋盛芥蓝。

# 花椰菜

十字花科芸薹属一、二年生草本植物，学名 *Brassica oleracea* L. var. *botrytis* L.，别名花菜、菜花、椰菜花，染色体数 $2n=2x=18$。

花椰菜原产地中海沿岸，广州已有百年以上的栽培历史，目前年种植面积约 1 300 公顷。

植物学性状：根系较发达。茎较短缩。叶互生，长椭圆形，叶缘波状，具叶耳或无，绿色，被蜡粉，茎中上部叶较小且无柄，长圆形至披针形，抱茎。花球由短缩、肉质的主花茎及其上多级侧花茎组成，白色。总状花序，花冠黄色。虫媒花。长角果。种子近圆形，红褐色或黑褐色，千粒重 3.5~4.5 克。

类型：品种主要从国内外引进，按成熟期可分为早熟、中熟和晚熟 3 种类型。早熟类型：定植到采收 70 天以内，品种主要有广泰 45、白狮王 65、五山 65 天松花菜、GL-65 松花等；中熟类型：定植到采收 70~100 天，品种主要有喜雪、华艺菜花王 80、利卡等；晚熟类型：定植到采收长达 100 天以上，广州地区一般不适宜种植。

栽培环境与方法：花椰菜属低温长日照蔬菜，喜光照、冷凉，半耐寒，不耐高温、干旱，不耐霜冻。营养生长温度为 8~24℃，而花球生育适温为 15~18℃，形成花球除了必需的低温感应条件外，还要求长日照条件，有些品种在短日照下不能形成花球。晚熟品种对春化低温要求严格，温度高于 25℃时，结出的花球小而且松散。广州栽培季节一般为 7 月至翌年 4 月，最适播期为 8—10 月，育苗移栽。

收获：待花球充分膨大、周边开始松散时即可采收。

病虫害：病害主要有炭疽病、软腐病、霜霉病、黑腐病等，虫害主要有黄曲条跳甲、小菜蛾、菜青虫、斜纹夜蛾、甜菜夜蛾、蚜虫等。

营养及功效：富含维生素、矿物质、食用纤维素和胡萝卜素。性温，具清热解渴、利尿通便、健胃助消化等功效。

# Cauliflower

*Brassica oleracea* L. var. *botrytis* L., an annual or biennial herb. Family: Cruciferae, Synonym: Vegetable flower. The chromosome number: $2n=2x=18$.

Cauliflower has been cultivated for more than 100 years in Guangzhou. At present, the cultivated area is about 1300 hm$^2$ per annum.

Botanical characters: Vigorous root system. Stem dwarfed. Leaves alternate, oblong. Leaf margin undulate. Leaf blade, green in color and wax-covered, with an ear or none. Few Leaf on the upon middle of the stalk, without leaf petioles, long round to lanceolate. Curd, the edible part, white in color, constituted by shortened and succulent main scape and many lateral scape. Raceme, entomophilous flower with yellow corolla, Silique, with approximate round, red brown or dark brown seeds. The weight per 1000 seeds is 3.5~4.5 g.

Types: Cauliflower cultivars, mainly introduced from other regions or abroad, can be classified into early, medium and late maturing types according to plant maturity. Early maturing type: less 70 days from sowing to harvest. The cultivars include Guangtai 45, Baishiwang 65, Wushan 65 tian songhua, GL-65 songhua, etc. Medium maturing types: 70~100 days from sowing to harvest, and the cultivars include Xixue, Huayicaihuawang 80, Lika, etc. Late maturing type: more than 100 days from sowing to harvest. It is unsuitable for cultivation in Guangzhou.

Cultivation environment and methods: Cauliflower is photophilous, not tolerant to drought, high temperature and frost. The suitable temperatureis 8~24℃ for vegetative growth, and 15~18℃ for crud development. The crud is small and loose above 25℃. Both low temperature and long day are required for crud formation. In some cultivars, the crud could not develop under short-day condition. Vernalization of the late maturing cultivars is sensitive to low temperature. Cauliflower is usually cultivated from July to April the following year in Guangzhou. Seedling transplanting. The most suitable sowing date is August to October.

Harvest: Cruds could be harvested when cruds are fully expanding and with initial loosing around.

Nutrition and efficacy: Cauliflower is rich in vitamins, minerals, edible fibers and carotene.

## 广泰 45

Guangtai 45 cauliflower

**品种来源**

广东省良种引进服务公司引进推广的杂种一代。

**分布地区** | 全市各区。

特　　征 | 株高52厘米，开展度75厘米。叶片长椭圆形，叶长45厘米，叶宽28厘米，绿色，有蜡粉；叶柄宽，白绿色。花球半圆形，横径15.1厘米，白色，结球紧实，单球重450克。

特　　性 | 早熟，播种至初收55天，延续采收5~7天。生长势强，耐寒，耐病毒病、黑腐病、霜霉病，抗病抗逆性较强。纤维少，商品综合性状好，品质优。每公顷产量15~18吨。

栽培要点 | 播种期7—9月。宜育苗，苗期可用纱网覆盖遮阳降温。幼苗2叶1心时，及时疏苗；幼苗5~8片真叶或苗龄30~40天时及时定植。株行距（50~60）厘米×（60~65）厘米。基肥施有机肥和硼砂。生长前期水肥要充足，封行前后结合中耕培土追施复合肥、钾肥，现蕾后施重肥，增施磷、钾肥。

## 白狮王 65　Baishiwang 65 cauliflower

> 品种来源

广州市兴田种子有限公司引进推广的杂种一代。

📍 分布地区｜全市各区。

特　　征｜株高 55 厘米，开展度 50 厘米。叶片长椭圆形，叶长 45 厘米，宽 30 厘米，绿色，有蜡粉；叶柄宽，白绿色。花球半圆形，横径 18 厘米，白色，结球紧实，单球重 600 克。

特　　性｜早中熟，播种至初收 65 天，延续采收 10~15 天。生长势强，耐寒，耐病毒病、黑腐病、霜霉病，抗病抗逆性较强。纤维少，商品综合性状好，品质优。每公顷产量 18~25 吨。

栽培要点｜播种期 7 月中旬至 9 月。其余参照广泰 45。

## GL-65 松花　GL-65 songhua cauliflower

> 品种来源

广东省良种引进服务公司引进推广的杂种一代。

📍 分布地区｜全市各区。

特　　征｜株高 55 厘米，开展度 78 厘米。叶片长椭圆形，叶长 48 厘米，宽 28 厘米，绿色，有蜡粉；叶柄宽，白绿色。花球半圆形，横径 19.5 厘米，白色，结球紧实，单球重 700 克。

特　　性｜早中熟，播种至初收 65 天，延续采收 10~15 天。生长势强，耐寒，耐病毒病、黑腐病、霜霉病。纤维少，商品综合性状好，品质优。每公顷产量 22~30 吨。

栽培要点｜播种期 8 月底至 10 月。采前若不覆叶遮球，能使花梗更绿。其余参照广泰 45。

## 五山 65 天松花菜　Wushan 65 tian songhua cauliflower

> 品种来源

广州华绿种子有限公司从台湾引进推广的杂交一代。

📍 分布地区｜全市各区。

特　　征｜株高 60 厘米，开展度 60 厘米。叶片长椭圆形，叶长 30 厘米，宽 25 厘米，绿色，有蜡粉；叶柄宽，白绿色。花球偏平圆形，松大，花层薄，花蕾乳白色，花梗淡绿色，球径约 25 厘米，单球重 1 200 克。

特　　性｜早中熟，播种至初收 65 天，延续采收 10~15 天。生长势强，耐寒，抗逆性强，耐病毒病、黑腐病、霜霉病。纤维少，商品综合性状好，品质优。每公顷产量 22~25 吨。

栽培要点｜播种期 8 月至 10 月中旬，株行距 50 厘米 ×75 厘米。采前若不覆叶遮球，能使花梗更绿。其余参照广泰 45。

## 喜雪 Xixue cauliflower

**品种来源**

广东省良种引进服务公司从荷兰引进推广的杂种一代。

分布地区｜全市各区。

特　　征｜株高 65 厘米，开展度 85 厘米。叶片长椭圆形，叶长 52 厘米，宽 32 厘米，绿色，有蜡粉；叶柄宽，白绿色。花球半圆形，横径 16.5 厘米，白色，结球紧实，单球重 1 200 克。

特　　性｜中熟，播种至初收 80 天，延续采收 10~15 天。生长势强，耐寒，抗逆性强，耐病毒病、黑腐病、霜霉病。纤维少，商品综合性状好，品质优。每公顷产量 22~30 吨。

栽培要点｜播种期 9—10 月。其余参照广泰 45。

## 华艺菜花王 80

Huayi caihuawang 80 cauliflower

**品种来源**

广州市华艺种苗行有限公司 2011 年从日本引进推广的杂种一代。

分布地区｜全市各区。

特　　征｜株高约 80 厘米，开展度 60 厘米。叶片长椭圆形，叶长约 70 厘米，宽 20~22 厘米，绿色，有蜡粉；叶柄宽，白绿色。花球半圆形，横径约 18 厘米，白色，结球紧实，单球重 1 500~2 000 克。

特　　性｜中熟，播种至初收 80 天，延续采收 10~15 天。生长势强，纤维少，自覆性好，耐寒，耐病毒病、黑腐病、霜霉病。商品综合性状好，品质优。每公顷产量 22~30 吨。

栽培要点｜播种期 9 月下旬至 11 月下旬。其余参照广泰 45。

## 利卡 Lika cauliflower

**品种来源**

美国先正达种子有限公司推广的杂交一代。

分布地区｜全市各区。

特　　征｜株高 70~80 厘米，开展度 60~70 厘米。叶片长椭圆形，叶长 50~60 厘米，宽 20~25 厘米，绿色，有蜡粉；叶柄长 10 厘米，白绿色。花球半圆形，横径约 20 厘米，白色，结球紧实，单球重 1 500 克。

特　　性｜中晚熟，定植至初收 95 天，延续采收 10~15 天。生长势强，耐寒，耐病毒病、黑腐病、霜霉病。纤维少，商品综合性状好，品质优。每公顷产量 24~36 吨。

栽培要点｜播种期 8—9 月。其余参照广泰 45。

# 青花菜

十字花科芸薹属一、二年生草本植物，学名 Brassica oleracea L. var. italica Plenck.，别名西兰花、绿菜花、木立花椰菜、茎椰菜、意大利芥蓝，染色体数 $2n=2x=18$。

青花菜原产意大利，广州于 20 世纪 30 年代引进栽培，80 年代以后，冬种青花菜生产得到快速发展，目前年种植面积约 1 360 公顷。

**植物学性状**：株型直立或半直立。根系较发达。茎较短缩。叶互生，多为阔卵形至椭圆形，基部具耳状裂片或深裂，叶面平滑，绿色或深绿色，被蜡粉。花球由短缩、肉质主花茎及其上侧花茎和花蕾组成，绿色。总状花序，花冠多为黄色。虫媒花。长角果。种子近圆形，棕褐色或黑褐色，千粒重 4~5 克。

**类型**：品种主要从国内外引进，按花球色泽分有绿花与紫花 2 种类型，其中以绿花类型较为普遍；按成熟期可分为早熟、中熟和晚熟 3 种类型。

1. **早熟类型** 适播期 7—9 月，播种至采收 80~100 天，耐热能力较强，对低温较敏感，苗期遇低温容易先期抽薹，不适宜冬种。品种主要有珠绿等。

2. **中熟类型** 适播期 8—11 月，播种至采收 100~130 天，耐寒能力较强。品种主要有曼陀绿、炎秀、格福等。

3. **晚熟类型** 播种至采收 130 天以上，耐寒能力强，不耐热，完成春化要求较长的低温时间。广州地区一般不适宜种植。

**栽培环境与方法**：属低温长日照蔬菜，喜光照、冷凉、耐寒，花球形成适温 15~18℃。形成花球除了必需的低温感应条件外，还要求长日照条件，有些品种在短日照下不能形成花球。育苗移栽。广州栽培季节一般为 8 月至翌年 4 月，最适播期为 8—10 月。

**收获**：当手感花蕾粒子开始有些松动或花球边缘的花蕾粒子略松散时以及花球表面紧密、平整、无凹凸时为采收适期。选择晴天的清晨或傍晚采收。

**病虫害**：病害主要有黑腐病、软腐病、霜霉病、炭疽病等，虫害主要有黄曲条跳甲、小菜蛾、菜青虫、斜纹夜蛾、甜菜夜蛾、蚜虫等。

**营养及功效**：富含蛋白质、维生素、矿物质、食用纤维素、糖、胡萝卜素、抗坏血酸、花青素和萝卜硫素等。性温，具健胃助消化、抗癌、增强肝脏的解毒能力、提高机体免疫力等功效。

# Broccoli

*Brassica oleracea* L. var. *italica* Plenck., an annual or biennial herb. Family: Cruciferae. Synonym: Sprouting broccoli, Asparagus broccoli, Italy kale. The chromosome number: $2n=2x=18$.

Broccoli was introduced to Guangzhou since the 1930s. The production of broccoli increased gradually since the 1980s. At present, the cultivation area is about 1360 $hm^2$ per annum.

Botanical characters: Plant erect or semi-erect. Vigorous root system. Stem dwarfed. Leaves alternate, broad oval to elliptical, with ear lobed or parted at the leaf base. Leaf surface smooth, green or deeply green in color and wax-covered. Curd green, the edible part, is constituted by shortened, succulent main scape and many lateral scape and flower buds. Raceme, entomophilous flower with yellow corolla. Silique, with approximate round, brown or dark brown seed. The weight per 1000 seeds is 4~5 g.

Types: Broccoli cultivars, which mainly introduced from other regions or abroad, can be classified into green crud and purple crud types according to crud color, and the green curd type is more popular. According to the maturity, it can be classified into early, medium and late maturing types.

1. Early maturing type: It has strong heat-tolerance. Sowing dates: July to September, 80~100 days from sowing to harvest. It is sensitive to low temperature and unsuitable for cultivation in winter. The seedlings are easy to bolt under low temperature. The cultivars include Zhulü, etc.

2. Medium maturing type: It has strong cold-tolerance. Sowing dates: August to November, 100~130 days from sowing to harvest. The cultivars include Mantuolü, Yanxiu, Gefu, etc.

3. Late maturing type: It has strong cold tolerance but no heat tolerance. It needs more than 130 days from sowing to harvest. Longer period at low temperature is required for vernalization. It is unsuitable for cultivation in Guangzhou.

Cultivation environment and methods: Broccoli is photophilous, cold-toleranct. The suitable temperature for crud development is 15~18 ℃. Both low temperature and long day are required for crud formation. In some cultivars, the crud could not develop under short-day condition. Broccoli is usually cultivated from August to April the following year in Guangzhou. Seedling transplanting. The most suitable sowing date is August to October.

Harvest: It's time to harvest when the bud grains begin loose, as well as the crud surface is tight and flat. Early morning or nightfall is the best time for harvest.

Nutrition and efficacy: Broccoli is rich in proteins, vitamins, minerals, edible fibers, sugars, carotene, ascorbic acid, cyanidin and sulforaphen.

## 珠绿

Zhulü broccoli

> 品种来源

广东省良种引进服务公司从日本引进推广的杂种一代。

📍 分布地区 | 全市各区。

特　征 | 株高55.0厘米，开展度90.0厘米，叶数27片。叶片椭圆形，叶长35厘米，宽25厘米，绿色，有蜡粉；叶柄长15.0厘米，白绿色。主花球大，半圆形，横径15~20厘米，绿色，蕾粒较细，结球紧实，单球重300~450克。

特　性 | 早熟，播种至初收90天。生长势强，抗病、抗逆性较强。质清香味甜，纤维少，商品性好，品质优。

栽培要点 | 适播期8月至10月中旬。宜育苗。幼苗2叶1心时及时疏苗，幼苗5~8片真叶或苗龄30~40天时及时定植。株行距（45~50）厘米×（55~60）厘米。一般全期追肥3~4次，前期主要结合浇水追施薄肥，封行时结合中耕除草培土，在行距间开浅沟追肥1次，每667米²施入复合肥20~25千克，现蕾时株距间穴施复合肥和钾肥，叶面喷施硼肥。若收侧球，在采收主球后再及时追肥1次。

## 曼陀绿

Mantuolü broccoli

> 品种来源

美国先正达种子有限公司引进的杂交一代。

📍 分布地区 | 全市各区。

特　征 | 株高55.0厘米，开展度95.0厘米，叶数27片。叶片椭圆形，叶长35厘米，宽25厘米，绿色，有蜡粉；叶柄长15.0厘米，白绿色。主花球大，半圆形，横径15~20厘米，绿色，蕾粒较细，结球紧实，单球重350~500克。

特　性 | 中熟，播种至初收100天。生长势强，耐涝，耐寒，耐病毒病、黑腐病、软腐病、霜霉病。质清香味甜，纤维少，商品性好，品质优。每公顷产量15~22吨。

栽培要点 | 适播期8月下旬至10月。其余参照珠绿。

## 炎秀

Yanxiu broccoli

### 品种来源

高华种子有限公司引进的杂交一代。

📍 分布地区｜全市各区。

特　　征｜株高 60.0 厘米，开展度 100.0 厘米，叶数 27 片。叶片椭圆形，叶长 38 厘米，宽 27 厘米，绿色，有蜡粉；叶柄长 15.0 厘米，白绿色。主花球大，半圆形，横径 15~20 厘米，绿色，蕾粒较细，结球紧实，单球重 350~500 克。

特　　性｜中熟，播种至初收 100~105 天，以采收顶球为主。生长势强，耐涝、耐寒，耐病毒病、黑腐病、软腐病、霜霉病。质清香味甜，纤维少，商品综合性状好，品质优。

栽培要点｜适播期 8 月下旬至 10 月。其余参照珠绿。

## 格福

Gefu broccoli

### 品种来源

广州市兴田种子有限公司引进推广的杂种一代。

📍 分布地区｜全市各区。

特　　征｜株高 60.0 厘米，开展度 100.0 厘米，叶数 28 片。叶片椭圆形，叶长 35 厘米，宽 25 厘米，绿色，有蜡粉；叶柄长 15.0 厘米，白绿色。主花球大，高圆形，横径 15~18 厘米，深绿色，蕾粒细，结球紧实，单球重 350~500 克。

特　　性｜中晚熟，秋种定植后 75~80 天采收。生长势强，耐寒，耐病毒病、黑腐病、软腐病、霜霉病。质清香味甜，纤维少，商品性好，品质优。

栽培要点｜适播期 8 月底至 10 月中旬。其余参照珠绿。

# 西兰薹

十字花科芸薹属一、二年生草本植物,学名 Brassica oleracea var. italica×alboglabra,别名小西兰花、青花笋、芦笋西兰薹,染色体数 $2n=2x=18$。

西兰薹是由西兰花与芥蓝杂交选育而成的一种新型蔬菜,最先由美国育成。广州最早在2006年引进,近年发展较快,目前年种植面积约100公顷。

**植物学性状**:根系较发达。茎较短缩,侧芽萌发力强。叶互生,多为阔卵形至椭圆形,基部具耳状裂片或深裂,叶面平滑或微皱,绿色或深绿色,被蜡粉;主要以侧枝抽生的肥嫩花茎(即菜薹)供食。菜薹色绿翠美,肉质脆嫩,风味香甜。菜薹收获之后,可长出下一级侧薹,一般可抽生3~4级菜薹。总状花序,花冠黄色或白色。虫媒花。长角果。种子近圆形,棕褐色或黑褐色。千粒重4.0~5克。

**类型**:按成熟期可分为早熟和中熟2种类型。早熟类型:定植至采收54~60天,如SBB-025、翠钰1号等;中熟类型:定植至采收60~90天,如芊秀、秀丽、翠钰2号等。

**栽培环境与方法**:西兰薹喜光照、冷凉,较耐寒,花球形成适温15~22℃。广州适播期为7—11月,育苗移栽,10月至翌年4月收获。

**收获**:一般当侧薹长至20~30厘米、花球稍松散时,在第2~4片叶的上方收割。

**病虫害**:病害主要有炭疽病、软腐病、霜霉病、黑腐病等,虫害主要有黄曲条跳甲、小菜蛾、菜青虫、斜纹夜蛾、甜菜夜蛾、蚜虫等。

**营养及功效**:富含蛋白质、维生素、食用纤维素、胡萝卜素、花青素、萝卜硫素及多种矿物质等。性温,具健胃助消化、抗癌、增强肝脏的解毒能力、提高机体免疫力等功效。

# Broccolini

*Brassica oleracea* var. *italica* × *alboglabra*, an annual or biennial herb. Family: Cruciferae. Synonym: Small broccoli, Sweet baby broccoli. The chromosome number: $2n=2x=18$.

Broccolini, a hybridized vegetable between broccoli and Chinese kale, was introduced to Guangzhou at 2006. At present, the cultivation area is about 100 hm$^2$.

Botanical characters: Vigorous fibrous root system. Stem dwarfed. Leaves alternate, broad oval to elliptical, with ear lobed or parted at the leaf base. Leaf surface smooth or slight wrinkled, green or deeply green in color and wax-covered. Lateral flower stalks, the edible part in green, tender, fragrant and sweet. After the first flower stalk is harvested, the next branch develop. There are usually three to four branch of flower stalk. Raceme entomophilous flower, with yellow or white corolla. Silique, with approximate round seed, brown or dark brown. The weight per 1000 seeds is 4.0~5.0 g.

Types: According to plant maturity, broccolini can be classified into early and medium maturing types. Early maturing type: 54~60 days from transplanting to harvest. The cultivars include SBB-025, Cuiyu No.1, etc. Medium maturing type: 60~90 days from transplanting to harvest. The cultivars include Qianxiu, Xiuli, Cuiyu No.2, etc.

Cultivation environment and methods: Broccolini is photophilous and cold-tolerant. The suitable temperature for crud development is 15~22 ℃. The most suitable sowing date is July to November. Seedling transplanting. The harvest date is from October to April the following year.

Harvest: The flower stalk is harvested when the lateral flower stalks are 20~30 cm in length, and the cruds are slightly loose.

Nutrition and efficacy: Broccolinis is rich in proteins, vitamins, edible fibers, carotene, cyanidin, sulforaphen and minerals, etc.

# 芊秀

Qianxiu broccolini

**品种来源**

广州市番禺区绿色科技发展有限公司2006年从日本引进的杂交一代。

**分布地区**｜全市各区。

特　　征｜株高70.0厘米，开展度90.0厘米，叶数27片左右。叶片椭圆形，叶长35厘米，宽25厘米，绿色，有蜡粉；叶柄长15.0厘米，白绿色。薹长12~18厘米，粗1.5~2.0厘米，单薹重35~50克，绿色。

特　　性｜中熟，定植至初收60~70天，以采收侧花薹为主，采收期60~85天。生长势强，分枝力强，侧薹多。耐涝，耐寒，耐病毒病、黑腐病、软腐病、霜霉病。肉质脆嫩，风味香甜，纤维少，商品性好，品质优。每公顷产量22~35吨。

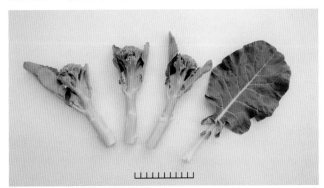

栽培要点｜播种期7月中旬至11月，宜育苗，幼苗5~8片真叶或苗龄30~40天时定植。株行距（40~45）厘米×50厘米，每667米$^2$用种量约15克，定植2 500~3 000株。施足基肥，一般全期需追肥4次：第1次追肥在植株开始迅速生长时；第2次在封垄时结合中耕除草培土，在行距间开浅沟施入复合肥；第3次在一级侧薹始收期，在株间开穴施入复合肥和钾肥；第4次在二级侧薹采收期间施入复合肥和钾肥，以后相隔20~25天追肥一次。淋水要充足。合理选留侧芽，主茎现蕾时及时打顶芽，以促进侧枝早日抽生。每个侧枝采收菜薹后，还应保留2~4片叶（芽）。当侧薹长至20~30厘米时，在第24片叶的上方收割菜薹。注意打顶和收割的刀口斜切，预防积水引起腐烂。

## 秀丽

Xiuli broccolini

> 品种来源

广州华绿种子有限公司2010年从日本引进的杂交一代。

📍 分布地区 | 全市各区。

特　　征 | 株高65.0厘米,开展度85.0厘米。叶数27片左右。叶片椭圆形,叶长35厘米,宽25厘米,绿色,有蜡粉;叶柄长14.0厘米,白绿色。薹长12~18厘米,粗1.5~2.0厘米,单薹重35~50克,绿色。

特　　性 | 中熟,定植至初收60~70天,以采收侧花薹为主,采收期60~85天。生长势强,分枝力强,侧薹多。耐涝,耐寒,抗逆性强,耐病毒病、黑腐病、软腐病、霜霉病。肉质脆嫩,风味香甜,纤维少,商品性好,品质优。每公顷产量22~35吨。

栽培要点 | 参照芥秀。

## 翠钰2号

Cuiyu No.2 broccolini

> 品种来源

广州市农业科学研究院2012年育成的杂交一代。

📍 分布地区 | 全市各区。

特　　征 | 株高60厘米,开展度80.0厘米。叶数25片。叶片椭圆形,叶长30厘米,宽24厘米,绿色,有蜡粉;叶柄长17.0厘米,白绿色。薹长15~20厘米,薹粗1.5~2.0厘米,单薹重40~55克,绿色。

特　　性 | 早中熟,定植至初收50~55天,以采收侧花薹为主,采收期65~90天。生长势强,分枝力强,侧薹多。耐涝,耐寒,耐病毒病、黑腐病、软腐病、霜霉病。菜薹含抗癌物质萝卜硫素,水焯后色泽翠绿,肉质脆嫩,风味香甜,纤维少,商品性好,品质优。每公顷产量24~37吨。

栽培要点 | 参照芥秀。

# 结球甘蓝

十字花科芸薹属二年生草本植物，学名 *Brassica oleracea* L. var.*capitata* L.，别名椰菜、卷心菜、洋白菜、包菜、圆白菜、包心菜、莲花白等，染色体数 $2n=2x=18$。

结球甘蓝起源于地中海沿岸，自16世纪开始从俄罗斯或东南亚传入中国。广州栽培200多年，目前年栽培面积约3 000公顷。

**植物学性状**：茎短缩，叶丛着生于短缩茎上。叶片椭圆形、倒卵圆形或近三角形，绿色、深绿色或紫色，被蜡粉；叶柄绿白色或紫红色。叶球心形、圆形或扁圆形，黄白色、绿白色或紫红色。幼苗在0~10℃通过春化，在长日照和适温下抽薹、开花、结果。复总状花序。异花授粉。长角果。种子圆球形，红褐色或黑褐色，千粒重4克左右。

**类型**：依植株色泽可分为绿色和紫色两种；依叶球形状和成熟期的迟早，可分为以下3种类型：

1. 尖球类型　叶球顶部尖，近似心脏形，多为早熟和早中熟品种，定植至采收50~70天。

2. 圆球类型　叶球圆球形，多为早熟和早中熟品种，定植至采收50~70天。

3. 扁圆球类型　叶球扁圆形，多为中熟或晚熟品种，定植至采收70~100天。

**栽培环境与方法**：喜温和湿润、充足的光照。较耐寒，生长适温15~20℃。叶球膨大期如遇30℃以上高温，肉质易纤维化。广州地区秋、冬季栽培为主。育苗移栽。叶球开始膨大时加大追肥量和均匀浇水，接近采收时，则要停止浇水。

**收获**：叶球充分膨大变紧实后收获。

**病虫害**：病害主要有黑腐病、软腐病、细菌性黑斑病等，虫害主要有小菜蛾、菜粉蝶、斜纹夜蛾等。

**营养及功效**：富含叶酸、维生素U、维生素C、粗纤维等。味甘，性平，具防衰老、抗氧化、预防感冒和便秘等功效。

# Cabbage

*Brassica oleracea* L. var. *capitata* L., a biennial herb. Family: Cruciferae. Synonym: White cabbage. The chromosome number: $2n=2x=18$.

Cabbage was introduced to China from Russia or Southeast Asia in 16th century. It has been cultivated for more than 200 years in Guangzhou. At present, the cultivation area is about 3000 $hm^2$ per annum.

Botanical characters: Stem dwarfed, with clustered leaves. Leaf, with green-white or purple-red petiole, elliptical, oval or approximate triangular in shape, green, dark green or purple in color and wax-covered. Leaf head, heart, round or flat round in shape, and yellow-white, green-white or purple-red in color. Raceme, cross-pollination. Silique, with round seed, red brown or dark brown. The weight per 1000 seeds is about 4.0 g.

Types: According to the color, cabbage can be classified into green and purple types. According to the shape of leaf head and maturity, cabbage can be classified into cuspidal, round and flat round types.

1. Cuspidal leaf head type: Leaf head is approximate heart in shape. Most of the cultivars are early or early-medium maturity. The growth period is 50~70 days from transplanting to harvest.

2. Round leaf head type: Leaf head is round in shape. Most of the cultivars are early or early-medium maturity. The growth period is 50~70 days from transplanting to harvest.

3. Flat round leaf head type: Leaf head is flat round in shape. Most of the cultivars are medium or late maturity. The growth period is 70~100 days from transplanting to harvest.

Cultivation environment and methods: Cabbage, the thermophilic, hydrophilous and photophilous vegetable, has cold tolerance. The suitable temperature for plant growth is 15~20℃. The fleshy leaf would be fibrillate above 30℃ during leaf head growing. It is usually cultivated in autumn and winter in Guangzhou. Seedling transplanting. Large fertilization and uniform irrigation should be applied during the leaf head growth stage. Irrigation should be stopped before harvest.

Harvest: The leaf head is harvested when it is fully and tightly.

Nutrition and efficacy: Cabbage is rich in proteins, vitamins, and edible fibers.

## 京丰 1 号

Jinfeng No.1 cabbage

**品种来源**

中国农业科学院蔬菜研究所和北京市农业科学院 1973 年育成的杂种一代。

**分布地区** | 全市各区。

特　　征 | 收获期植株开展度 40~50 厘米，外叶 12~14 片。成叶近圆形，叶色绿，背面灰绿色，蜡粉中等。叶球扁圆形。单球重 2.0 千克左右。

特　　性 | 扁圆球类型。中晚熟，定植到初收 70~80 天。田间表现耐热，高温下结球较紧实，不易开裂，抗病性强，冬性较强，不易抽薹。球叶纤维较粗。每公顷产量 60~90 吨。

栽培要点 | 按 1.8 米左右包沟起畦种植，适施基肥，加强前期管理。8—10 月播种育苗，6~8 片真叶时定植，株行距 35 厘米 ×40 厘米，每 667 米² 种植 2 500~2 800 株。

# KK 甘蓝

KK cabbage

### 品种来源

引自日本。

📍 分布地区 | 全市各区。

特　　征 | 收获期植株开展度 40~50 厘米。外叶深绿色，蜡质少，球叶淡绿色。单球重 1.6~1.8 千克。
特　　性 | 扁圆球类型。中晚熟，定植到初收 55~65 天。田间表现耐热，高温下结球较紧实，不易开裂，抗黑腐病和软腐病较强，冬性较强，不易抽薹。球叶纤维较粗。每公顷产量 60~90 吨。
栽培要点 | 7 月中旬至 10 月播种育苗。其余参照京丰 1 号。

# 元绿甘蓝

Yuanlü cabbage

### 品种来源

广东省良种引进服务公司引进推广。

📍 分布地区 | 全市各区。

特　　征 | 收获期植株开展度 35~40 厘米。外叶深绿，球叶有光泽，蜡粉少。单球重约 1 千克。
特　　性 | 圆球类型。中早熟，从定植到初收 60 天左右。田间表现抗逆性强。叶球紧实，耐裂，耐贮运。球叶质地脆嫩，风味品质优良。每公顷产量 60~90 吨。
栽培要点 | 8 月下旬至 12 月播种育苗，株行距 25 厘米 ×35 厘米，每 667 米$^2$ 种植 4 500~5 000 株。其余参照京丰 1 号。

## 中甘 11 号

Zhonggan No.11 cabbage

> 品种来源

中国农业科学院蔬菜花卉研究所育成。

📍 **分布地区** | 全市各区。

特　　征 | 植株开展度 35~40 厘米。幼苗期真叶卵圆形。叶球近圆形，深绿色，蜡粉中等，球内中心柱长 6~7 厘米，单球重约 1 千克。
特　　性 | 早中熟，定植至初收 60 天左右。球叶质地脆嫩，风味品质优良。抗寒性较强，不容易先期抽薹，抗干烧心病。每公顷产量 60~90 吨。
栽培要点 | 8 月下旬至 12 月播种育苗，株行距 30 厘米 ×35 厘米，每 667 米$^2$ 种植 4 500~5 000 株。采收期遇雨易裂球。结球紧实后应及时收获。其余参照京丰 1 号。

## 紫甘 1 号

Zigan No.1 cabbage

> 品种来源

北京市农林科学院蔬菜研究中心育成。

📍 **分布地区** | 全市各区稀有种植。

特　　征 | 植株开展度 35~40 厘米。叶球紫色，紧实。
特　　性 | 圆球类型，中早熟，从定植到初收需 60 天左右。单球重约 1.5 千克。每公顷产量 60~90 吨。

栽培要点 | 起畦种植，1.8 米包沟，施足基肥，加强前期管理。广州地区 8—9 月播种育苗，6~7 片真叶时定植，株行距 30 厘米 ×35 厘米，每 667 米$^2$ 种植 3 000~3 500 株。

# 抱子甘蓝 Brussels sprout

十字花科芸薹属甘蓝种二年生草本植物，为甘蓝种中腋芽能形成小叶球的变种，学名 *Brassica oleracea* L. var. *gemmifera* Zenk.，别名芽甘蓝、子持甘蓝、椰菜仔，染色体数 $2n=2x=18$。

抱子甘蓝原产于地中海沿岸，我国于 20 世纪 80 年代开始引入栽培。

**植物学性状**：根系发达。植株高大，茎直立。顶芽不断生长形成同化叶，叶深绿色，叶缘上卷，叶面皱缩，叶柄较长。在茎周围每个叶腋处自下而上不断长出外部深绿色、紧实、横径 2~4 厘米的小叶球，犹如子附母怀，故称"抱子甘蓝"，该小叶球为食用器官。总状花序，花冠多为黄色。异花授粉。长角果。种子球形，灰棕色，千粒重约 2.8 克。

**类型**：按植株高矮可分为高生种和矮生种两种类型，一般矮生种品质较好；按成熟期可分为早熟、中熟和晚熟 3 种类型。

**栽培环境与方法**：属长日照蔬菜，喜冷凉气候，耐寒，不耐热，对光照要求不严格，高温和强光不利于芽球的形成。生长发育适温为 18~22℃，结球期适温为 12~15℃。广州最适播期为 8—9 月，11 月至翌年 3 月收获，每株可收 40~100 个，每 667 米² 产量 1 000~1 200 千克。

**收获**：叶球亮绿、紧实时可采收，一般自下而上依次采收，用刀沿茎将小叶球割下。

**病虫害**：病害主要有黑腐病、软腐病、立枯病、炭疽病、霜霉病、花叶病等，虫害主要有蚜虫、黄曲条跳甲、小菜蛾、菜青虫、斜纹夜蛾、甜菜夜蛾等。

**营养及功效**：富含蛋白质、维生素 C 和微量元素硒。味甘，性平，具清热止痛、强壮筋骨、健肤美容等功效。

*Brassica oleracea* L. var. *gemmifera* Zenk., a bennial herb which is the varieties of cabbage with small leaf heads formatted by axillary buds. Family: Cruciferae. Synonym: Sprouted cabbage, Young cabbage. The chromosome number: $2n=2x=18$.

Brussels sprout was introduced to Guangzhou in the 1980s.

**Botanical characters**: Vigorous root system. Plant large, with erect stem. Top buds grow gradually and develop to assimilated leaves. Leaf, deeply green in color and ladled or round in shape, leaf margin curl upward, leaf surface wrinkled, with long petiole. Small leaf heads, the edible parts, borne on leaf axiles from bottom to top, tight and deeply green in color and 2~4 cm in diameter. Raceme, cross-pollination, usually with yellow corolla. Silique, with round seed, grey brown. The weight per 1000 seeds is about 2.8 g.

**Types**: According to the plant height, brussels sprout can be classified into short and tall types, and the short type has a higher quality. According to the maturity, it can be classified into early, medium and late maturing types.

**Cultivation environment and methods**: Brussels sprout, the thermophilic and long-day vegetable, is resistant to cold but not heat. It's not beneficial to leaf head formation under high temperature and high light condition. The suitable temperature is 18~22 ℃ for plant growth, and 12~15 ℃ for leaf head formation. Sowing from August to September. Harvesting from November to March the following year. There are 40~100 leaf heads grown on the same plant, and 15000~18000 kg per ha.

**Harvest**: Brussels sprout is harvested from bottom to top when the leaf head is tight and bright green. The leaf head would be easily cracked, and the quality decreased if delays harvesting.

**Nutrition and efficacy**: Brussels sprout is rich in proteins, vitamins, edible fibers.

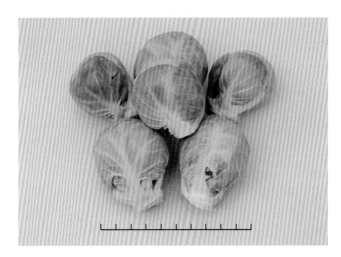

# 羽衣甘蓝 Kale

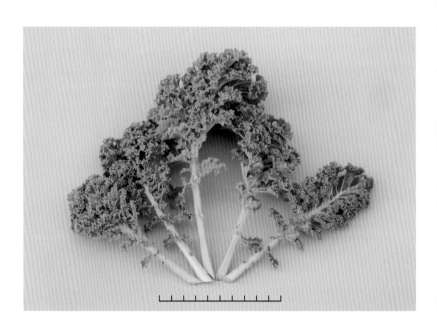

十字花科芸薹属二年生或多年生草本植物，学名 *Brassica oleracea* L. var. *acephala* DC.，别名菜用羽衣甘蓝、叶牡丹、花包菜，染色体数 $2n=2x=18$。

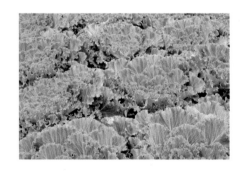

**植物学性状**：茎短而坚硬。叶呈长椭圆形，叶缘羽状深裂。菜用类型叶色有灰绿、浅黄绿、紫红（心叶）等色。复总状花序，花黄色。种子圆球形，黄褐色至棕褐色。

**栽培环境与方法**：喜冷凉气候，生长适温 20~25℃。广州地区的生长期为 9 月至翌年 4 月。

**收获**：采收时可以连续不断地剥取植株不断长出的嫩叶。

**营养及功效**：羽衣甘蓝富含 $\beta$- 胡萝卜素、维生素 K、维生素 C 和钙。

*Brassica oleracea* L. var. *acephala* DC., a biennial or perennial herb. Family: Cruciferae. Synonym: Curly kale, Borecole. The chromosome number: $2n=2x=18$.

**Botanical characters:** Stem dwarfed and hard. Leaf oval in shape, with the pinnate margine and parted, grey green, light yellow green, or purple red in color. Compound raceme, with yellow corolla. Silique, with round seed, yellow brown to brown.

**Cultivation environment and methods:** Kale favours cold weather, with the suitable temperature is 20~25℃. The cultivation season in Guangzhou is September to April the following year.

**Harvest:** The tender leaf is continous harvested as the edible part.

**Nutrition and efficacy:** Kale is rich in beta carotene, vitamin K, vitamin C, and calcium.

# 三、芥菜类
# MUSTARD VEGETABLES

- 叶用芥菜
- 结球芥菜
- 根用芥菜

# 叶用芥菜 Leaf mustard

十字花科芸薹属一年生或二年生草本植物，学名 *Brassica juncea* (L.) Coss. var. *foliosa* Bailey，别名芥菜、青菜、苦菜、春菜等，染色体数 $2n=4x=36$。

叶用芥菜起源于中国，广州地区已有千年以上栽培历史，目前年种植面积约 3 000 公顷。

**植物学性状**：直根系。茎短缩。叶长椭圆形或长倒卵形，叶面平滑或皱缩，叶缘有锯齿或深裂，幼苗期有粗毛，中肋宽，扁平或近圆形；总状花序，花冠黄色。异花授粉。种子圆形或椭圆形，红褐色或褐色，千粒重 1.8~2.0 克。

**类型**：20 世纪 70 年代以前有三月青、南风芥、油叶、葫芦芥、高脚芥、马登芥、迟芥菜、黄尾及乌尾大芥菜 9 个品种，70 年代后有些品种被淘汰，目前主要有特选客家芥、竹芥、水东芥菜等。

**栽培环境与方法**：喜冷凉、湿润的环境。如果仅以小株食用为主的芥菜，在广州周年均可播种，如南风芥多用直播。大叶芥如客家芥、竹芥、水东芥等，可在 10—11 月播种，12 月至翌年 2 月收获，以育苗移栽为主。

**收获**：一般在播后 30~60 天、植株抽薹前采收。

**病虫害**：病害主要有病毒病、软腐病、菌核病、霜霉病等，虫害主要有黄曲条跳甲、小菜蛾、菜青虫、斜纹夜蛾、菜蚜等。

**营养及功效**：富含硫葡萄糖苷、维生素、磷、钙等。性温，味辛，具宣肺豁痰、利气温中、解毒消肿、开胃消食、温中利气、明目利膈等功效。

*Brassica juncea* (L.) Coss. var. *foliosa* Bailey, an annual or biennial herb. Family: Cruciferae. Synonym: Indian Mustard, Mutard green, Chinese mustard, White mustard. The chromosome number: $2n=4x=36$.

The leaf mustard is originated in China and has been grown for more than 1000 years in Guangzhou. At present, the cultivated area is about 3000 $hm^2$ per annum.

**Botanical characters**: Leaf mustard, with tap root system. Stem dwarfed. Leaf oblong or long obvate, leaf surface smooth or wrinkle, leaf margin serrate or parted, with thick hairs when young, midrid broad, flat or nearly round. Inflorescene: a compound raceme with yellow corolla, cross-pollination. Seeds round or oval, red-brown or brown. The weight per 1000 seeds is 1.8~2.0 g.

**Types**: There were nine cultivars before the 1970s, namely March Green, Nanfeng, Oiled-leaf, Squash, Long-petiole, Madeng, Late, Yellow leaf and Black leaf. Some cultivars have been eliminated after the 1970s. At present, there are three cultivars, Hakka mustard, Zhu mustard and Shuidong mustard.

**Cultivation environment and methods**: The leaf mustard grows in a cold and humid environment. The small mustard cultivar Nanfeng could be sown throughout the year in Guangzhou. The Large mustard cultivars, i. e. Hakka mustard, Zhu mustard and Shuidong mustard, are cultivated mainly with seedling transplanting. Sowing dates: October to November, Harvesting dates: December to February the following year.

**Harvest**: Leaf mustards are harvested before plant bolting in 30~60 days after sowing or transplanting.

**Nutrition and efficacy**: Leaf mustard is rich in glucosinolates, vitamins, phosphorus, calcium, etc.

## 南风芥

Nanfeng mustard

**品种来源**

引自佛山。

**分布地区** | 花都、白云、增城、从化等区。

特　　征 | 株高29~42厘米，开展度36~45厘米。叶片长椭圆形，长28.2厘米，宽19.4厘米，浅绿色，叶面有光泽，叶缘锯齿状；叶柄扁平，长7厘米，宽3.1厘米，厚0.5厘米，浅白色，背部淡绿色。单株重100~160克。

特　　性 | 早熟，播种至初收40~60天。耐热，耐风雨。纤维少，质脆嫩，味微苦，品质好。每公顷产量12~16吨。

栽培要点 | 播种期4—9月，适播期5—8月。直播，间苗1~2次，株行距9厘米×12厘米。高温期间用遮阳网覆盖栽培，适施基肥，勤追薄肥。注意及时间苗，防治跳甲、菜青虫为害。收获期5—11月。

## 特选客家芥

Texuan hakka mustard

**品种来源**

广州市农业科学研究院从地方品种中提纯选育而成。

**分布地区** | 全市各区。

特　　征 | 株高70.5~80.5厘米，开展度85厘米。叶长椭圆形，长78.5厘米，宽35.7厘米，绿色，有光泽，叶缘锯齿状；叶柄扁狭，长8.2厘米，宽7.2厘米，厚0.8厘米，淡绿色。单株重1 500~2 000克。

特　　性 | 早熟，播种至初收40~60天。耐热，耐风雨。纤维少，质脆嫩，味微苦，品质好。每公顷产量25~30吨。

栽培要点 | 播种期4—9月。直播，间苗1~2次。高温期间用遮阳网覆盖栽培，适施基肥，勤追薄肥。注意及时间苗，防治黄曲条跳甲、菜青虫为害。

## 竹芥

Zhu mustard

### 品种来源

蔡兴利菜种行有限公司。

📍 **分布地区** | 全市各区。由于良好的栽培条件和技术，从化吕田采用竹芥311在冬季生产出优质产品，称为"吕田大芥菜"。

**特　　征** | 株高75~88厘米，开展度60~70厘米。叶宽椭圆形，长78~85厘米，宽39~45厘米，浅绿色，叶面平滑，叶脉明显，叶缘微波状，基部锯齿状；叶柄长23厘米，宽6.8厘米，厚0.7厘米，中肋横切面半月形，背面具有纵槽线，绿白色。单株重2 000~2 500克。

**特　　性** | 中晚熟，播种至初收90~110天。耐风雨，耐热。无苦味，肉质嫩，纤维少，品质优。每公顷产量40~50吨。

**栽培要点** | 周年均可播种，适播期7月下旬至8月。其余参照特选客家芥。

## 水东芥菜

Shuidong mustard

### 品种来源

引自电白。

📍 **分布地区** | 花都、白云、增城、从化等区。

**特　　征** | 株高50~63厘米，开展度54~61厘米。叶片宽椭圆形，长54厘米，宽30.4厘米，青绿色，叶面平滑，叶缘锯齿状，心叶向内卷；叶柄短，长7厘米，宽3.6厘米，厚1.2厘米，浅白色，背部青绿色。单株重1 000~1 200克。

**特　　性** | 早熟，播种至初收45~60天。植株脆嫩，无渣，无苦味，菜形美观，色油绿，有光泽，商品性好，品质优。抗性强，耐热，耐雨水，不易抽薹。每公顷产量35~45吨。

**栽培要点** | 周年均可播种。其余参照特选客家芥。

# 结球芥菜 Head mustard

十字花科芸薹属二年生草本植物，学名 *Brassica juncea* (L.) Coss.var. *capitata* Bailey，别名包心芥、潮州大芥菜，染色体数 $2n=4x=36$。

直根系。茎短缩。基叶阔矩圆形，平展生长，叶面皱缩，叶缘波状，浅绿色，幼苗期有粗毛，叶柄短，中肋宽厚，具沟，绿白色。心叶结球，叶球扁圆形、圆形或尖圆形。复总状花序，花冠浅黄色。种子圆形或椭圆形，红褐色或褐色，千粒重1.5~1.6克。

广州已有千年以上栽培历史，但种植面积不大，主要分布在白云、增城、番禺及从化等地。20世纪70年代以前栽培有鸡心和哥苈大芥菜两个品种，80年代后引入大坪埔品种。除作鲜菜外，主要腌渍咸酸菜。

*Brassica juncea* (L.) Coss. var. *capitata* Bailey, a biennial herb. Family: Cruciferae. Synonym: Large Mustsrd. The chromosome number: $2n=4x=36$.

Head mustard, with taproot system. Stem dwarfed. Basal leaf broad-long round and spreading, leaf surface wrinkled, leaf margin undulate, with thick hairs and slight green. Petiole short, midrid broad and thick, with grooves and green-white in color. Inner leaves forming a flat found, round or acuminate-round head. Inflorescence: a compound raceme, light yellow corolla. Seed round or oval, brown red or brown. The weight per 1000 seeds is 1.5~1.6 g.

Head mustard has been in cultivation for more than 1000 years in Guangzhou. But its distribution is limited, mainly in Baiyun, Zengcheng, Panyu and Conghua district. Two cultivars Chicken-Heart and Geli head mustard existed before the 1970s. Cultivar Dapingpu was introduced after the 1980s. It is used to process into sour vegetable or as fresh vegetable for cooking.

## 大坪埔 11 号包心芥菜

Dapinpu No.11 head mustard

**品种来源** | 汕头市白沙蔬菜原种研究所育成。

**分布地区** | 南沙、花都、增城、从化等区。

**特　征** | 株高47厘米，开展度65厘米。植株半披生，叶葵扇形，叶色浅绿，叶长44厘米，宽41厘米；叶柄厚阔，肉厚。叶球圆形、大而结实，高19厘米，横径12.7厘米。单球重1 000~1 500克。

**特　性** | 中迟熟，定植60天后可收获。侧芽萌发力弱，根系发达，耐抽薹，抗病毒病。口感清脆，品质优。每公顷产量40~50吨。

**栽培要点** | 播种期7—11月，适播期8月下旬至9月。育苗移栽，每667米$^2$用种量100克，苗期25~30天，11月至翌年1月采收。适施基肥，追肥4~5次。保持土壤湿润，及时防治病虫害。

# 根用芥菜 Root mustard

十字花科芸薹属二年生草本植物，学名 *Brassica juncea* (L.) Coss.var. *megarrhiza* Tsen et Lee，别名大头菜、冲菜，染色体数 $2n=4x=36$。

直根系，直根膨大形成肉质根。茎短缩，叶片簇生于短缩茎上，长倒卵形，绿色，叶缘缺刻或羽状深裂，叶面和背脉具稀疏刺毛。肉质根扁圆形、圆锥形或短圆柱形，皮厚，肉质致密，有芥辣味。复总状花序，花黄色。种子近圆形、红褐色，千粒重 1.5~1.6 克。

唐代已有广州栽培根用芥菜的记载，品种多从省外引进，20 世纪 70 年代以前有细苗、粗苗、乌苗及黄苗（荷塘冲菜）等 4 个品种，其中乌苗品种因种源及生产需求被淘汰。目前栽培面积不大，白云区、番禺区、增城区以栽培光头芥菜为主。一般秋季播种，冬春收获，主要用作腌渍咸菜。

*Brassica juncea* (L.) Coss. var. *megarrhiza* Tsen et Lee, a biennial herb. Family: Cruciferae. The chromosome number: $2n=4x=36$.

Root mustard, has a succulent taproot. Stem dwarfed. Leaves clustered on dwarf stem, long obovate and green in color. Leaf margin lobbed or pinnately-parted, leaf surface and back veins with sparse stiff hairs. The succulent root flat round, coniform or short cylindrical, firm textured, thick shin, with a strong pungent odor. Inflorescence: a compound raceme, yellow corolla. Seed nearly round, brown red. The weight per 1000 seeds is 1.5~1.6 g.

Root mustard has been in cultivation for more than 1000 years in Guangzhou. The cultivars in Guangzhou were mainly introduced from the other regions. 4 cultivars: Small, Thick, Black and Yellow leaf root mustards were cultivated before the 1970s. The Black leaf was eliminated due to the changes of germ resource and producing need. Its distribution is limited, mainly in Baiyun, Panyu and Zengcheng district. Sowing season: Autumn. Harvesting season: winter and spring the following year. It is used to process into sour vegetables.

## 光头芥菜

Guangtou root mustard

**品种来源** | 从日本引进。

**分布地区** | 白云、番禺、增城等区。

**特　征** | 植株直立，株高 52~64 厘米，开展度 45~55 厘米。叶片簇生，倒卵圆形，长 60 厘米，宽 26 厘米，深绿色。叶面皱缩，叶缘波状缺刻，基部深裂；叶柄青绿色，长 10~14 厘米。肉质根圆锥形，高 10~14 厘米，横径 8~10 厘米，皮淡绿色，土层部为白色，表皮光滑，两侧有须根。单个重 400~500 克。

**特　性** | 生长期 80~90 天，耐寒。纤维少，组织致密，质嫩，有辣味，品质好。用于鲜食或加工腌渍。每公顷产量 25~30 吨。

**栽培要点** | 播种期 9—10 月。苗期 25~30 天，具有 4~5 片真叶、根部稍膨大时即可定植，移栽的株行距 20 厘米 ×40 厘米。每公顷施腐熟基肥 22 吨，追肥 4~5 次。保持土壤湿润，及时防治病虫害。收获期 12 月上旬至翌年 1 月上旬。

# 四、绿叶菜类
## GREEN VEGETABLES

- 蕹菜
- 莴苣
- 苦荬菜
- 苦苣
- 菠菜
- 茼蒿
- 芹菜

- 苋菜
- 落葵
- 叶菾菜
- 枸杞
- 千宝菜
- 银丝菜
- 芝麻菜

- 番杏
- 冬寒菜
- 香麻叶
- 益母草
- 藤三七
- 菊花脑
- 一点红

- 马齿苋
- 鱼腥草
- 紫背菜
- 珍珠菜
- 人参菜
- 无瓣簿菜
- 刺芫荽

- 车前草
- 野苋
- 番薯叶
- 观音菜

# 蕹菜

旋花科番薯属一年生草本植物，学名 *Ipomoea aquatica* Forsk.，别名通菜、空心菜，染色体数 $2n=2x=20$。

蕹菜原产中国，公元 3 世纪已普遍种植。广州市各区广泛种植，年栽培面积 3 000 公顷左右。

植物学性状：根系浅，易生不定根。茎扁圆形或近圆形，中空，绿色或绿白色。叶互生，卵形或披针形，全缘，绿色，具长柄。聚伞花序，腋生，花冠漏斗形，白色或浅紫色。蒴果，卵形，含种子 2~4 粒。种子近圆形，皮厚，黑褐色。

类型：依其繁殖方法分为种子繁殖和无性繁殖 2 种类型。种子繁殖类型茎叶粗大，以水生为主，有青梗大叶、白梗大叶等品种；无性繁殖类型茎叶细小，以旱生为主，在广州不开花结籽，用扦插或分株繁殖，较耐寒，早熟，有丝蕹品种。

栽培环境与方法：栽培方法有旱地栽培、浅水栽培和浮水栽培，以浅水栽培为主。

采收：当嫩梢长至 25~30 厘米时即可采摘上市。

病虫害：病害主要有根腐病、白锈病、轮斑病等，虫害主要有螨类、红蜘蛛、菜蛾、斜纹夜蛾等。

营养及功效：富含蛋白质、脂肪、碳水化合物、胡萝卜素、维生素、微量元素、粗纤维素等。味甘，性平，具凉血止血、清热利湿等功效。

# Water spinach

## 白梗柳叶

### White petiole willow leaf water spinach

*Ipomoea aquatica* Forsk., an annual herb. Family: Convolvulaceae. Synonym: Kang kong, Swamp cabbage. The chromosome number: $2n=2x=20$.

Water spinach, originated in China, had been cultivated since the third century. And it is wildly cultivated in Guangzhou. At present, the cultivation area is about 3000 $hm^2$ per annum.

Botanical Characters: Shallow root system, it root adventitiously easily. Stem hollow, flat round or approximate round, green or green white in color. Leaves alternate, ovate or lanceolate in shape and green in color, with entire margin and long petiole. Cymes, axillary, with funnel-shaped corollas, white or light purple in color. Capsule, ovate, with two to four seeds. Seed, approximate round and black brown, with thick seed coat.

Types: According to the propagating method, it can be classified into seed and vegetative propagation types. Seed propagation type: Both stem and leaf are large, and primarily aquatic. The cultivars include green-petiole large-leaf, white-petiole large-leaf water spinach etc. Vegetative propagation type: Both stem and leaves are small, and primarily terrestrial. It is propagated with cutting or branched plant, could not bloom in Guangzhou. The cultivars have cold tolerance and early maturity, include filiform-leaf water spinach.

Cultivation environment and methods: Water spinach can be cultivated in dry land, shallow water or floating, primarily in shallow water.

Harvest: Water spinach can be harvested When the shoots grew to 25~30 cm.

Nutrition and efficacy: Water spinach is rich in proteins, fats, carbohydrates, carotene, vitamins, minerals, crude fibres, etc.

> 品种来源

地方品种。

📍 分布地区｜全市各区。

特　　征｜植株直立，株高40厘米，开展度25厘米。茎细长，横径0.8厘米，白色，节间长4厘米。叶柳叶形，长12厘米，宽1.5~2厘米，浅绿色；叶柄长6厘米，白色。单株重12克。

特　　性｜播种至初收30~40天，延续采收期长。生长旺盛，分枝少。耐热。纤维少，口感爽脆。每公顷全期产量约70吨。

栽培要点｜适播期2—3月。水田育苗，株行距15厘米×15厘米。耐肥，宜勤施薄肥。收获期4—6月，每隔25天采收1次。

## 白梗大叶 White petiole large leaf water spinach

**品种来源**

地方品种。

📍 分布地区｜全市各区。

特　　征｜株高45厘米，开展度35厘米。茎粗大，横径1.3厘米，白色，节间长5厘米。叶长16厘米，宽5.5厘米，浅绿色，卵形，基部心形，叶脉明显；叶柄长10厘米，白色。单株重25克。

特　　性｜播种至初收60~70天，延续采收120天。生长旺盛，分枝少。耐热。纤维少，质柔软，品质优。每公顷全期产量约80吨。

栽培要点｜适播期为3—4月。水田育苗，株行距15厘米×15厘米。耐肥，宜勤施薄肥。收获期5—6月，每隔25天采收1次。

## 青梗柳叶 Green petiole willow leaf water spinach

**品种来源**

地方品种。

📍 分布地区｜全市各区。

特　　征｜植株直立，株高50厘米，开展度25厘米。茎细长，横径1厘米，绿色，节间长5厘米。叶片窄长，长15厘米，宽1.5~2厘米，绿色；叶柄长8厘米，绿色。单株重12克。

特　　性｜播种至初收25~30天，延续采收100天。生长快速，适应性强。耐热。口感爽脆，品质优。每公顷全期产量约60吨。

栽培要点｜适播期为2—3月。水田育苗，株行距15厘米×15厘米。耐肥，宜勤施薄肥。收获期4—6月，每隔25天采收1次。

## 青梗大叶 Green petiole large leaf water spinach

**品种来源**

地方农家品种。

📍 分布地区｜全市各区。

特　　征｜株高45厘米，开展度40厘米。茎较细长，横径0.8厘米，绿色，节间长4厘米。叶长16厘米，宽5厘米，浓绿色，基部心形，叶脉明显；叶柄长11厘米，绿色。单株重18克。

特　　性｜播种至初收40天，延续采收160天。生长旺盛，抗逆性强。较耐热。每公顷全期产量约80吨。

栽培要点｜播种期1—3月。直播或育苗移栽，播后薄膜覆盖防寒，旱作或水田栽培，株行距15厘米×15厘米。

## 青白柳叶

Green white willow leaf water spinach

> 品种来源

农家品种。

📍 分布地区 | 全市各区。

特　　征 | 植株直立，株高 45 厘米，开展度 32 厘米。茎粗大，横径 1.2 厘米，白色，节间长 4.5 厘米。叶柳叶形，长 14 厘米，宽 2~3 厘米，浅绿色；叶柄长 7.5 厘米，白色。单株重 15 克。

特　　性 | 播种至初收 60~70 天，延续采收 120 天。生长旺盛，分枝少。耐热。纤维少，质柔软，品质优。每公顷全期产量约 70 吨。

栽培要点 | 播种期 3—4 月。水田育苗，株行距 15 厘米×15 厘米。耐肥，宜勤施薄肥。收获期 5—6 月，每隔 25 天采收 1 次。

## 丝蕹

Filiform leaf water spinach

> 品种来源

农家品种，别名细通。

📍 分布地区 | 全市各区。

特　　征 | 株高 30 厘米，开展度 30 厘米。茎细小，横径 0.6 厘米，绿色，节间长 2.8 厘米，较密。叶长条形，长 15 厘米，宽 1~1.5 厘米，深绿色；叶柄长 6.5 厘米，绿色。单株重 8 克。

特　　性 | 播种至初收 50~60 天，延续采收半年以上。以旱地种植为主，也可水田种植。抗逆性强，较耐寒，耐风雨。一般不开花结实。质脆，味浓，品质优。每公顷全期产量约 38 吨。

栽培要点 | 分株繁殖。采用上一年宿根长出的侧芽作为幼苗。3 月定植，株行距 12 厘米×12 厘米。苗期勤施薄施追肥，每次收获后追肥。旱地种植在高温季节勤淋水、灌水。

# 莴苣

菊科莴苣属一、二年生草本植物，学名 Lactuca sativa L.，别名千斤菜等，染色体数 $2n=2x=18$。

莴苣原产于东亚和地中海沿岸。在隋代传入我国。全国各地均有栽培。广州是我国栽培莴苣最早的地区之一，年栽培面积约 10 000 公顷，可周年生产供应。

植物学性状：根系浅而密集。茎短缩。叶互生，近圆形或倒卵形等，叶面光滑或皱缩，叶缘波状或浅裂、全缘或有缺刻，绿色、黄绿色、紫色或紫红色。头状花序，花黄色。瘦果披针形，黑褐色或银白色，含种子1粒，成熟时顶端具伞状冠毛，能随风飞散。种子成熟后有休眠期，千粒重 0.8~1.2 克。

类型：按产品器官的不同，可分为叶用莴苣和茎用莴苣（莴笋）两种类型，含有4个变种：

1. 皱叶莴苣　学名 Lactuca sativa L. var. crispa L., 叶片有深裂，叶面皱缩，心叶抱合或不抱合。品种有意大利生菜、玻璃生菜、奶油生菜和红叶生菜等。

2. 长叶莴苣　学名 Lactuca sativa L. var. longifolia Lam., 叶狭长，直立，故也称直立莴苣，一般不结球。品种有罗马生菜等。

3. 结球莴苣　学名 Lactuca sativa L. var. capitata L., 叶全缘，有锯齿或浅裂，叶面平滑或皱缩，外叶开展，绿色，心叶抱合成叶球。叶球圆形、扁圆形或圆锥形。品种主要从国外引进，有万利包心生菜等。

4. 茎用莴苣　学名 Lactuca stative L. var. asparagina Bailey，即莴笋，叶片披针形、长卵圆形、长椭圆形等，叶色淡绿、绿、深绿或紫红。茎部肥大，茎的皮色有浅绿、绿或带紫红色斑块，茎的肉色有浅绿、翠绿及黄绿色。广州主栽品种为白皮尖叶莴笋。

栽培环境与方法：喜较冷凉的气候，发芽适温为 15~20℃，高于 25℃时发芽不良，夏季播种时种子须进行低温处理。育苗移栽。喜潮湿、微酸性土壤，适宜的土壤 pH 为 6.0 左右，pH＜5 或 pH＞7 时生育不良。

收获：叶片已充分生长或叶球已成熟要及时采收。散叶莴苣一般在定植后 30~40 天采收，结球莴苣一般在定植后 50~60 天采收。

病虫害：病害主要有霜霉病、菌核病、灰霉病、软腐病和病毒病等，虫害主要有蚜虫及螨类等。

营养及功效：富含蛋白质、脂肪、碳水化合物、各种维生素、矿物质和微量元素。味甘、苦，性凉，具利五脏、通经脉、清胃热等功效。

# Lettuce

*Lactuca sativa* L., an annual or biennial herb. Family: Compositae. The chromosome number: $2n=2x=18$.

Lettuce was introduced to China during Sui Dynasty. Guangzhou is one of the original regions for lettuce cultivation in China. At present, the cultivation area is about 10000 $hm^2$ per annum in Guangzhou. It is cultivated all the year round.

Botanical characters: Root shallow and dense. Stem dwarfed. Leaves alternate, approximate round or oval. Leaf surface smooth or slight wrinkled, green, yellow green, purple or purple red in color. Leaf margin, entire, undulate or robed. Capitulum, with yellow flowers. Achene lanceolate, black brown or grey white, with one seed and umbrella-shaped cap hairs at the terminal during maturing period. Because the seed has dormancy. The weight per 1000 seeds is 0.8~1.2 g.

Types: According to the edible part, lettuce can be classified into leaf and stem types, including four varieties.

1. Wrinkled leaf lettuce: *Lactuca sativa* L. var. *crispa* L., Leaf deeply cracked and wrinkled. The inner leaves form leaf head or not. The cultivars include Yidali, Boli, Naiyou, Hongye, etc.

2. Long leaf lettuce: *Lactuca sativa* L. var. *longifolia* Lam. Leaf long-narrow, usually no leaf head. The cultivars include Luoma, etc.

3. Leaf head lettuce: *Lactuca sativa* L. var. *capitata* L. Leaf margin entire, serrate or shallow cracked. Leaf surface, smooth or slight wrinkled. The outer leaves spread, green in color. The inter leaves form a leaf head, round, flat round or coniform in shape. The cultivars include Wanli, etc.

4. Stem lettuce: *Lactuca sativa* L. var. *asparagina* Bailey. Leaf lanceolate or oral, light green, green, deeply green or purple red in color. Stem succulent, light green, green with purplered spots, and flesh light green, bluish green or yellow green. The cultivars include Baipi jianye, etc.

Cultivation environment and methods: Lettuce is the chimonophilous vegetable. The suitable temperature is 15~20 ℃ for seed germination. The seed germination is inhibited above 25 ℃, and no germination above 30 ℃. The seed should be treated with low temperature for summer sowing. Seedling transplanting. The suitable soil pH is 6.0. Lettuce growth is inhibited below 5.0 or above 7.0.

Harvest: It's time to harvest when the leaf grows fully or the leaf head is mature. Growing period from sowing to harvest: romaine lettuce 30~40 days and leaf head lettuce 50~60 days.

Nutrition and efficacy: Lettuce is rich in proteins, fats, carbohydrates, vitamins and minerals.

## 皱叶莴苣 Wrinkled leaf lettuce
### 玻璃生菜
Boli lettuce

> 品种来源

地方品种。

📍 分布地区｜全市各区。

特　　征｜株高25厘米，开展度30厘米。叶簇生，近圆形，较薄，长23厘米，宽20厘米，黄绿色，有光泽，叶面皱缩，叶缘波状，心叶抱合；叶柄扁宽，长1厘米，白色。单株重400~500克。

特　　性｜定植后35~40天收获。较耐寒，不耐热，易感菌核病。叶质脆嫩，纤维少，品质优。每公顷产量30~45吨。

栽培要点｜播种期9月至翌年3月，育苗移栽，苗期30天左右。其余参照意大利生菜。

## 意大利生菜（全年生菜）
Yidali lettuce

> 品种来源

高华种子有限公司从意大利引进。

📍 分布地区｜全市各区。

特　　征｜株高30厘米，开展度35厘米。叶簇生，近圆形，长23厘米，宽18厘米，黄绿色，有光泽，叶面皱缩，叶缘波状，心叶微弯；叶柄长2厘米，宽1.8厘米，白色。单株重400~600克。

特　　性｜定植后30~40天收获。较耐寒和耐热，抗逆性和抗病性较强，适应性广。叶质软滑，品质中等。每公顷产量35~50吨。

栽培要点｜广州地区基本上周年均可播种，育苗移栽，苗期25~30天，5~6片真叶时定植，株行距25厘米×30厘米，每公顷用种量约0.5千克。宜施用有机质作基肥，封行前勤施薄肥，封行后少淋水，以防腐烂。

## 奶油生菜 Naiyou leaf lettuce

**品种来源** 20世纪80年代从国外引进。

**分布地区** | 全市各区。

特　　征 | 株高15~20厘米，开展度30~35厘米。叶簇生，椭圆形，长18~20厘米，宽12~15厘米，黄绿色，微皱，柔软，全缘，心叶抱合或不抱合；叶柄长1~2厘米，宽1.5~2.5厘米。单株重300~400克。

特　　性 | 定植后35~40天收获。较耐寒，不耐热，抗逆性和抗病性一般。叶厚，纤维少，味清香，品质优。每公顷产量25~35吨。

栽培要点 | 播种期9月至翌年3月，育苗移栽，苗期30~35天。其余参照意大利生菜。

## 红叶生菜 Red leaf lettuce

**品种来源** 20世纪80年代从国外引进。

**分布地区** | 全市各区。

特　　征 | 株高20~25厘米，开展度25~30厘米。叶簇生，卵圆形，长18~20厘米，宽15~18厘米，紫绿色或紫红色，叶面多皱，叶缘波状浅裂，心叶不抱合；叶柄长1~2厘米，宽1~1.5厘米。单株重350~450克。

特　　性 | 定植后35~40天收获。较耐寒，不耐热，较抗霜霉病和病毒病。叶厚，味清香，品质中等。每公顷产量30~35吨。

栽培要点 | 参照奶油生菜。

## 长叶莴苣 Long leaf lettuce
## 罗马生菜 Luoma lettuce

**品种来源** 20世纪80年代从欧洲引进。

**分布地区** | 全市各区。

特　　征 | 株高25~30厘米，开展度20~25厘米。叶片直立，倒卵形，长20~23厘米，宽15~20厘米，深绿色，有光泽，叶面平滑，刺毛少，全缘无锯齿，后期心叶半抱合；叶柄长3~4厘米，宽1.5~2.5厘米。单株重450~600克。

特　　性 | 定植后30~40天收获。耐寒性好，抽薹较晚，抗病性和抗逆性较强，适应性较广。叶厚，较爽脆，味清香，品质好。每公顷产量40~50吨。

栽培要点 | 秋、冬、春植为主。其余参照意大利生菜。

## 尖叶无斑油荬菜

Jianye wuban lettuce

**品种来源**

广州市农业科学研究院育成。

📍 分布地区 | 全市各区。

特　征 | 株高33厘米，开展度25~30厘米。叶披针形，先端尖，较直立，长30厘米，宽6厘米，青绿色。

特　性 | 播种至初收60~70天。生长势强。香脆爽口，品质优。每公顷产量30~45吨。

栽培要点 | 播种期9—12月。苗期25~30天。株行距23厘米×26厘米。适施基肥，追肥4~5次。保持土壤湿润。整株采收或分期剥叶采收。

## 香水油荬菜

Xiangshui lettuce

**品种来源**

地方品种。

📍 分布地区 | 全市各区。

特　征 | 株高20厘米，开展度50~60厘米。叶披针形，长20~25厘米，宽6~8厘米，青绿色，有光泽，微皱；叶柄不明显。

特　性 | 播种至初收70~80天。生长势强，耐寒，耐湿。叶薄，质软，具香味，品质优。每公顷产量20~30吨。

栽培要点 | 播种期7—12月，适播期9—11月。苗期25天左右，株行距21厘米×24厘米。适施基肥，追肥4~5次。9月至翌年5月采收。

## 结球莴苣 Leaf head lettuce
### 万利包心生菜
Wanli head lettuce

**品种来源**

20世纪80年代从国外引进。

📍 分布地区｜全市各区。

特　征｜株高20厘米，开展度38厘米。叶长21厘米，宽22厘米，黄绿色，叶面微皱，叶缘锯齿状；叶柄扁短，长1厘米，宽2厘米，黄白色。叶球近圆形，紧实，高12厘米，横径13厘米，黄白色。单球重400~550克。

特　性｜早熟，定植后40~50天收获。不耐寒，较耐热，较耐菌核病，适应性一般。质脆，纤维少，品质优。每公顷产量35~45吨。

栽培要点｜广州地区基本上周年均可播种，适播期8月至翌年2月，夏播前期须用遮阳网覆盖。育苗移栽，苗期25~30天，株行距夏植20厘米×25厘米，春、秋植25厘米×30厘米。施足基肥，前期追肥2~3次，定植后约25天施1次重肥。封行前保持土壤湿润，结球期宜用沟灌，注意防治软腐病和斜纹夜蛾。

## 茎用莴苣（莴笋） Stem lettuce
### 白皮尖叶莴笋
Baipi jianye lettuce

**品种来源**

20世纪60年代从四川引进。

📍 分布地区｜全市各区。

特　征｜株高50~60厘米，开展度35~40厘米。叶披针形，长25厘米，宽7厘米，绿色，有光泽，叶面微皱，叶缘有不规则尖齿；叶柄长约1.5厘米，宽约2.5厘米，绿白色。肉质茎长25~30厘米，横径5~7厘米，绿白色。种子黑褐色。

特　性｜定植后50~60天收获，延续采收30天。生长势较强，耐寒，耐风雨。质脆，微甜，纤维少，品质优。每公顷产量30~35吨。

栽培要点｜播种期9月至翌年2月。育苗移植，株行距25厘米×30厘米。施足基肥，定植后5天薄施速效肥，全期追肥4~5次。封行前勤淋水。采收幼嫩植株或肉质茎供食。

# 苦荬菜 Common sowthistle

菊科苦荬菜属一、二年生草本植物，学名 *Sonchus oleraceus* L. 别名苦荬，染色体数 $2n=4x=32$。

广州栽培历史悠久，栽培普遍，其适应性广，抗性强，收获期长。

**植物学性状**：须根系。茎短缩，有乳汁。叶互生，长椭圆形至披针形，全缘至羽状深裂，基部具披针形叶耳，抱茎。叶色浅绿色至深绿色或红色等，味微苦或无。头状花序，花舌状，黄色。瘦果卵形或椭圆形，白色。种子褐色。

**类型**：品种有尖叶苦荬、花叶苦荬等。

**栽培环境与方法**：一般8—12月播种育苗，10月至翌年5月收获。可直播或育苗移栽。

**收获**：一般植株约10片真叶或株高约20厘米时开始采收，可多次采收嫩茎叶。每10~15天采收1次。

**病虫害**：病害主要有霜霉病、叶斑病，虫害主要有蚜虫、红蜘蛛。

**营养及功效**：苦荬菜茎叶脆嫩，稍有苦味，具清热解毒、凉血止血等功效。

*Sonchus oleraceus* L., an annual or biennial herb. Family: Compositae. Synonym: Annual sowthistle, Smooth sowthistle. The chromosome number: $2n=4x=32$.

Bitter sowthistle has long been cultivated in Guangzhou and distributed widely. It has strong adaptability, abotic tolerance and long harvesting period.

**Botanical characters**: Fibrous root system. Stem dwarfed, with latex. Leaves alternate, oval to lanceolate, with entire or pinnate-cracked margin and lanceolate auricle at the base, clasp around the stem, light green, deeply green or red in color, and tasted bitter or not. Capitulum, with ligulate and yellow flowers. Achene, ovate to oval, white. Seed brown.

**Types**: The cultivars include Acute leaf bitter sowthistle, Parted leaf bitter sowthistle, etc.

**Cultivation environment and methods**: Bitter sowthistle is usually sowed during August to December, and harvested during October to May the following year. Seed sowing or seedling transplanting.

**Harvest**: The harvest begins when the common sowthistle plant has 10 true leaves or 20 cm in height. The tender leaves and stems are harvested every 10~15 days.

**Nutrition and efficacy**: Bitter sowthistle is rich in vitamin C, proteins, fats, carbohydrates.

## 尖叶苦荬

Acute leaf common sowthistle

**品种来源**

地方品种。

📍 分布地区｜全市各区。

特　　征｜株高40厘米，开展度35~40厘米。叶长披针形，先端渐尖，长35厘米，宽8厘米，绿色，叶面平滑，叶缘有不规则小尖齿。

特　　性｜播种至初收60天，延续采收180天。生长势强。耐寒，耐湿。质脆，味较苦，品质较优。每公顷产量15~25吨。

栽培要点｜播种期8—12月。育苗移植，苗期约30天。株行距30厘米×40厘米。适施基肥，勤追薄肥。采收期10月至翌年3月。

## 花叶苦荬

Parted leaf common sowthistle

**品种来源**

地方品种。

📍 分布地区｜全市各区。

特　　征｜株高45厘米，开展度60厘米。叶戟形，长43厘米，宽10厘米，绿色，薄被蜡粉，叶缘深裂。

特　　性｜播种至初收70~80天，延续采收5~6个月。生长势与抗逆性强，耐寒，耐湿。有苦味，品质中等。每公顷产量20~30吨。

栽培要点｜参照尖叶苦荬。

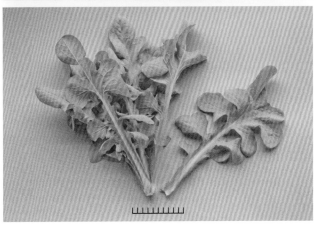

# 苦苣 Endive

菊科菊苣属一、二年生草本植物，学名 *Cichorium endivia* L.，别名细菊、花叶生菜等，染色体数 $2n=4x=32$。

苦苣原产于印度和欧洲南部，在唐代或其后由中亚细亚通过"丝绸之路"传入我国。广州年栽培面积约100公顷。

**植物学性状**：根系发达。茎短，根出叶互生于短缩茎上。叶片较大，长倒卵形、长椭圆形或长卵圆形，叶面皱缩或平展，叶缘深裂或全缘；外叶绿色，心叶浅黄色至黄白色，叶背面稍具茸毛。头状花序，花冠淡紫色。种子较小，钟状，灰白色，千粒重约1.6克。

**类型**：按叶片不同形态，可分为阔叶（大叶）、碎叶（细叶）2个变种：

1. 阔叶苦苣（大叶苦苣） 学名 *Cichorium endivia* L. var. *latifolia* Hort.。叶浓绿色，叶面平，全缘稍具毛刺，稍具苦味，耐热性好，抗寒性强，适应性广。近年来从美国引进的大叶苦苣即为此类。

2. 碎叶苦苣（细叶苦苣） 学名 *Cichorium endivia* L. var. *crispa* Hort.。叶翠绿色，叶缘具深缺刻，多皱褶，呈鸡冠状（碎叶状）。微苦，耐热性较好，适应性较强。近年来从荷兰引进的细叶苦苣即为此类。

**栽培环境与方法**：喜冷凉湿润环境，发芽适温为15~20℃，30℃以上高温抑制发芽。耐寒性、耐热性和耐旱性均较强，不易感染病虫害。苦苣叶片柔嫩，含水量高，又以叶片为食用部分，因此整个生长期间需有均匀而充足的水分和氮肥供应。宜选择有机质丰富、土层疏松、保水保肥力强的黏壤土或壤土栽培。可直播也可育苗移栽，多采用育苗移栽。忌连作，其前作以葱蒜类、豆类蔬菜为佳。

**收获**：播后60~90天，叶片长至30~40厘米，宽达8~10厘米，叶簇仍处旺盛生长时应适时采收。

**病虫害**：病害主要有白粉病、软腐病、菌核病、褐斑病；虫害主要有蚜虫。

**营养及功效**：富含维生素、矿物质、甘露醇、生物碱等，具清热、凉血、解毒、消炎和预防贫血等功效。

*Cichorium endivia* L., an annual or biennial herb. Family: Compositae. The chromosome number: $2n=4x=32$.

Endive was native to India and sourthern Europe, and introduced to China from Central Asia after Tang Dynasty. At present, the cultivation area is about 100 $hm^2$ per annum in Guangzhou.

Botanical characters: Vigorous root system. Stem dwarfed. Leaves alternate, born on the dwarfed stem, oval, long ellipsoid or lanceolate, wrinkle or smooth, with deeply cracked or entire leaf margin. Outer leaves green. Inner leaves light yellow to yellow white in color, with downy hair on the leaf back blade. Capitulum, with light purple corolla. Seed is little and bell-shaped, grey white in colour. The weight per 1000 seeds is 1.6 g.

Types: According to the shape of leaf, the cultivars can be classified into broad leaf and narrow leaf types.

1. Broad leaf type: *Cichorium endivia* L. var. *latifolia* Hort. Leaf deeply green in color, with smooth leaf surface. Leaf margin entire, with burr. It has strong tolerance to heat and cold. The cultivars include Broad leaf endive, introduced from the United States.

2. Narrow leaf type: *Cichorium endivia* L. var. *crispa* Hort. Leaf bluish green in color. Leaf margin deeply cracked and rugose in shape. It has strong tolerance to heat. The cultivars include narrow leaf endive, introduced from the Netherland.

Cultivation environment and methods: Endive favours cool weather. The suitable seed germination temperature is 15~20℃. The seed can not germinate above 30℃. The crop has stronger tolerance to cold, heat, drought, diseases and insect. The tender leaves are edible parts. Adequate irrigation and fertilization especially nitrogen fertilization are required for leaf growth. Seed sowing or seedling transplanting. It should not be continuous cropped.

Harvest: It's time to harvest 60~90 days after sowing, when the leaves are 30~40 cm in length and 8~10 cm in width.

Nutrition and efficacy: Endive is rich in vitamins, minerals, mannitol, alkaloid.

## 美国大叶苦苣

Broad leaf endive

**品种来源**

沈阳市农友种子商行从美国引进。

📍 分布地区｜全市各区。

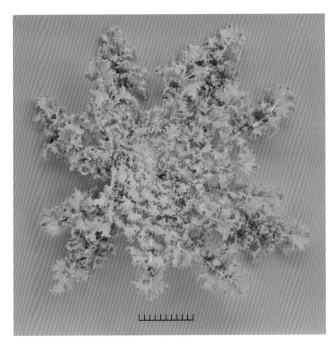

特　　征｜叶簇半直立，株高约25厘米，开展度约35厘米，呈莲座状。叶色浓绿，叶片椭圆形，叶缘缺刻多，深裂，叶长约30厘米，宽约9厘米，外叶105~135片。单株重0.5~1千克。

特　　性｜生育期55天左右。耐热性好，耐寒性强，抗病，适应性广。味清香，稍具苦味，品质佳。每公顷产量35~50吨。

栽培要点｜广州地区基本上周年均可播种。直播或育苗移栽，株行距30厘米×30厘米。每公顷用种量约0.5千克。宜施用有机质作基肥，封行前勤施薄肥，封行后少淋水，以防腐烂。

## 荷兰细叶苦苣

Narrow leaf endive

**品种来源**

沈阳市农友种子商行从荷兰引进。

📍 分布地区｜全市各区。

特　　征｜叶簇直立，株高约28厘米，开展度33厘米左右。外叶浅绿色，心叶黄绿色，叶缘缺刻深，叶长约35厘米，宽约7厘米，外叶105~120片。单株重0.4~0.8千克。

特　　性｜生育期55天左右。耐寒性和耐热性好，抗病性和抗虫性强，适应性广。质脆嫩，清香可口，稍具苦味，品质极佳。每公顷产量35~50吨。

栽培要点｜广州地区基本上周年均可播种。直播或育苗移栽，株行距30厘米×30厘米。每公顷用种量约0.5千克。宜施用有机质作基肥，封行前勤施薄肥，封行后少淋水，以防腐烂。

# 菠菜 Spinach

藜科菠菜属一、二年生草本植物，学名 *Spinacia oleracea* L.，别名菠棱菜、赤根菜、角菜、波斯草等，染色体数 $2n=2x=12$。

菠菜原产波斯，唐代开始传入中国。《唐会要》（8世纪后期）卷一百《尼波罗国》记载："贞观二十一年，遣使献菠棱菜、浑提葱。"广州地区栽培历史悠久，年栽培面积约 1 400 公顷。

**植物学性状**：主根较粗大，须根系。茎短缩。叶簇生在短缩茎上，椭圆形或戟形，浓绿色，基部有裂片；叶柄长，绿色或浅绿色。雌雄异株，少数雌雄同株，雌花簇生于叶腋，雄花为穗状花序，黄绿色。风媒花。异花授粉。播种用的种子为果实，具棱刺或无刺，果皮坚硬，灰褐色，千粒重 8.5~12.5 克。

**栽培环境与方法**：耐寒性较强。种子萌发需敲破果皮后浸种催芽，高温条件下需要搭遮阳网栽培。8月至翌年2月播种，10月至翌年4月收获。

**收获**：叶片肥大、鲜嫩时整株采收。

**病虫害**：病害主要有病毒病、霜霉病、炭疽病及斑点病等；虫害主要有潜叶蝇、美洲潜叶蝇、蚜虫等。

**营养及功效**：富含维生素A、维生素C及多种矿物质。味甘，性凉，对于胃肠障碍、便秘、痛风、皮肤病、各种神经疾病、贫血有一定食疗效果。菠菜草酸含量较高，一次食用不宜过多。

*Spinacia oleracea* L., an annual or biennial herb. Family: Chenopodiaceae. The chromosome number: $2n=2x=12$.

Spinach was native to persia, and introduced to China in Tang Dynasty. It has long been cultivated in Guangzhou. At present, the cultivation area is about 1400 $hm^2$ per annum.

**Botanical characters:** Fibrous root system with strong main root. Stem dwarfed. Leaves, clasping around the dwarf stem, oval or hastate, with lobes at the base, and deeply green. Petiole long, green or light green. Flower, mainly dioecious. Female flowers cluster on the leaf axil, while male flower is spike, yellow green in color. Anemophilous flower. The seed sowed is fruit, with thorn or not, and hard seed coaty, grey brown in color. The weight per 1000 seeds is 8.5~12.5 g.

**Cultivation environment and methods:** Spinach has strong cold tolerance. The seed coat should be broke and presoaked for germination. Shade net is required for cultivation under high temperature. Sowing date: August to February the following year, and harvesting date: October to April the following year.

**Harvest:** The whole plant could be harvested when the leaves are fleshy and tender.

**Nutrition and efficacy:** Spinach is rich in vitamin A, vitamin C and minerals.

## 新时代 A 级菠菜

Xinshidai A spinach

**品种来源**

广东省良种引进服务公司的杂种一代。

📍 **分布地区** | 全市各区。

特　　征 | 株高 13 厘米,开展度 35~46 厘米。叶戟形,以六角形为主,长 32 厘米,宽 10 厘米,叶片碧绿;叶柄浅绿色,具长梗、中凹槽,长 18 厘米。单株重 160 克。

特　　性 | 中熟,播种至初收 45 天左右。较耐热、耐湿,冬性强,晚抽薹,可抗霜霉病生理小种 1~7 型。纤维少,味甜,品质优,商品性佳。每公顷产量 18~22 吨。

栽培要点 | 播种期 10—12 月,生长适温为 15~25℃,低于 3℃或高于 28℃对生长有不利影响。直播,每 667 米$^2$ 用种量 2~5 千克。适施基肥,前期少量多次追肥,后期增加施肥量。分 2~3 次收获。注意防治霜霉病。

## 亨达利菠菜

Hengdali spinach

**品种来源**

广州市兴田种子有限公司从荷兰引进的杂种一代。

📍 **分布地区** | 全市各区。

特　　征 | 株高 18 厘米,开展度 35~44 厘米。叶戟形,以三角形为主,长 36 厘米,宽 11 厘米,叶片碧绿;叶柄浅绿色,具长梗、中凹槽,长 22 厘米。单株重 150 克。

特　　性 | 中熟,播种至初收 43 天左右。较耐热,耐湿,耐抽薹。味甜,品质好。每公顷产量 18~22 吨。

栽培要点 | 参照新时代 A 级菠菜。

# 茼蒿　Garland chrysanthemum

菊科茼蒿属一、二年生草本植物，学名 *Chrysanthemum coronarium* L. var. *spatisum* Bailey，别名塘蒿、蓬蒿、春菊等，染色体数 $2n=2x=18$。

茼蒿原产于地中海地区，广州栽培历史悠久，年栽培面积约 500 公顷。

**植物学性状**：须根系。茎短缩。叶互生，披针形，二回羽状深裂，叶厚，基部耳状抱茎，绿色。头状花序，舌状花，黄色或白色。自花授粉。瘦果近三角形，有明显棱沟，褐色，千粒重 1.8~2.2 克。

**类型**：依叶片大小可分为大叶、中叶和小叶（细叶）茼蒿。

**栽培环境与方法**：喜冷凉湿润气候，较耐寒，耐热性较差，在 10~30℃均可生长。广州地区 9—12 月播种育苗，11 月至翌年 3 月收获，撒播或条播。生长期间需保持土壤湿润，以施用氮肥为主。

**收获**：一般播种后 40~50 天可采摘嫩叶或嫩梢。

**病虫害**：病害主要有褐斑病、芽枯病等，虫害主要有蚜虫等。

**营养及功效**：富含维生素 A、钙、钾等，具有特殊香味。味甘、辛，性平，具调胃健脾、降压补脑等功效。

*Chrysanthemum coronarium* L. var. *spatisum* Bailey, an annual or biennial herb. Family: Compositae. Synonym: Tanghau. The chromosome number: $2n=2x=18$.

Garland chrysanthemum, the crop originated from the Mediterranean, has long been planted in Guangzhou. At present, the cultivation area is about 500 $hm^2$ per annum.

Botanical characters: Fibrous root system. Stem dwarfed. Leaves alternate, lanceolate, bipinnatipartite, thick and green, with ear-shaped embraced stem at the base. Capitulum, ligulate flowers, yellow or white in color. Self pollination. Achene, approximate triangular and brown in color, with obvious ribbed groove. The weight per 1000 seeds is 1.8~2.2 g.

Types: According to leaf size, it can be classified into broad leaf, medium leaf and narrow leaf types.

Cultivation environment and methods: Garland chrysanthemum favours cool and moisture weather. It has strong cold tolerance but poor heat tolerance, and could grow under 10~30℃. Sowing date: September to December. Broadcast or row sowing. Harvesting date: November to March the following year. For optimal growth, garland chrysanthemum requires a constant and abundant supply of moisture and nitrogen throughout its growth.

Harvest: The tender leaf or shoot could be harvested about 40~50 days after sowing.

Nutrition and efficacy: Garland chrysanthemum is rich in vitamin A, calcium and potassium.

## 大叶茼蒿

Large leaf garland chrysanthemum

**品种来源**

地方品种。

📍 分布地区｜全市各区。

特　　征｜株高21厘米，开展度28厘米。茎青绿色。叶长倒卵形，深裂，叶片大，长18厘米，宽8厘米，绿色，叶面微皱；叶柄长1.4厘米，抱茎，浅绿色。

特　　性｜播种至初收直播30~40天，移植60~70天。耐热性、耐寒性较好，多雨季节不易腐烂。爽脆味香，品质优。每公顷产量约23吨。

栽培要点｜9月至翌年1月播种，苗期25~30天，定植株行距12厘米×15厘米。直播每667米²用种量750克，育苗移栽每667米²用种量150~200克。

## 小叶茼蒿

Small leaf garland chrysanthemum

**品种来源**

地方品种。

📍 分布地区｜全市各区。

特　　征｜株高20厘米，开展度18厘米。茎直立，青绿色。叶倒卵形，深裂，长15厘米，宽4~5厘米，绿色，叶面微皱，叶缘缺刻较深；叶柄长1.7厘米，抱茎，浅绿色。

特　　性｜播种至初收直播30~35天，移植60~70天。分枝中等，抗逆性、适应性较强，立春前后抽薹。香味较淡，品质优。每公顷产量约20吨。

栽培要点｜播种期9—12月，直播或育苗移栽，移栽苗期25~30天，株行距10厘米×20厘米。勤施薄肥。收获期11月至翌年2月。

## 广良803中叶茼蒿

Guangliang 803 mid-leaf garland chrysanthemum

**品种来源**

广东省良种引进服务公司的一代杂交种。

📍 分布地区｜全市各区。

特　　征｜株高35~40厘米，开展度18~20厘米。茎直立，青绿色。叶倒卵形，深裂，长12~15厘米，宽4.5~6.0厘米，绿色，叶面微皱；叶柄长1.5厘米，抱茎，浅绿色；节间短，侧枝多。

特　　性｜播种至初收直播30~35天，移植60~70天。生长势旺盛，分枝力强。立春前后抽薹。抗逆性、适应性较强。香味浓郁，品质优。每公顷产量约25吨。

栽培要点｜播种期9月至翌年1月。其余参照大叶茼蒿。

# 芹菜

伞形花科芹菜属二年生草本植物，学名 *Apium graveolens* L. var. *dulce* DC.，别名芹、药芹、苦堇、堇菜等，染色体数 $2n=2x=22$。

芹菜原产地中海沿岸及瑞典等地的沼泽地带，于汉代由高加索传入我国。广州栽培历史悠久，公元6世纪初期已有记述，目前年栽培面积约600公顷。

**植物学性状**：根系浅。茎短缩。叶着生在短缩茎的基部，二回羽状复叶，每叶有2~3对小复叶和1片尖端小叶，小叶卵圆形，2裂或3裂，叶缘锯齿状，深绿色、绿色或淡绿色。叶柄长而肥大，并有突起纵棱，绿色、淡绿色、黄绿色或绿白色。伞形花序，花小，黄白色。虫媒花。异花授粉。双悬果，圆球形，含1~2粒种子。种子细小，椭圆形，褐色，千粒重约0.4克。

**类型**：按叶柄的形态可分为叶柄较细的本地芹菜和叶柄宽厚的西洋芹菜（西芹）2种类型；按叶柄颜色可分为白梗类型和青梗类型。

1. **白梗类型** 植株较矮小。叶片较小，淡绿色；叶柄较细，黄白色或白色。香味浓，质地较细嫩，易软化。抗病性差。目前主栽品种有黄叶香芹（白梗芹菜）等。

2. **青梗类型** 植株较高大。叶片较大，绿色至深绿色；叶柄较粗，淡绿色至绿色。香味较浓，产量高。目前主栽品种有青梗芹菜、西芹等。其中西芹叶柄更宽、厚，较短，纤维少，纵棱突出，多实心。

**栽培环境与方法**：喜冷凉湿润，较耐寒。日温23℃、夜温18℃、地温13~23℃的条件下最适生长。长日照下植株表现直立性，短日照下植株成开张性。直播和育苗移栽均可。在广州地区以冬春生产为主，7—12月播种，10月至翌年4月收获。

**收获**：芹菜的收获期不是十分严格，一般播种至采收需80~140天。

**病虫害**：病害主要有斑枯病、菌核病、叶斑病、病毒病、假黑斑病等，虫害主要有斜纹夜蛾、甜菜夜蛾、蚜虫、粉虱等；生理病害主要有烧心、空心、叶柄开裂、叶柄弯曲等。

**营养及功效**：富含蛋白质、碳水化合物、矿物质、维生素等多种营养物质和挥发性芳香油。味甘，性凉，具平肝清热、祛风利湿、除烦消肿、凉血止血、解毒宣肺、健胃利血、清肠利便、润肺止咳、降低血压、健脑镇静等功效。

# Celery

*Apium graveolens* L. var. *dulce* DC., a biennial herb. Family: Umbelliferae. Synonym: Kan tsoi, Honkan. The chromosome number: $2n=2x=22$.

Celery was introduced to China in Han Dynasty. It has long been cultivated in Guangzhou. At present, the cultivation area is about 600 $hm^2$ per annum.

Botanical characters: Shallow root system. Stem dwarfed, with numerous grooves. Leaf bipinnate, with 2~3 pairs leaflets, ovate, dual or tris lobed, deeply green, green or light green in color, with serrate leaf margin. Petiole, the main edible part, long, succulent, and green, light green, yellow green or green white in color, with longitudinal arris. Umbel, entomophilous, small and yellow white in color. Cremocarp, spherical in shape and with 1~2 seeds. Seed, little, oval and brown in color. The weight per 1000 seeds is 0.4 g.

Types: According to the petiole shape, celery can be classified into local celery (thinner petiole) and the West celery ( thicker petiole). According to the petiole color, celery can be classified into white petiole and green petiole types.

1. White petiole type: Small plant. Leaf smaller and light green. Petiole thinner and yellow white or white in color, tender with aroma. It has poor diseases resistant. The cultivars include Yellow leaf celery (white petiole celery), etc.

2. Green petiole type: Large plant. Leaf larger and green to deeply green in color, Petiole thicker and light green to green in color. The cultivars include Green petiole celery, etc. Xiqin celery has wide, thick and short petiole, obvious arris.

Cultivation environment and methods: Celery favours cool weather. The suitable temperature is 23 ℃ (day), 18 ℃ (night), and 13~23 ℃ (soil). The plant behaves upright under long-day, while spread under short-day. Celery is mainly cultivated in winter and spring in Guangzhou. Sowing dates: July to December. Direct sowing or transplanting. Harvested date: October to April the following year.

Harvest: Harvest occurs about 80~140 days after sowing.

Nutrition and efficacy: Celery is rich in proteins, carbohydrates, minerals, vitamins and aromas.

## 黄叶香芹（白梗芹菜）

White petiole celery

**品种来源**

地方品种。

📍 分布地区｜白云、花都、增城等区。

特　　征｜株高70厘米左右，开展度25~35厘米。奇数羽状复叶。小叶长7厘米，宽7厘米，浅绿色，叶缘有深缺刻；叶柄长45厘米，宽1厘米，绿白色。单株重150~190克。

特　　性｜早熟，播种至收获90~100天。生长势强，耐寒，易抽薹。质脆，香味浓，品质优。每公顷产量60~80吨。

栽培要点｜播种期8—12月。育苗移植，苗期40天左右，早播须用遮阳网覆盖防晒，株行距10厘米×12厘米，浅植。施足基肥，追肥勤施薄施，生长前期多淋水，后期注意排水。株高30厘米时，用稻草屏或遮阳网遮光软化。11月至翌年2月采收。

## 青梗香芹

Green petiole celery

> 品种来源

引自上海，栽培已数十年。

📍 分布地区｜白云、从化等区少量种植。

特　　征｜株高 65 厘米左右，开展度 20~30 厘米。奇数羽状复叶。小叶长 5 厘米，宽 5 厘米，深绿色，叶缘有深缺刻；叶柄长 30~40 厘米，宽 1 厘米，青绿色。单株重 200~250 克。

特　　性｜晚熟，播种至收获 130~150 天。生长势强，冬性强，迟抽薹。质脆，香味浓，品质优。每公顷产量 60~75 吨。

栽培要点｜播种期 10—12 月。苗期 50 天左右，株行距 13 厘米 ×16 厘米。翌年 1—4 月采收。其余参照黄叶香芹。

## 荷兰西芹

Helan xiqin celery

> 品种来源

20 世纪 90 年代从荷兰引进。

📍 分布地区｜南沙、番禺等区。

特　　征｜株型紧凑，株高 75 厘米，开展度 45 厘米。奇数羽状复叶。小叶卵圆形，长 8.2 厘米，宽 7.7 厘米，绿色，叶缘有深缺刻；叶柄第一节长 30~35 厘米，基部宽 3.5 厘米，浅绿色。单株重 0.8~1.0 千克。

特　　性｜早熟，定植至收获 70~80 天。耐热性、耐寒性较好，冬性强，较耐抽薹，抗病能力强。纤维少，品质优。每公顷产量 80~90 吨。

栽培要点｜播种期 7—12 月，宜育苗移栽，每 667 米$^2$ 用种量 10~20 克，苗期 50~60 天，早播可用遮阳网覆盖降温防雨。株行距（20~23）厘米 ×（20~23）厘米，每 667 米$^2$ 种植 1.0 万 ~1.3 万株。施足基肥，生长期间合理追肥并注意补充钙元素和硼元素。注意病虫害防治。11 月至翌年 4 月收获。

# 苋菜

# Edible amaranth

苋科苋属一年生草本植物，学名 *Amaranthus mangostanus* L., 别名米苋、赤苋、刺苋、青香苋等，染色体数 $2n=2x=34$。

广州栽培历史悠久，各区均有栽培，年栽培面积约 200 公顷。

**植物学性状**：根系发达。茎肥大直立，分株少，茎色有绿、白、红。叶互生，全缘，平滑或皱形，叶形有披针形、近圆形、卵形，叶色有绿、黄绿、紫红色及彩色（绿色与紫红色镶嵌）。花顶生或腋生，短穗状花序，花小。胞果矩圆形。种子粒小，扁圆形，黑色有光泽，千粒重 0.7 克。

**类型**：按叶型与色泽分，主要栽培的类型与品种有尖叶绿苋、圆叶绿苋、花红苋和红苋等。

**栽培环境与方法**：喜温暖，耐热，耐湿，耐旱，不耐寒，对低温很敏感，10℃以下种子发芽困难，20℃以下生长缓慢。采用种子直播。当幼苗 2 片真叶时进行第 1 次追肥，每隔 10 天左右追肥一次。生长期间保持土壤湿润，田间不能积水。

**收获**：一般采收幼苗供食用，在植株高 15 厘米左右时开始采收。

**病虫害**：病害主要有白锈病，虫害主要有食叶跗线螨、小地老虎、斜纹夜蛾、蚜虫等。

**营养及功效**：富含铁、钙和维生素 K。味甘，性凉，具清热解毒、明目利咽等功效。

*Amaranthus mangostanus* L., an annual herb. Family: Amaranthaceae. The chromosome number: $2n=2x=34$.

Edible amaranth, the crop has long been cultivated in Guangzhou and wildly in its each district. At present, the cultivation area is about 200 $hm^2$ per annum.

**Botanical characters**: Root vigorous. Stem erect, succulent, few branched and green, white or red in color. Leaves alternate, lanceolate, approximate round or ovate in shape, and green, yellow green, purple red or colorful spotted in color, with entire margin and smooth or wrinkle surface. Spike short, terminal or axillary, small flower. Utricle, square round. Seed, little and flat-round, black and grossy. The weight per 1000 seeds is 0.7 g.

**Types**: According to the shape and color of leaf, edible amaranth can be classified into Green acute leaf, Green round leaf, Central red leaf and Red leaf types.

**Cultivation environment and methods**: Edible amaranth favours warm weather. It has strong tolerance to heat, wet, drought, while is sensitive to low temperature. Seed germinates difficultly below 10℃, and plant grows slowly below 20℃. Mainly direct sowing. Fertilizers should be used about 10 days at each interval, when the plant has second true leaf.

**Harvest**: Young seedlings, the edible parts, could be harvested when the plants are 15 cm in height.

**Nutrition and efficacy**: Edible amaranth is rich in iron, calcium, vitamin K.

## 尖叶绿苋

Green acute leaf edible amaranth

**品种来源**

地方品种。

📍 分布地区｜全市各区。

特　　征｜株高35~45厘米，开展度28厘米。叶长卵形，先端尖，长13厘米，宽5厘米，绿色，全缘；叶柄长5厘米，青绿色。单株重20~30克。

特　　性｜播种至初收40~50天。耐热力较强。品质中等。每公顷产量11~15吨。

栽培要点｜春播期1—4月，采收期3—6月；秋播期7—9月，采收期9—11月。多混沙撒播，每667米$^2$用种量6~7.5千克，全期追肥5~6次。注意防治斜纹夜蛾、小地老虎等。

## 圆叶绿苋

Green round leaf edible amaranth

**品种来源**

地方品种。

📍 分布地区｜全市各区。

特　　征｜株高25厘米，开展度22厘米。茎绿色。叶卵圆形，长13厘米，宽11厘米，叶面微皱，绿色；叶柄长5厘米。单株重26克。

特　　性｜播种至初收30~40天，延续采收20~30天。抽薹较迟，耐热性强，耐寒性弱。品质优。每公顷产量15~18吨。

栽培要点｜播种期3—8月，采收期4—9月。直播，每667米$^2$用种量5~6千克。全期追肥5~6次。注意防治斜纹夜蛾、小地老虎等。

## 红苋

Red edible amaranth

**品种来源**

地方品种。

📍 分布地区｜全市各区。

特　　征｜株高27厘米，开展度20~25厘米。叶近圆形、心形或叶面微皱，紫红色，全缘；叶柄长4厘米，紫红色。单株重20~25克。

特　　性｜播种至初收25~35天，延续采收15~25天。抽薹较迟，较耐热。品质优。每公顷产量15~20吨。

栽培要点｜播种期3—8月，采收期4—10月。全期追肥5~6次。注意防治斜纹夜蛾、小地老虎等。

## 花红苋

Central red edible amaranth

**品种来源**

地方品种。

📍 分布地区｜全市各区。

特　　征｜株高25~30厘米，开展度18~20厘米。叶卵圆形至圆形，叶长8厘米，宽6~7厘米，叶面微皱，叶边绿色，叶脉附近红色、紫红色，全缘；叶柄长4~5厘米，红绿色或青绿色。单株重约30克。

特　　性｜播种至初收35天，延续采收15~20天。较耐热。品质中等。每公顷产量15~20吨。

栽培要点｜播种期1—8月，适播期3—4月，采收期4—9月。早播薄膜覆盖防寒。宜混细沙撒播，每667米$^2$用种量1—2月12千克，3—4月7~8千克。全期追肥5~6次。注意防治斜纹夜蛾、小地老虎等。

# 落葵

落葵科落葵属一年生缠绕草本植物，学名 *Basella* spp.，别名木耳菜、潺菜、藤菜，染色体数红花落葵 $2n=4x=48$，白花落葵 $2n=5x=60$。

落葵原产于我国和印度，广州栽培历史悠久，300 多年前已有记述，目前年栽培面积约 300 公顷。

**植物学性状**：须根系。茎绿色或略带紫红色。单叶互生，全缘，叶片卵形或近圆形，顶端渐尖。叶形较宽大。穗状花序腋生，长 3~15 厘米；花无花瓣，淡红色或淡紫色，卵状长圆形，全缘，顶端钝圆，内折，下部白色，连合成筒；雄蕊着生于花被筒口，花丝短，基部扁宽，白色，花药淡黄色；柱头椭圆形。浆果圆球形，直径 5~6 毫米，熟后红色至紫黑色，内含 1 粒种子。种子近圆形，紫黑色，千粒重 25 克。

**类型**：按照花的颜色分为白花、红花两类。白花落葵又叫白落葵，茎淡绿色，叶绿色，品种有青梗藤菜。红花落葵茎和花淡紫色至紫红色，叶片深绿色，品种有红梗藤菜。

**栽培环境与方法**：喜温暖湿润和半阴环境，不耐寒，怕霜冻，耐高温多湿，宜在肥沃疏松和排水良好的沙壤土中生长。春、夏、秋季栽培，2—5 月播种。通常采用播种繁殖，也可用扦插繁殖。种子可用 50℃温水搅拌 15 分钟，然后在常温水中浸泡 4~6 小时，搓洗干净后，在 30℃左右的条件下保温催芽。穴播，每穴 3~4 粒种子，播后覆盖 1.5~2 厘米厚的细土。4 叶期定苗，每穴留 2~3 株。定苗后，加强肥水管理，保持土壤湿润，每采收 1 次追 1 次肥水。

**收获**：在株高 25~30 厘米时采收嫩茎叶，保留茎基部 3~4 片叶，长出侧枝陆续采收。

**病虫害**：在持续骤雨高湿时易发生褐斑病和灰霉病，褐斑病俗称"蛇眼病"，主要为害叶片，灰霉病侵染叶、叶柄、茎和花序；虫害主要有蚜螨、小地老虎等。

**营养及功效**：含粗脂肪、粗蛋白、粗纤维、多种维生素及钾、钙、镁、磷等矿物质。味甘，性寒，具滑肠利便、清热解毒等功效。

# Malabar spinach

*Basella* spp., an annual herb. Family: Basellaceae. Synonym: Ceylon, Indian spinach, Malabar nightshade. The chromosome number: *Basella rubra* L. $2n=4x=48$, *Basella alba* L. $2n=5x=60$.

Malabar spinach, originated from China and Indian, has long been cultivated in Guangzhou. At present, the cultivation area is about 300 $hm^2$ per annum.

Botanical characters: Fibrous root system. Stem, fleshy and green or light purple red in color. Leaves alternate, ovate or approximate round, broad, with entire margin. Spike, axillary, 3~15 cm in length. Flower, light red or light purple in color and ovate to long round in shape, without petal, with margin entire and folded. Stamen, white and flat broad at the base, with short filaments and light yellow anthers. Stigma, oval. Bacca, spherical in shape and 0.5~0.6 cm in diameter, red to purple black in color, with one seed. Seed, approximate round and purple black. The weight per 1000 seeds is 25 g.

Types: According to the flower color, malabar spinach can be classified into white flower and red flower types. White flower type: Stem, light green. Leaf green. Cultivars include Green malabar spinach, etc. Red flower type: Stem and flower are light purple to purple red in color. Leaf, deeply green. Cultivars include Red malabar spinach, etc.

Cultivation environment and methods: Malabar spinach favours warm, moist, shady weather. It has tolerance to heat and wet, and no tolerance to cold and frost. Fertile, porous sandy soil with well drainage is desirable for malabar spinach production. Malabar spinach is usually cultivated in spring, summer and autumn. Sowing dates: February to May. Malabar spinach is usually cultivated by direct seeding, and also by cuttage. The seeds should be stirred in 50 ℃ water for 15 minutes and soaked for 4~6 hours in water under room temperature, then scrubbed and forced germination under 30 ℃. Hole sowing, with 3~4 seeds for each hole. After sowing, 1.5~2 cm soil covering is required. Field planting should be done at the fourth leaf stage, 2~3 seedlings in each hole. A constant and abundant supply of moisture throughout the growth is required. Fertilization and irrigation apply after harvest each time.

Harvest: Tender stems and leaves could be harvested when the plants are 25~30 cm in height, while 3~4 leaves should be kept at the base for continuous harvest.

Nutrition and efficacy: Malabar spinach is rich in fat, protein, crude fiber, Vitamin and mineral.

## 青梗藤菜

Green petiole malabar spinach

> 品种来源

地方品种。

📍 **分布地区** ｜ 全市各区。

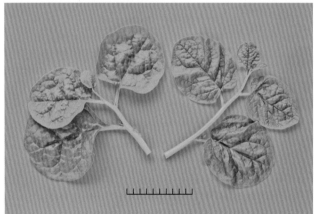

特　　征 ｜ 植株蔓生。株高可达200~300厘米。茎圆形，肉质，绿色。叶互生，卵圆形，长14厘米，宽10厘米，肥厚，光滑，绿色，全缘；叶柄长2.7厘米，绿色。

特　　性 ｜ 播种到初收60~70天。生长势强，易生侧枝。耐高温多雨，骤雨易罹患蛇眼病。质嫩滑，品质中等。每公顷产量23~38吨。

栽培要点 ｜ 播种期2—6月，适播期3—4月。苗期30~35天。株行距17厘米×17厘米。生长期勤追肥，保持土壤湿润。收获期5—8月。主茎长约30厘米时开始收获，以后陆续采收叶腋长出的侧枝。

## 红梗藤菜

Red petiole malabar spinach

> 品种来源

地方品种。

📍 **分布地区** ｜ 全市各区。

特　　征 ｜ 植株蔓生。株高可达200~300厘米。茎圆形，肉质，紫红色。叶卵圆形，长12厘米，宽8~10厘米，肥厚，光滑，深绿色，全缘；叶柄长3厘米，紫绿色。

特　　性 ｜ 播种至初收60~70天，延续采收约70天。生长势强，易生侧枝，耐高温，耐湿，骤雨易罹患蛇眼病。质嫩滑，品质中等。每公顷产量23~38吨。

栽培要点 ｜ 菜农习惯与青梗藤菜混种，以增加商品美观性。栽培管理参照青梗藤菜。

# 叶菾菜 Swiss chard

藜科甜菜属二年生草本植物，学名 *Beta vulgaris* L. var. *cicla* L.，别名莙荙菜、猪姆菜，染色体数 $2n=2x=18$。

**植物学性状**：根系较发达。茎短缩。叶卵形或长卵形，表面平滑或皱缩，浅绿色或深绿色，有光泽。叶柄肥厚多肉，白色、浅绿色或红色。青梗叶菾菜和白梗叶菾菜花腋生，小，绿色，单生或2~3朵聚生成圆锥花序。球果，果皮革质，褐色，内含种子数粒。

**栽培品种**：广州栽培历史悠久，百年前已称之为"常蔬"，各地均有栽培。广州的叶菾菜依其叶柄色泽，分为青梗、白梗和红梗3种类型，以青梗种栽培较普遍。

**栽培环境与方法**：喜冷凉湿润气候，生长适温15~25℃，耐寒，耐热，耐瘠薄，耐盐碱。青梗叶菾菜和白梗叶菾菜在秋、冬季栽培，红梗叶菾菜适应性广，在广州地区不抽薹，一年四季都可栽培。可直播或育苗移栽。

**病虫害**：病害主要有褐斑病、病毒病等，虫害主要有蚜虫、小地老虎、潜叶蝇等。

**营养及功效**：富含还原糖、粗蛋白、粗纤维及维生素等，可煮食、凉拌或炒食。叶菾菜味甘、苦，性寒、滑，具解风热毒、止血生肌、理脾除风等功效。

## 青梗莙荙菜

Green petiole swiss chard

**品种来源**

地方品种。

**分布地区** | 全市各区。

**特　　征** | 株高50~60厘米，开展度30~40厘米。叶卵形，长40厘米，宽15厘米，绿色，微皱，全缘；叶柄长11厘米，宽3厘米，厚0.5厘米，浅绿色。

**特　　性** | 播种至初收约85天。抽薹较早。耐热力中等。质柔软。每公顷产量32~40吨。

**栽培要点** | 播种期9—12月，收获期11月至翌年3月。苗期35天，株行距18厘米×23厘米。施足基肥，定植后勤追肥。

## 白梗莙荙菜

White petiole swiss chard

**品种来源**

地方品种。

📍 **分布地区** | 全市各区。

**特　　征** | 株高60厘米，开展度65厘米。叶椭圆形，长40厘米，宽15厘米，较厚，深绿色，微皱，全缘；叶柄长12厘米，宽4厘米，厚1.0厘米，白色。

**特　　性** | 播种至初收90~100天。生长势强，抽薹较迟。耐热性较弱。味淡，品质中等。每公顷产量约40吨。

**栽培要点** | 播种期9—12月，收获期11月至翌年3月。苗期30~40天。株行距20厘米×24厘米。其余参照青梗莙荙菜。

## 红梗莙荙菜

Red petiole swiss chard

**品种来源**

引自北京。

📍 **分布地区** | 全市各区有少量栽培。

**特　　征** | 叶长卵圆形或卵圆形，叶片肥大，叶面有皱褶，鲜红色或紫红色，叶缘无缺刻，叶柄断面呈"凹"字状，肥厚，肉质鲜红色或紫红色。整株有很高的观赏价值，可盆栽观赏兼食用，亦可作室外摆花。

**特　　性** | 播种至初收55~65天。广州地区不抽薹，耐寒性、耐热性较强。味淡，品质中等。每公顷产量约40吨。

**栽培要点** | 周年均可播种。苗期30~40天，株行距20厘米×24厘米。定植后25天左右可陆续采收外叶。

# 枸杞

*Chinese wolfberry*

茄科枸杞属多年生灌木，但菜用枸杞一般作一年生栽培，学名 *Lycium chinense* Mill.，染色体数 $2n=2x=24$。

广州栽培历史悠久，公元 11 世纪后期已有栽培。

**植物学性状**：根系浅。枝柔软，分枝力强，茎节有刺或无。叶互生，卵形或阔卵圆形。花紫色，1~4 朵同生于一叶腋。浆果卵形或长椭圆形，橙红色或大红色。在广州一般不开花结籽。

**栽培品种**：广州的枸杞品种主要有细叶枸杞和大叶枸杞，其中大叶枸杞以前以圆叶为主，后来经过品种改良后，现在主栽的大叶枸杞品种以长卵叶为主。

**栽培环境与方法**：喜温暖，但不耐炎热，生长适温 15~25℃。生长发育需要充足的阳光。不耐涝。对土壤的适应性强。插条繁殖，扦插期 8 月。采收期 11 月至翌年 4 月。

**收获**：一般扦插后 50~60 天开始采收，可采收嫩枝、嫩叶或嫩梢，先采收生长最旺的枝条，留下其余的枝条继续生长，以后分次采收。

**病虫害**：病害主要有病毒病、霉斑病、白粉病等，虫害主要有斜纹夜蛾、螨虫、蓟马、蚜虫等。

**营养及功效**：嫩茎叶含多种维生素和氨基酸，可凉拌、炒食、作汤料。枸杞嫩茎叶味甘，性平，有清肝明目功效。

# 大叶枸杞

Large leaf Chinese wolfberry

### 品种来源

地方品种。

分布地区 | 全市各区。

特　　征 | 株高65~75厘米，开展度55厘米。茎横径0.7厘米，青绿色。叶互生，长卵形，叶肉较薄，长8厘米，宽5厘米，绿色，叶背浅绿色。节无刺或具小软刺。

特　　性 | 插条至初收约60天，延续采收6个月。易生侧枝。耐寒，耐风雨，不耐热，遇高温易发病或落叶。味较淡。每公顷产量60~70吨。

栽培要点 | 9月插条栽植，插条长10~15厘米，具2~3节。7—8月插条须用遮阳网降温。株行距12厘米×18厘米。前期可间种叶菜，注意淋水，保持土壤湿润。勤追薄肥。注意综合防治蚜虫和蓟马。收获期10月至翌年4月，约15天收获1次，每次收割时留2~3条嫩枝，全期收10余次。

# 细叶枸杞

Small leaf Chinese wolfberry

### 品种来源

地方品种。

分布地区 | 从化、增城、白云等区。

特　　征 | 株高70厘米，开展度25厘米。嫩茎青绿色，收获时青褐色，横径0.6厘米。叶互生，卵形，长5厘米，宽3厘米，叶肉较厚，绿色，叶背浅绿色。节有刺或无。

特　　性 | 插条至初收50~60天，延续采收约5个月。易生侧枝。耐寒，耐风雨，不耐热，高温易发病或落叶。味浓，质优。每公顷产量53~60吨。

栽培要点 | 参照大叶枸杞。

## 千宝菜 Senposai

十字花科芸薹属一、二年生草本植物，学名 *Brassica juncea* (L.) Czern. var. *japonica* Hort.。

原产日本，1991年引进中国，目前在山东、云南、湖南、浙江、广东等地少量栽培。

**植物学性状**：茎在营养生长期为短缩茎、绿色，生殖生长期抽生花薹。单叶互生，叶片浓绿色，圆形，全缘，肉厚，光滑，有褶皱。叶柄长，淡绿色，叶翼2~4枚。复总状花序，完全花，花冠黄色，花瓣十字形排列。长角果。种子粒大，圆形，黑褐色。

**栽培环境与方法**：耐热，耐寒，较耐阴，但不耐涝，要求光照充足。生长适温为20~25℃，可耐-5~-3℃的低温。多采用直播。植株具真叶6~8片时整株采收供食用。可炒食、凉拌、作火锅配菜，也可作色拉生食等。

**病虫害**：病害主要有叶斑病和炭疽病，虫害主要有蚜虫（春季）、黄曲条跳甲（夏季）、菜青虫（秋季）等。

## 银丝菜 Jade silk vegetable

十字花科芸薹属一年生或二年生草本植物，学名 *Brassica juncea* var. *multisecta* L. H. Bailey.，别名京水菜、水晶菜。

引自日本，广州各地零星种植。

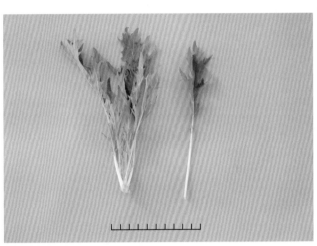

**植物学性状**：植株较直立，外形介于不结球小白菜和花叶芥菜之间。茎在营养生长期为短缩茎，基生叶簇丛生于短缩茎上，并具有很强的分蘖能力。叶绿色，羽状深裂，有光泽，质薄，光滑；叶柄长，横切面呈楔形至半圆形，有浅沟，银白色，故有"银丝"之称。花黄色。异花授粉。长角果。种子圆形，黑色至红褐色。

**栽培环境与方法**：喜冷凉气候，耐寒力较强，属长日照蔬菜，喜湿但不耐涝。生长适温为18~25℃。广州地区适播期为9月至翌年5月。多采用撒播。苗高20~30厘米时可一次性采收，也可陆续采收。

**病虫害**：高温高湿易发软腐病；虫害主要有黄曲条跳甲、蚜虫及潜叶蝇等。

**营养及功效**：富含矿质营养，质柔软，清香，口感好，风味类似不结球小白菜，可炒食、作火锅配菜、腌渍等。

## 芝麻菜 Eruca sativa

十字花科芝麻属一年生草本植物，学名 *Eruca sativa* Mill.，别名臭菜、紫花南芥（芸芥）、德国芥菜等。

原产于欧洲北部和亚洲西部及北部等，我国华北和西北地区都有野生种分布，栽培面积不大。

**植物学性状**：植株高 20~90 厘米。茎直立，圆形，有细茸毛，上部常分枝。叶片深绿色，簇生基生叶和茎下部叶呈羽状分裂，叶缘波状，具叶柄；茎上部叶近长椭圆形，具钝齿，无柄，下部具 1~3 对裂片；茎顶部叶宽椭圆形或长卵形，常为羽状深裂。顶生总状花序，花瓣黄色、乳黄色或渐转为白色，有紫纹，短倒卵形，基部有窄线形长爪。长角果圆柱形，种子近球形或卵形，黄棕色或淡褐色。

**栽培环境与方法**：喜冷凉湿润的气候，耐寒性较强。属短日照蔬菜，抗盐碱，耐旱涝，生长适温为 15~25℃。广州地区可周年直播种植，最适播期为 9 月至翌年 4 月。幼苗供菜用。

**病虫害**：抗性强，病害少，虫害主要有黄曲条跳甲、小菜蛾等。

**营养及功效**：含特殊芳香及钾、钙、硒等物质。味甘，有浓郁的芝麻香。种子可入药。具有下气行水、祛痰定喘作用。

## 番杏 New zealand spinach

番杏科番杏属一年生肉质草本植物，学名 *Tetragonia expansa* Murray，别名新西兰菠菜、洋菠菜等。

原产于澳大利亚、新西兰、智利及东南亚等地，主栽区分布在热带和温带地区，中国东南沿海地区于 20 世纪 20 年代引进栽培，广州地区引种于 80 年代。

**植物学性状**：根系发达。植株高 30~50 厘米。茎初期直立，长出分枝后匍匐生长。叶片卵状菱形或卵状三角形，互生，绿色，肉厚而软，嫩叶上有银色茸毛，边缘波状。叶柄肥粗。温度较高时，每个叶腋由下而上着生黄色小花，花单生或簇生。坚果菱角形，褐色，每果含种子数粒。

**栽培环境与方法**：喜凉爽湿润，较耐阴，耐寒，较耐热，不耐涝。属长日照蔬菜，对光照条件要求不严格。生长适温为 20~25℃。以直播为主，也可育苗移栽，浸种 24 小时后在 25℃ 条件下催芽，种子稍膨胀裂开时播种，也可干种直播。采收 8~12 厘米长嫩梢供食用。

**病虫害**：抗病虫能力强。病害主要有病毒病等，虫害主要有甜菜夜蛾等。

**营养及功效**：含蛋白质、维生素 A、铁、钙、磷脂、番杏素等成分。质柔嫩，味清淡。味甘，性平，全株可入药，具有清热解毒、祛风消肿、凉血利尿功效。

## 冬寒菜 Malva verticillata

锦葵科锦葵属一年生或二年生草本植物，学名 Malva Verticillata L.，别名冬葵、滑菜、冬苋菜等。

原产中国，我国各地均有小面积栽培。

**植物学性状**：茎直立，采收植株高 30~60 厘米，留种植株高达 100 厘米左右，采摘后分枝较多。叶大肥厚，互生，圆扇形，掌状浅裂，叶面微皱褶，叶缘波状，茎叶被有白色茸毛，叶柄长。花小，浅红色或白色，簇生于叶腋。果实为蒴果，扁圆形。种子细小，黄白色，肾形，扁平。按梗颜色分主要有紫梗和白梗两个类型。

**栽培环境与方法**：喜冷凉湿润气候，耐寒性较强，不耐高温。属短日照蔬菜。生长发育适温为 15~20℃。直播或育苗移栽。在广州地区以春、秋、冬季栽培为主，8月底至翌年3月可播种，可穴播、撒播或育苗移栽。当株高 15~20 厘米时开始采摘顶梢，茎基部留 1~2 节。

**病虫害**：病害主要有炭疽病、根腐病等，虫害主要有小地老虎、斜纹夜蛾和蚜虫等。

**营养及功效**：富含胡萝卜素、维生素 C 和钙等，口感柔滑。味甘，性寒，全株可入药，具有清热润燥、利尿除湿、滑肠解毒等功效。

## 香麻叶 Sesame leaves

椴树科黄麻属一年生草本植物，学名 Corchorus olitorius L.，别名食(叶)用黄麻、甜麻叶。

原产非洲，我国南北各地均有种植，主要分布于长江以南及台湾等地区，广东地区栽培历史较早。

**植物学性状**：茎直立，绿色或红色，较多分枝。叶互生，卵形或披针形，先端渐尖，边缘锯齿状，最下部 2 齿伸长为尾状裂片，具叶柄。聚伞花序，顶生及腋生，花黄色，完全花。蒴果长圆筒形，茶褐色。种子小，千粒重约 1.8 克。

**栽培环境与方法**：喜高温湿润气候，不耐寒，低温生长缓慢或停止，遇霜会枯死，耐旱性强，不耐涝。短日照蔬菜。生长适温为 22~30℃。在广州地区可春、夏、秋种植，多采用直播。可采收嫩茎叶供食用。

**病虫害**：病害少，虫害有斜纹夜蛾、蚜虫等。

**营养及功效**：富含胡萝卜素、维生素 B、维生素 C、磷、钾、钙、铁等，味微甘香。具有止血、解暑和清热等功效。

## 益母草 Leonurus artemsisia

唇形科益母草属一年生或二年生草本植物，学名 Leonurus japonicus Houtt.，别名益母艾、红花艾、云母草、茺蔚等。

原产于中国，亚洲东部、非洲、美洲各地区等均有分布。

**植物学性状**：植株直立。茎青绿色，方柱形，有4钝棱，微具纵沟。叶片对生，形状变化较大，基生叶圆心形，下部茎叶卵形或近圆形，茎中部叶菱形，较小，上部叶羽状深裂或浅裂成3片，花序上的叶片线形或披针形。下部叶具叶柄，向上叶柄渐短至无柄。轮伞状花序腋生，花粉红色至淡紫红色或白色。种子小，灰棕色至灰褐色。按花瓣颜色可分为红花益母草和白花益母草两个类型。

**栽培环境与方法**：喜温暖湿润气候，喜阳光，较耐阴，怕涝。属短日照蔬菜。生长发育适温为22~30℃。广东可春、夏、秋种植，多采用直播。可采收嫩茎叶供食用。

**病虫害**：病害主要是白粉病，虫害主要是蚜虫。

**营养及功效**：含有硒、锰等多种微量元素。味辛，性凉，具有活血调经、利尿消肿等功效。

## 藤三七 madeira—vine

落葵科落葵薯属多年生宿根稍带木质的缠绕藤本植物，学名 Anredera cordifolia (Ten.) Steenis，别名田三七、洋落葵等。

原产巴西，我国云南、广东、四川、湖北等地均有栽培，广州市郊区有少量种植。

**植物学性状**：根系发达，定植半年后，在地下可形成瘤块状块茎。茎圆形，基部簇生，初为绿色，老熟后变棕褐色。叶绿色，呈心形，互生，光滑无毛，有短柄。叶腋能长出瘤块状珠芽。叶腋抽生穗状花序，花小，数十朵至200余朵，下垂，花冠5瓣，花白色，花谢后变黑褐色，不容易脱落。花虽两性，但通常不结实，无种子。

**栽培环境与方法**：喜温暖湿润气候，生长适温为17~25℃，耐寒性较强，能耐0℃以上的低温。在35℃以上的高温下，生长不良。耐阴，扦插繁殖。可搭架栽培，也可铺地栽培。适宜在遮阳或在半遮蔽的地方栽培。

**病虫害**：病害有蛇眼病，虫害少。

**营养及功效**：富含蛋白质、碳水化合物、维生素、胡萝卜素等，具有滋补、壮腰健膝、消肿散瘀及活血等功效。

## 菊花脑 *Chrysanthemum nankingense*

菊科菊属多年生宿根性草本植物，学名 *Chrysanthemum nankingense* H. M.，别名菊花叶、路边黄、黄菊仔等。

原产中国，江苏、湖南、贵州等地均有栽培，南京地区栽培历史悠久，已成为地方特色蔬菜品种，广州各区均有少量种植。

**植物学性状**：根系发达。植株直立，高30~60厘米。茎纤细，半木质化，具地下匍匐茎，分枝性强，无毛或茎上部有细毛。叶片互生，卵圆形，绿色，无毛或叶脉处有稀疏细毛，羽状缺刻。头状花序，总苞半球形。舌状花黄色。果实为瘦果，11~12月成熟。种子小，灰褐色。

**栽培环境与方法**：周年可生长，生长适温15~25℃，冬季低温生长缓慢。适应性强，耐贫瘠，耐旱，忌涝。为短日照植物。多采用种子繁殖，也可分株繁殖和扦插繁殖。可采收嫩梢供食用。

**病虫害**：夏、秋季容易发生根腐病，虫害主要为蚜虫。

**营养及功效**：含有丰富的蛋白质、维生素A和钾等矿物质，具有浓郁的菊香味，食之清凉，味苦、辛，性凉，有清热解毒、调中开胃、降血压等功效。

## 一点红 *Emilia sonchifolia*

菊科一点红属一年生或多年生植物，学名 *Emilia sonchifolia* (L.)DC，别名清香菜、羊蹄菜、红背叶、叶下红等。

原产中国，广东、广西、台湾等地均有分布，广州各区有零星种植。

**植物学性状**：根系浅，侧根多。株高20~50厘米。茎直立，横切面为圆形。单叶互生，叶稍肉质，下部叶卵形，上部叶披针形，羽状分裂或具锯齿，基部抱茎，叶柄具叶翼。叶面灰绿色，叶背常为紫红色，密布细茸毛。头状花序，具长柄，通常2~3分枝，总苞绿色，圆柱形，花紫色。果实为瘦果，矩圆形，有棱，冠毛白色。种子细小，披针形，浅褐色，表皮有条纹。采收种子应在早上进行，并用纱网罩住晒干。

**栽培环境与方法**：喜温暖湿润气候，生长适温20~30℃，冬季生长缓慢。直播和移植均可。可采收嫩梢供食用。

**病虫害**：抗逆性强，较少发生病虫害。

**营养及功效**：含丰富的矿物质和氨基酸，质地爽脆，味道清香。具清热解毒、利尿凉血、活血散瘀功效。

# 马齿苋 *Portulaca oleracea*

马齿苋科马齿苋属一年生肉质草本植物，学名 *Portulacea oleracea* L.，别名瓜子菜、马齿菜、五行草、酸苋、猪母乳、地马菜、长寿菜等。

广州各区田间、路旁、庭园很多地方都有野生种。

**植物学性状**：茎平卧或斜倚，伏地铺散，多分枝，圆柱形，淡绿色或带暗红色。叶互生，有时近对生，叶片扁平，肥厚，倒卵形，似马齿状，顶端圆钝或平截，也有微凹，基部楔形，全缘，叶面暗绿色，叶背淡绿色或带暗红色，中脉微隆起；叶柄粗短。花无梗，常3~5朵簇生枝端，午时盛开；子房无毛，花柱比雄蕊稍长，柱头4~6裂，线形。蒴果卵球形，长约5毫米，盖裂；种子细小，偏斜球形，黑褐色，有光泽。

**栽培环境与方法**：喜温暖湿润气候，适应性较强，耐旱，耐贫瘠。多采用播种及压条两种方法繁殖。可采收嫩茎叶供食用。

**病虫害**：病害主要有病毒病、白粉病及叶斑病，虫害少。

**营养及功效**：含有丰富的去甲基肾上腺素和钾盐，以及二羟乙胺、苹果酸、维生素 $B_1$、维生素 $B_2$ 等，具有解毒、消炎、利尿、止痛等功效。

# 鱼腥草 *Houttuynia cordata*

三白草科蕺菜属多年生宿根草本植物，学名 *Houttuynia cordata* Thunb.，别名蕺菜、侧耳根等。

原产亚洲、北美，我国长江流域以南各省均有分布。

**植物学性状**：根系较发达。植株高30~50厘米。茎细长，匍匐生长，上部直立，茎上有节，每节都可萌芽，地下茎白色，地上部紫色。叶互生，绿色，心脏形，顶端短渐尖，基部心形，两面有时除叶脉被毛外余均无毛，背面常呈紫红色。叶柄无毛。托叶下部与叶柄合生成鞘，稍抱茎。夏季开花，穗状花序，花小而密，白色，无花被。蒴果近球形，广州地区不结实。植株具有特殊鱼腥气味。

**栽培环境与方法**：喜温暖潮湿，怕强光，较耐寒，忌干旱，生长发育适温为16~25℃。一般于春季采用老熟地下根茎（即母根）进行繁殖，冬季植株地上部渐干枯，翌年春季又开始萌发，在广州地区可越冬。可采收嫩茎叶及地下根茎供食用。

**病虫害**：病害主要有茎腐病、白绢病、叶斑病。虫害主要有螨类（红蜘蛛）、斜纹夜蛾等。

**营养及功效**：富含蛋白质、脂肪、糖类、维生素C及钙、铁等，还含挥发油和黄酮类等成分。全株可入药，具清热解毒、利尿消肿等功效。

## 紫背菜 Gynura

菊科菊三七属多年生宿根性草本植物，学名 Gynura bicolor DC.，别名红背菜、血皮菜等。

原产中国，广东、海南、福建、四川等地均有分布，广州各区有少量栽培。

**植物学性状**：株高约50厘米，分枝性强。全株肉质，茎近圆形，直立，绿色中带紫红色，嫩茎紫红色，被茸毛。单叶互生，5叶为一叶序，叶宽披针形，先端尖，叶缘锯齿状，叶面绿色，略带紫色，叶背紫红色，表面蜡质，有光泽。头状花序，在花序梗上呈伞房状排列，花筒状，黄色，两性花，在广州很少结籽。

**栽培环境与方法**：抗逆性强，耐旱，耐热，耐贫瘠，耐阴。生长适温为20~30℃，可耐3℃的低温，在广州露地栽培可安全过冬，喜充足光照。采用扦插繁殖，可采收嫩梢供食用。

**病虫害**：病害主要有病毒病、菌核病等，虫害主要有蚜虫、B-生物型烟粉虱。

**营养及功效**：维生素C和铁含量较高，还含有黄酮苷。具治咳血、血崩、痛经、血气亏、支气管炎、中暑、阿米巴痢疾等功效。

## 珍珠菜 Ghostplant wormwood

菊科蒿属多年生草本植物，学名 Artemisia lactiflora Wall. ex Dc.，别名鸭脚艾、白苞蒿、珍珠花菜等。

原产中国，分布于华南地区，广东潮汕地区盛产，广州各区有少量栽培。

**植物学性状**：植株直立，株高40~60厘米，茎紫色，分枝能力强。茎基部有较强的不定根形成能力。茎生叶具长柄，广卵形，1次或2次羽状深裂，裂片卵形，顶端略钝或渐尖，边缘具不规则锯齿。茎生叶向上渐小，具短柄至无柄，羽状分裂，通常掌状深裂，裂片顶端最大，侧裂1~2对，有锯齿或粗齿。叶片深绿色。头状花序为多数，在枝端排列成短或长的复总状花序，花浅黄色，外层雌性，内层两性。瘦果圆柱形。

**栽培环境与方法**：喜温暖，有很强的耐高温能力，在35~38℃高温下生长良好，在广州露地栽培也能安全越冬。耐旱，耐涝，耐盐碱能力强。分株繁殖或扦插繁殖。采收嫩梢供食用。

**病虫害**：病害主要有菌核病，虫害主要有斑潜蝇等。

**营养及功效**：含有特殊的芳香物质及多种氨基酸、钾、钙等，属高钾低钠食品，常食对预防高血压、保护心脏和血管健康有益。

## 人参菜 Panicled fameflower

马齿苋科土人参属多年生草本植物，学名 Talinum paniculatum (Jacq.) Gaertn.，别名土人参、绿兰菜、玉兰菜、玉参、飞来菜等。

原产热带美洲，20世纪初引入我国台湾，目前广东、广西、海南等地均有栽培，广州各区均有少量栽培。

**植物学性状**：植株高 30~60 厘米。直根系，主根圆柱形，微弯曲，肉质。茎直立或半平卧，茎圆柱形，肉质，花茎三角形。叶片互生或近对生，肉质，倒卵形或椭圆形，基部尖，无叶柄，先端尖，全缘，无茸毛，叶脉不明显。3~7 对叶片后即抽生花序，花序顶生，为散房状聚伞圆锥花序，常二歧分枝。花两性，花萼宿存，花冠 5 瓣，桃红色。蒴果椭圆形至球形，褐色，具 2~3 裂片，成熟时弹裂。种子近球形，棕褐色。

**栽培环境与方法**：耐热，喜光，耐阴。可采用种子繁殖，也可扦插繁殖。夏季高温时可遮阳栽培。可采收嫩梢供食用。

**病虫害**：抗性强，较少发生病害，虫害主要有蚜虫、小地老虎、斜纹夜蛾等。

**营养及功效**：含有丰富的蛋白质、脂肪、钙、维生素等营养物质。具清热解毒、畅通乳汁、补中益气功效。

## 无瓣蔊菜 Indian rorippa

十字花科蔊菜属一年生草本植物，学名 Rorippa dubia (Pers.) Hara，别名塘葛菜、野油菜等。

分布于东南亚各国，我国长江以南各省区有分布。广州郊区有少量种植，以野生为主。

**植物学性状**：直根系，主根细长，侧根多。植株较柔弱，光滑无毛，直立或呈铺散状分枝，浅绿色或浅紫红色，高 10~30 厘米。单叶互生，叶大而薄。基生叶与茎下部叶倒卵形或倒卵状披针形，长 3~8 厘米，宽 1.5~3.5 厘米，多数呈大头羽状分裂；茎上部叶卵状披针形或长圆形，边缘具波状齿，短柄或无柄。总状花序，花小，多数，萼片 4 片，无花瓣或不完全花瓣，雄蕊 6 枚。长角果线形，种子每室 1 行，多数，细小，褐色，近卵形。

**栽培环境与方法**：喜温暖湿润气候，生长适温 20~30℃，冬天生长缓慢。每年 3—11 月均可采用直播繁殖。

**病虫害**：很少发生病害，虫害主要有黄曲条跳甲、小菜蛾和菜粉蝶幼虫等。

**营养及功效**：营养丰富，不仅含有蛋白质、氨基酸、多种维生素、有机酸、有机糖类、胡萝卜素、纤维素，以及钾、钙、镁、铁、磷、锌、锰、铜等营养元素，还含有蔊菜素、蔊菜酰胺、黄酮化合物和生物碱类等药用成分。具解表健胃、止咳化痰、清热解毒、散热消肿等功效。

# 刺芫荽 Foecid eryngo

伞形科刺芹属多年生草本植物，学名 *Eryngium foetidum* L.，别名野芫荽、假芫荽、刺芹。

我国广东、广西、云南等地均有分布，广州郊区零星种植。

**植物学性状**：株高 20~60 厘米。叶自根出，基生叶披针形，革质，边缘有硬骨质刺状齿，基部渐窄，无叶柄。花梗直立、粗壮，二歧分枝，其上有茎生叶。多数头状花序组成聚伞花序，总苞片叶状，开展且反折，边缘或有 1~2 疏生尖刺；花小，白色或淡绿色。双悬果球形或卵形。种子细小。

**栽培环境与方法**：喜温喜湿，耐热，耐阴，生长适温 15~35℃。可采用育苗移植，也可分株繁殖。可搭遮阳网遮阴栽培。

**病虫害**：病害主要有白粉病，虫害主要有蚜虫。

**营养及功效**：富含胡萝卜素、维生素 $B_2$、维生素 C 等，有特殊香味。嫩茎叶用于菜用，全草用于煲汤或药用，具芳香健胃、祛风清热、消炎等功效。

# 车前草 Plantain herb

车前科车前属多年生草本植物，学名 *Plantago asiatica* L.，别名钱贯草、猪肚草、车轮草、车前、田灌草等。

各地均有野生种或少量栽培。常见于路旁、沟边、田埂、荒地等。

**植物学性状**：须根系。叶卵形、宽卵形或长椭圆匙形，叶片与叶柄略等长，薄纸质，五出脉，叶基下延至叶柄，全缘、波浪缘或浇齿缘皆有，对生。穗状花序，花茎具有棱角与疏毛。蒴果，黑褐色。有车前、大车前和平车前 3 种类型。

**栽培环境与方法**：喜温暖湿润气候，较耐寒。气温 20~24℃最适宜播种。多采用种子直播繁殖。

**病虫害**：病害主要有叶斑病、根腐病、霜霉病，虫害主要有蚜虫、蝼蛄等。

**营养及功效**：含车前苷、熊果酸、维生素 C、维生素 B 等，通常拔取成熟植株用于煲汤，也可晒干备用。具清热利尿、渗湿止泻、明目、祛痰等功效。

## 野苋 Wild amaranth

苋科苋属一年生草本植物，学名 *Amaranthus viridis* L.，别名马屎苋、细苋、白苋。

各地均有分布，多处于野生状态，生长在丘陵、平原地区的路边、河堤、沟岸、田间、地埂等处。

**植物学性状**：茎直立或伏卧。叶互生，全缘，有柄。叶腋无刺。花单性或杂性，雌雄同株或异株，排成无梗的花簇，生于叶腋，或组成腋生或顶生的穗状花序。胞果卵球形，种子扁球形，黑色或褐色，平滑有光泽。

**栽培环境与方法**：耐旱，耐热，喜肥，喜阳，生命力强，不耐寒，不耐涝，容易栽培。在高温短日照条件下易开花结籽。可采用种子直播繁殖。保持肥水充足时产品较鲜嫩。

**病虫害**：病害主要有锈病，虫害主要有蚜虫等。

## 番薯叶 Sweet potato leaves

旋花科番薯属多年生草本植物，学名 *Ipomoea batatas* Lamk.，别名地瓜叶。

番薯原产于热带亚热带地区，起源于南美洲、非洲。全国各地均有栽培。

**植物学性状**：地下部分具圆形、椭圆形或纺锤形的块根，一般作蔬菜的专用品种块根少或没有。茎平卧或上升，偶有缠绕，多分枝，圆柱形或具棱，绿色或紫色，被疏柔毛或无毛，茎节易生不定根。叶片形状、颜色常因品种不同而异，通常为宽卵形，叶片基部心形或近于平截，顶端渐尖，花序腋生，在气温高、日照短的地区常见开花，温度较低的地区很少开花。蒴果卵形或扁圆形，异花授粉，自花授粉常不结实，所以有时只见开花不见结果。

**栽培环境与方法**：广州地区生产季节为3—10月，当气温稳定在15℃以上即可种植。多采用茎蔓种植，也可采用薯块繁殖。可采摘茎尖和嫩叶供食用。

**病虫害**：病害少，虫害有斜纹夜蛾等。

## 观音菜 Wild chives

百合科葱属多年生草本植物，学名 *Allium hookeri* Thwaites，别名野茎菜、宽叶韭、神仙菜等。

观音菜分布于西藏、云南、四川、贵州、广东、广西等地区。广东龙门南昆山盛产，20世纪90年代逐渐成为广州地区的特色蔬菜品种之一。

**植物学性状**：须根少，吸收能力较弱。茎为假茎，柱状，由叶鞘抱合而成，能不断分蘖出植株。叶条状，绿色，肉质。四季可抽薹开花，伞形花序近球形，花白色，以初夏最多，但不结果。与普通韭菜很相似，但其叶片较宽，纤维含量稍多，具有韭、蒜、葱三种蔬菜的香味。

**栽培环境与方法**：喜冷凉湿润气候，不耐高温和低温，生长适温为15~20℃，气温高于32℃或低于5℃时生长缓慢。忌强光照，忌积水。多采用分株繁殖，宜采用遮阳网覆盖栽培。可采收假茎叶供食用。

**病虫害**：病害以灰霉病、根腐病为主，虫害少。

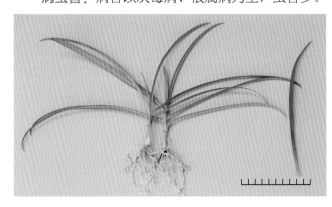

# 五、瓜类
# GOURDS

- 丝瓜
- 苦瓜
- 节瓜
- 冬瓜
- 瓠瓜
- 黄瓜
- 南瓜
- 越瓜
- 西葫芦
- 佛手瓜

# 丝瓜

葫芦科丝瓜属一年生攀援、短日性草本植物，染色体数 $2n=2x=26$。丝瓜有两个栽培种：普通丝瓜和有棱丝瓜。

丝瓜起源于亚洲热带地区，6世纪传入中国，中国古籍中记载的丝瓜主要是普通丝瓜，清代及民国年间修撰的两广地区一些县志中所载的丝瓜多指有棱丝瓜。有棱丝瓜主要分布在广东、广西、海南、福建、台湾等地，普通丝瓜则以长江流域栽培为多。

## 有棱丝瓜

学名 Luffa acutangula (L.) Roxb.，别名丝瓜、胜瓜、棱角丝瓜、棱丝瓜、角瓜。

有棱丝瓜在广州地区已有百余年栽培历史，传统栽培以长棒绿皮类型为主，近年来短棒花皮类型因具有品质优、抗逆性强、适应广的特点，栽培面积不断扩展，年栽培面积约1 500公顷。播种期2—8月，采收期4—11月。

**植物学性状**：根系发达，分布广泛。茎5棱，绿色，部分茎节墨绿色，主蔓长7~11米。叶心脏形至掌状浅裂，深绿色。雌雄同株异花，雌花单生，雄花为总状花序，偶有单生，花冠黄色，接近傍晚时分开花。果实棒形，具9~11棱，棱色绿色至墨绿色，果皮有皱纹，绿白色至墨绿色，嫩果供食。种子椭圆形，扁，表面有皱纹，黑色，千粒重110~185克。

**类型**：根据果形、皮色分为：

1. 长棒绿皮类型（又称长绿类型） 长棒形，皮色绿色至墨绿色，无花点至少花点。主栽品种有绿胜1号、绿胜3号、夏绿3号、雅绿二号、雅绿六号、雅绿八号、翠丰、满绿、夏晖3号、华绿宝丰等。

2. 短棒花皮类型（又称大肉瓜、麻皮瓜类型） 短棒形，皮色绿白色至绿色，多有花点。主栽品种有翠雅、夏胜2号、粤优2号、步步高、美味高朋、夏优4号、新秀五号等。

**栽培环境与方法**：喜温，喜光，耐热，耐湿。适宜温度：种子发芽为25~35℃，茎叶生长为20~35℃，开花结果为25~30℃。长日照下延迟花芽分化，短日照促进花芽分化。品种间对日照长短的反应有差异，栽培时应根据这一差异选择适宜品种。温暖、阳光充足、空气湿度较大、土壤水分充足、有机质丰富、土层深厚的环境条件对丝瓜生长发育最为理想。

**收获**：开花后7~12天就可采收。

**病虫害**：病害主要有猝倒病、霜霉病、疫病、白粉病、枯萎病、褐斑病、病毒病等，虫害主要有守瓜、美洲斑潜蝇、瓜实蝇、斜纹夜蛾、瓜绢螟、粉虱、蓟马等。

**营养及功效**：富含维生素、矿物质、碳水化合物、木糖胶等。具清热解暑功效。

# Loofah

Loofah (*Luffa*) an annual climbing gourd and short-day herb. Family: Cucurbitaceae. The chromosome number: $2n=2x=26$. There are two varieties: angular sponge gourd and sponge gourd.

Loofah, originated in tropical Asia, is introduced to China in the 6th century. Sponge gourd is recorded in Chinese ancient books, and angular sponge gourd is recorded in some books of Guangdong Province and Guangxi Province since Qing Dynasty. Angular sponge gourd is mainly distributed in Guangdong, Guangxi, Hainan, Fujian and Taiwan. Sponge gourd is mainly cultivated in the Yangtze River Basin.

## Angular sponge gourd

*Luffa acutangula* (L.) Roxb., synonym: Angled loofah, Ridged gourd, Angled gourd.

Angular sponge gourd has been cultivated more than 100 years in Guangzhou. The traditional cultivars are long rod green type. In recent years, the cultivation area of short rod spotted skin is continuous expanded, due to good quality, stress resistance and high adaptability. At present, the cultivation area is about 1500 hm² per annum in Guangzhou. Sowing dates: February to August. Havesting dates: April to November.

Botanical characters: Angular sponge gourd has a strong root system. Stem 5 edges, green in color or dark green in part. The main vine length can be 7~11 m. Leaf cordate or palmate, label, dark green in color. Monoecism, female flowers solitary, male flower in racemes, sometimes solitary, yellow corolla. The flower open at sundset. The fruit is cylindrical, with 9~11 edges, green or dark green in color. The skin has wrinkles, green white to dark green in color. Tender fruit is edible. Seeds oval, flat, with wrinkles, black in color. The weight per 1000 seeds is 110~185 g.

Types: According to fruit shape and color, angular sponge gourd in Guangzhou is divided into:

1. Long cylindrical and green type. Long cylindrical shaped, green or dark green in color, no or little spot. The main cultivars include: Lüsheng No.1, Lüsheng No.3, Xialü No. 3, Yalü No.2, Yalü No.6, Yalü No.8, Cuifeng, Manlü, Xiahui No.3, Hualübaofeng, etc.

2. Short cylindrical and spotted skin type. Short cylindrical shape, green white and green in color, with more spot. The main cultivars include: Cuiya, Xiasheng No.2, Yueyou No.2, Bubugao, Meiweigaopeng, Xiayou No. 4 and Xinxiu No.5, etc.

Cultivation environment and methods: The warm and sunny weather is favour for angular sponge gourd growth. Angular sponge gourd has strong tolerance to heat and wet. Suitable temperature: seed germination, 25~35 ℃, leaf and vine growth, 20~35 ℃, and flower and fruit development, 25~30 ℃. The differentiation of flower bud is delayed in the long-day and promoted in the short-day. There are differences in response to day-length in cultivars. Warm, sunny, moist weather, more organic matter, deep soil are optimal for angular sponge gourd cultivation.

Harvest: Angular sponge gourd fruit can harvest 7~12 days after anthesis.

Nutrition and efficacy: Angular sponge gourd is rich in vitamins, minerals, carbohydrates, xylans, etc.

# 有棱丝瓜 Angular sponge gourd

## 绿胜 1 号
Lüsheng No.1 angular sponge gourd

**品种来源**

广州市农业科学研究院育成的杂种一代，2001 年通过广东省农作物品种审定。

**分布地区** | 番禺、南沙、增城、花都、从化、白云等区。

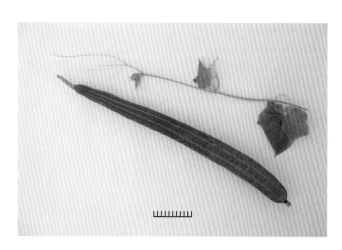

特　　征 | 叶深绿色。主蔓第 1 雌花节位第 7~22 节。瓜皮绿色，少花点，长绿类型，瓜长 50~60 厘米，横径约 4.5 厘米，具 10 棱，棱角墨绿色，光滑，棱沟较浅，单果重 350~450 克。

特　　性 | 生长势中等，分枝性较强。早熟，播种至初收春季约 45 天、秋季约 33 天，延续采收 40~60 天。主侧蔓结果，坐果性好。肉质紧实，品质优，味甜，肉质爽脆，耐贮运。耐寒性、耐热性较强，适宜春、夏、秋种植。每公顷产量 45~47.5 吨。

栽培要点 | 选择肥沃、土层深厚、前作为非瓜类的田块种植，施足基肥。播种期 2—8 月。每 667 米$^2$ 种植 500~1 300 株。搭平棚架、人字架等。出现雌花时及时追肥；早春栽培生长势较弱，应提早施肥促苗并摘除 1 米以下的雌花；初花期重施追肥，采收期勤追肥。生长前期需水较少，开花结果期需保证水分供应充足而均衡。前期摘除 1 米以下的雌花，中后期要摘除下部衰老的黄叶、病叶。保证充足的养分供应，及时采收，注意防治病虫害。

## 绿胜 3 号
Lüsheng No.3 angular sponge gourd

**品种来源**

广州市农业科学研究院育成的杂种一代，2008 年通过广东省农作物品种审定。

**分布地区** | 番禺、南沙、增城、花都、从化、白云等区。

特　　征 | 叶深绿色。主蔓第 1 雌花节位第 10~18 节。瓜皮深绿色，无花点，长绿类型，瓜条顺直，头尾匀称，长 45~55 厘米，横径 4.5~5 厘米，具 10 棱，棱角墨绿色，光滑，棱沟较浅，单果重 400~500 克。

特　　性 | 生长势较强，分枝性较强。中熟，播种至初收春季约 60 天、秋季约 44 天，延续采收 40~60 天。主侧蔓结果，坐果性好。肉质紧实，品质优，耐贮运。耐寒性较好，适宜春、秋种植。每公顷产量 37.5~52.5 吨。

栽培要点 | 选择肥沃、土层深厚、前作为非瓜类的田块种植，施足基肥。播种期 2—3 月、7—8 月。每 667 米$^2$ 种植 500~1 200 株。出现雌花后开始追肥，初花期应重施追肥，采收期勤追肥；及时采收。其余参照绿胜 1 号。

## 夏绿 3 号

Xialü No.3 angular sponge gourd

**品种来源**

广州市农业科学研究院育成的杂种一代，2006 年通过广东省农作物品种审定。

**分布地区** | 番禺、南沙、增城、花都、从化、白云等区。

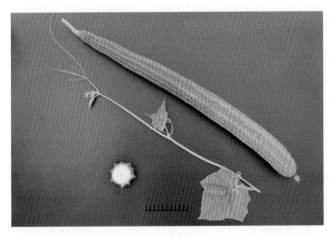

**特　　征** | 叶深绿色。主蔓第 1 雌花节位第 11~27 节。瓜皮绿色，少花点，长绿类型，头尾较均匀，瓜长 55~65 厘米，横径 4.5~5 厘米，具 10 棱，棱角墨绿色，棱沟较浅，单果重 400~500 克。

**特　　性** | 生长势较强，分枝性中等。中早熟，播种至初收 40~60 天，延续采收 40~60 天。主侧蔓结果。味甜，肉质爽脆，品质优。耐热，较耐雨水。适宜春、夏、秋种植，夏季种植优势明显。每公顷产量 45~52.5 吨。

**栽培要点** | 播种期 3—8 月，最适期 4 月和 7 月。每 667 米² 种植 500~1 200 株。夏季种植在出现雌花后方可开始追肥；勤引蔓，避免枝蔓过早爬上竹顶。其余参照绿胜 1 号。

## 雅绿二号

Yalü No.2 angular sponge gourd

**品种来源**

广东省农业科学院蔬菜研究所育成的杂种一代，2006 年通过广东省农作物品种审定。

**分布地区** | 从化、增城、番禺、南沙、白云等区。

**特　　征** | 叶深绿色。主蔓第 1 雌花节位第 9~22 节。瓜皮绿色，少花点，长绿类型，瓜长 55~65 厘米，横径 4.5~5 厘米，具 10 棱，棱角墨绿色，棱沟中等，单果重 400~550 克。

**特　　性** | 生长势较强，分枝性中等。中熟，播种至初收春季 40~55 天、秋季 40~50 天，延续采收 40~60 天。主侧蔓结果。味甜，肉质稍绵，品质良。较耐寒，适宜春、秋种植。每公顷产量 45~52.5 吨。

**栽培要点** | 播种期 2—4 月、7—8 月。每 667 米² 种植 500~1 200 株。其余参照绿胜 1 号。

## 雅绿六号 Yalü No.6 angular sponge gourd

**品种来源** 广东省农业科学院蔬菜研究所育成的杂种一代，2013年通过广东省农作物品种审定。

**分布地区**｜白云、从化、增城、番禺、南沙等区。

特　　征｜叶深绿色。主蔓第1雌花节位第6~14节，瓜皮深绿色至墨绿色，光泽性好，无花点，长绿类型，上部稍细，瓜长50~65厘米，横径4.5~5.5厘米，具10棱，棱角墨绿色，棱沟稍深，单果重400~550克。

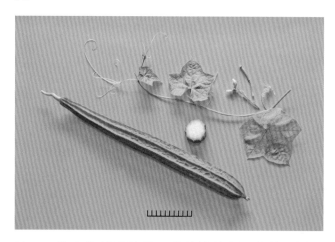

特　　性｜生长势较强，分枝性较强。中熟，播种至初收春季40~70天、秋季40~50天，延续采收40~60天。主侧蔓结果。肉质紧实，味甜，口感爽脆，品质优。较耐寒，抗病性中等。每公顷产量45~60吨。

栽培要点｜参照绿胜1号。

## 雅绿八号 Yalü No.8 angular sponge gourd

**品种来源** 广东省农业科学院蔬菜研究所育成的杂种一代。

**分布地区**｜白云、从化、增城、南沙、番禺等区。

特　　征｜叶深绿色。主蔓第1雌花节位第6~14节，瓜皮墨绿色，有光泽，无花点，长绿类型，头尾匀称，瓜长50~65厘米，横径5~5.5厘米，具10棱，棱角墨绿色、光滑，棱沟浅，单果重400~500克。

特　　性｜生长势较强，分枝性较强。中早熟，播种至初收40~60天，延续采收40~60天。主侧蔓结果。肉质紧实，味甜，口感爽脆，品质优。较耐热，抗病性、抗逆性中等。适宜春、夏、秋季种植，夏植优势较强。每公顷产量45~52.5吨。

栽培要点｜参照夏绿3号。

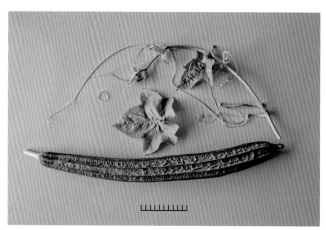

## 翠丰 Cuifeng angular sponge gourd

**品种来源** 广东省良种引进服务公司育成的杂种一代，2008年通过广东省农作物品种审定。

**分布地区**｜番禺、南沙、增城、从化等区。

特　　征｜叶深绿色。主蔓第1雌花节位第10~20节。瓜皮深绿色，无花点，长绿类型，头尾较匀称，瓜长50~60厘米，横径约4.5厘米，具10棱，棱角墨绿色，棱沟稍深，单果重约400克。

特　　性｜生长势强，分枝性较强。中熟，播种至初收44~68天，延续采收40~60天。主侧蔓结果。瓜身稍硬，肉质紧实，口感脆，味微甜，品质中等。每公顷产量37.5~52.5吨。

栽培要点｜参照绿胜1号。

## 满绿

Manlü angular sponge gourd

**品种来源** | 蔡兴利国际有限公司育成的杂种一代。

**分布地区** | 番禺、南沙、增城、从化等区。

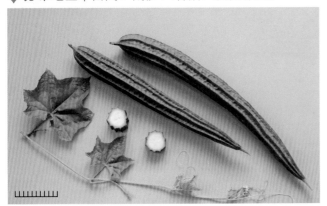

特　征 | 叶绿色。主蔓第 1 雌花节位第 15~20 节。瓜皮深绿色，无花点，长绿类型，瓜上部稍细，瓜长 50~55 厘米，横径约 4.5 厘米，具 10 棱，棱角墨绿色，棱沟较深，单果重 350~400 克。

特　性 | 生长势强，分枝性较强。中迟熟，播种至初收 40~60 天。肉质较紧实，味微甜，品质一般，瓜身较硬。耐秋风，较抗霜霉病，适宜春、秋种植，晚秋种植优势明显。每公顷产量 30~45 吨。

栽培要点 | 播种期春植 2—3 月，秋植 7—9 月，每 667 米² 种植 500~1 100 株。其余参照绿胜 1 号。

## 夏晖 3 号

Xiahui No.3 angular sponge gourd

**品种来源** | 广州市华艺种苗行有限公司育成。

**分布地区** | 全市各区。

特　征 | 叶深绿色，主蔓第 1 雌花节位第 10~26 节，瓜皮深绿色，稍有花点，长绿类型，头尾较均匀，瓜长 55~65 厘米，横径 4.5~5 厘米，具 10 棱，棱角墨绿色。棱沟较浅，单果重 400~500 克。

特　性 | 生长势较强，分枝性较强。中早熟，播种至初收 45~55 天，延续采收 45~75 天。主侧蔓结果。肉质紧实，口感爽脆，味甜，品质优。耐热，较耐雨水。适宜春、夏、秋种植。每公顷产量 45~60 吨。

栽培要点 | 参照夏绿 3 号。

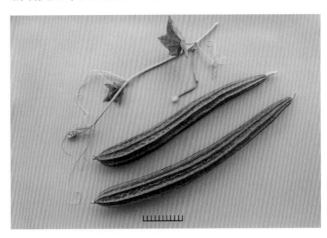

## 华绿宝丰

Hualübaofeng angular sponge gourd

**品种来源** | 广东华农大种业有限公司育成。

**分布地区** | 番禺、南沙、增城、从化等各区。

特　征 | 叶深绿色。主蔓第 1 雌花节位第 10~24 节。长绿类型，瓜皮深绿色，无花点，头尾均匀，瓜长 50~60 厘米，横径 4.5~5.5 厘米，具 10 棱，棱角墨绿色，棱沟浅，单果重 450~550 克。

特　性 | 生长势强，茎蔓粗壮，分枝性中等。中熟，播种至初收 41~60 天，延续采收 40~60 天。主侧蔓结果。肉质紧实，口感爽脆，味甜，品质优。耐热性较强，不耐雨水，适宜夏、秋种植。每公顷产量 37.5~45 吨。

栽培要点 | 每 667 米² 种植 500~1 100 株。注意防治疫病。其余参照夏绿 3 号。

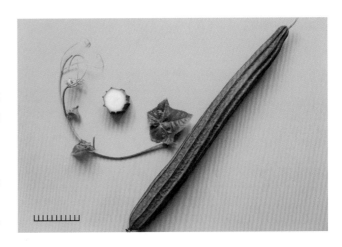

## 新秀五号 Xinxiu No.5 angular sponge gourd

**品种来源** ｜ 鹤山市沙坪恒丰种子经营部育成的杂种一代。

**分布地区** ｜ 番禺、南沙、增城、从化、白云等区。

**特　　征** ｜ 叶绿色。主蔓第1雌花节位第10~28节。瓜皮绿色，少光泽，少花点，长绿类型，瓜条端正，头尾较匀称，瓜长50~60厘米，横径4.5~5.3厘米，具10棱，棱角绿色间中有细条墨绿色，棱沟中等深，单果重400~500克。

**特　　性** ｜ 生长势强，分枝性中等。中熟，播种至初收38~55天，延续采收期较长，肉质紧实，味微甜，品质良。耐湿性强，耐热性强，较抗霜霉病。适宜春、夏、秋种植。每公顷产量45~60吨。

**栽培要点** ｜ 播种期2—8月。每667米$^2$种植500~1 100株。

## 夏胜2号 Xiasheng No.2 angular sponge gourd

**品种来源** ｜ 广州市农业科学研究院育成的杂种一代。

**分布地区** ｜ 从化、增城、花都、番禺、南沙等区。

**特　　征** ｜ 叶绿色。主蔓第1雌花节位第15~33节。大肉瓜类型，瓜形较匀称，瓜皮绿白色，有细花点，瓜长37~40厘米，横径4.5~5.3厘米，具10棱，棱角绿色，棱沟较浅，单果重300~350克。

**特　　性** ｜ 生长势较强，侧蔓较多。中熟，播种至初收45~59天，延续采收45~65天。主侧蔓结果，坐果能力强。味甜，肉质脆，品质优。抗逆性较强，适应性较广。每公顷产量45~67.5吨。

**栽培要点** ｜ 选择肥沃、土层深厚、前作为非瓜类的田块种植，施足基肥。播种期2—3月，秋植7—8月。每667米$^2$种植500~1 000株。摘除前期侧枝，后期适当疏叶。出现雌花后开始追肥，初花期重施追肥，采收期勤追肥。及时采收。注意防治病虫害。

## 翠雅 Cuiya angular sponge gourd

**品种来源** ｜ 广州市农业科学研究院育成的杂种一代。

**分布地区** ｜ 花都、增城、番禺、南沙、从化等区。

**特　　征** ｜ 叶绿色。主蔓第1雌花节位第12~28节。大肉瓜类型，瓜皮绿白色，有花点，瓜长30~35厘米，横径5~6厘米，具10棱，棱角绿色，单果重300~350克。

**特　　性** ｜ 生长势较强，侧蔓较多。中熟，播种至初收45~60天，延续采收45~60天。主侧蔓结果，连续坐果能力强。味甜，肉厚，质爽脆，品质优。适宜春、秋种植。每公顷产量45~67.5吨。

**栽培要点** ｜ 参照夏胜2号。

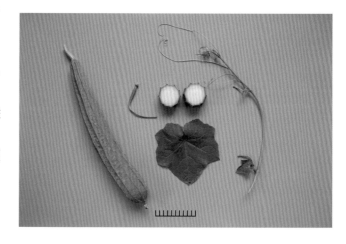

## 粤优 2 号

Yueyou No.2 angular sponge gourd

## 美味高朋

Meiweigaopeng angular sponge gourd

**品种来源**

广东省农业科学院蔬菜所育成的杂种一代，2012 年通过广东省农作物品种审定。

📍 **分布地区**｜番禺、南沙、从化、增城、花都等区。

**特　　征**｜叶绿色。主蔓第 1 雌花节位第 8~30 节。大肉瓜类型，瓜形较匀称，瓜皮绿白色，有花点，瓜长 40~47 厘米，横径约 5 厘米，具 10 棱，棱角绿色、光滑，棱沟较浅，单果重约 400 克。

**特　　性**｜生长势较强，侧蔓较多。中熟，播种至初收 40~60 天，延续采收 45~65 天。主侧蔓结果。品质优。耐寒性较强，耐旱性较强，较抗枯萎病。每公顷产量 45~60 吨。

**栽培要点**｜参照夏胜 2 号。

**品种来源**

广东省良种引进服务公司育成的杂种一代，2009 年通过广东省农作物品种审定。

📍 **分布地区**｜番禺、南沙、从化、增城、花都等区。

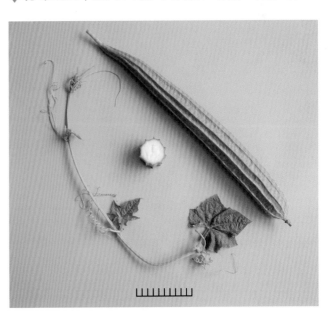

**特　　征**｜叶绿色。主蔓第 1 雌花节位第 11~28 节。大肉瓜类型，瓜条上部逐渐收缩变细，瓜皮绿白色，有少量细花点，瓜长 40~45 厘米，横径 4.5~5.0 厘米，具 10 棱，棱角浅绿色，棱沟中等深，单果重 350~410 克。

**特　　性**｜生长势强，分枝能力强。中迟熟，播种至初收 46~62 天，延续采收 45~65 天。主侧蔓结果。商品率高，肉质脆，味微甜，品质良。耐寒性较强，耐旱性较强，较抗霜霉病。播种期春植 2 月至 4 月初，秋植 7—9 月。每公顷产量 45~60 吨。

**栽培要点**｜适宜稀植，每 667 米$^2$ 种植 400~800 株，采用拱棚或平棚栽培。其余参照夏胜 2 号。

## 步步高

Bubugao angular sponge gourd

> 品种来源

广东省良种引进服务公司育成的杂种一代。

📍 分布地区 | 番禺、南沙、从化、增城、花都等区。

特　征 | 叶绿色。主蔓第1雌花节位第15~32节，大肉瓜类型，瓜条上部稍细，瓜皮绿白色，有细花点，瓜长25~30厘米，横径4.5~5厘米，具10棱，棱角绿色且有白斑，棱沟较浅，单果重270~350克。

特　性 | 生长势强，分枝能力强。中迟熟，播种至初收48~65天，延续采收45~65天。主侧蔓结果。肉质细密，味清甜，品质良。抗逆性和抗霜霉病能力较强。每公顷产量45~60吨。

栽培要点 | 参照美味高朋。

## 夏优4号

Xiayou No.4 angular sponge gourd

> 品种来源

汕头市白沙蔬菜原种研究所育成的杂种一代，2011年通过广东省农作物品种审定。

📍 分布地区 | 花都、增城、从化等区。

特　征 | 叶绿色。主蔓第1雌花节位第16~27节。大肉瓜类型，瓜条上部稍细，细花点，瓜长40~45厘米，横径4.5~5厘米，具10棱，棱角绿色，单果重300~350克。

特　性 | 生长势强，分枝性强。播种至初收40~50天，延续采收45~60天。主侧蔓结果，坐果性较强。味微甜，肉质较绵，品质良。耐寒性较强，耐热性较强，较抗霜霉病。播种期春植2—3月，秋植7—8月。每公顷产量45~52.5吨。

栽培要点 | 每667米$^2$种植500~1 100株。其余参照夏胜2号。

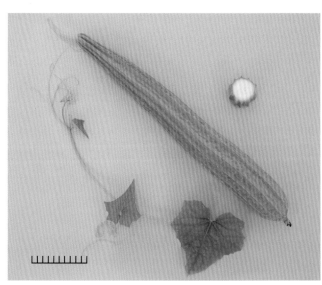

# 普通丝瓜

学名 *Luffa cylindrica* (L.) M. J. Roem.，别名水瓜、天罗瓜。

广州栽培历史悠久，公元 13 世纪已有记述，传统多为零星栽培，近年呈规模化种植，出现了水瓜专业村，如南沙区万顷沙镇同兴村。目前年栽培面积约 700 公顷。播种期春季 1—3 月，秋季 7—8 月，采收期 4—11 月。

**植物学性状**：根系发达，分布深广，易发生不定根。茎 5 棱，绿色，侧蔓多。叶掌状，叶裂较深，深绿色。雌雄同株异花，雌花单生，雄花为总状花序，花冠黄色，凌晨开花。果实圆筒形，表皮多被有茸毛和突起的瘤粒或深绿色条纹，绿色至深绿色，少数品种为白色。嫩果供食。种子椭圆形，扁，表面平滑，具翅状边缘，多为黑色，极少数为米黄色，千粒重 90~125 克。

广州地区水瓜栽培种类以中度水瓜为主，也有少数短度水瓜。

**栽培环境与方法**：喜温，喜光，耐热，耐湿，是最耐涝的瓜类蔬菜之一。适宜温度：种子发芽为 25~35℃，茎叶生长为 20~35℃，开花结果为 25~30℃。多数品种适宜春、秋季栽培，直播或育苗移植均可，出现雌花前适当控肥、控水，出现雌花后开始追肥。

**收获**：开花后 7~12 天可采收。

**病虫害**：病害主要有猝倒病、疫病、霜霉病、白粉病、枯萎病等，虫害主要有守瓜、美洲斑潜蝇、瓜实蝇、斜纹夜蛾、瓜绢螟、蓟马等。

**营养及功效**：富含维生素、矿物质、碳水化合物、木糖胶等。具清热解毒、止咳化痰等功效。

## Sponge gourd

*Luffa cylindrica* (L.) M. J. Roem. Family: Cucurbitaceae. Synonyms: Smooth loofah, Luffa gourd.

Sponge gourd has long been cultivated in Guangzhou. It was mentioneded in old literatures during 13th centuty. Sparsely cultivated in the past, and large-scaled cultivation began in recent years. At present, the cultivation area is about 700 $hm^2$ per annum in Guangzhou. Sowing dates: January to March, July to August. Harvesting dates: April to November.

Botanical characters: Sponge gourd has a strong root system. Stem 5 edges, green in color, vine with many laterals, leaf palmate. Leaf deep splitting, dark green in color. Monoecism, female flowers solitary, male flower in racemes, yellow corolla. The flower open before dawn. Cylinder fruit with villi and protuberance, or dark green stripes green to dark green, few white. Young fruit is edible. Seed oval, smooth, flat, with winged rim, black and few milk-yellow in color. The weight per 1000 seeds is 90~125 g.

In Guangzhou, the mainly cultivated type is medium body sponge gourd, and few short body sponge gourd.

Cultivation environment and methods: The warm and sunny weather is favour for sponge gourd growth. Sponge gourd has strong tolerance to heat and wet, especially to wet. Suitable temperature: seed germination, 25~35℃, leaf and vine growth, 20~35 ℃, and flower and fruit development, 25~30 ℃. Most of cultivars are suitable for cultivation in spring and autumn. Direct seeding or transplanting. Fertilization and irrigation should be properly controlled before the female flowers appearance Fertilization is enhanced after the female flowers appearance.

Harvest: Sponge gourd fruit can harvest 7~12 days after anthesis.

Nutrition and efficacy: Sponge gourd is rich in vitamins, minerals, carbohydrates, xylans, etc.

## 普通丝瓜 Sponge gourd

### 中度水瓜
Medium body sponge gourd

**品种来源**

地方品种。

📍 **分布地区** | 南沙、番禺、增城、从化等区。

特　　征 | 叶深绿色。主蔓第 1 雌花节位第 27~31 节。瓜皮深绿色，有光泽，有少量瘤状突起，瓜圆筒形，上下较匀称，瓜长 25~32 厘米，横径约 4.5 厘米，单果重 300~350 克。

特　　性 | 生长势强，分枝性强。中迟熟，播种至初收 45~75 天，延续采收 60~150 天。主侧蔓均可结果，坐果性好。肉质较紧实，品质较好。耐涝，较耐寒，较抗枯萎病。每公顷产量 37.5~75 吨。

栽培要点 | 选择肥沃、土层深厚、前作为非瓜类的田块种植，施足基肥。春播 1—3 月，秋播 7—8 月，春播为主。每 667 米² 种植 300~500 株。搭平棚架。生长前期控肥、控水，防止徒长，出现雌花时开始追肥，采收期及时追肥，开花结果期保证水分供应充足而均衡。上棚后摘除 1 米以下叶片，以利于通风透光。及时采收。因水瓜主产区多年连作导致枯萎病发生严重，可采用嫁接的方法减轻枯萎病的发生，同时注意防治白粉病、针蜂、美洲斑潜蝇、斜纹夜蛾等病虫害。

### 短度水瓜
Short body sponge gourd

**品种来源**

地方品种。

📍 **分布地区** | 南沙、番禺、增城、从化等区。

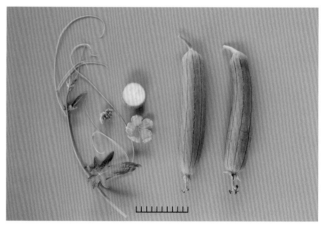

特　　征 | 叶深绿色。主蔓第 1 雌花节位第 25~35 节。瓜皮深绿色，有光泽，有少量瘤状突起，商品瓜圆筒形，上下匀称，瓜长 18~25 厘米，横径约 4.5 厘米，单果重 250~300 克。

特　　性 | 生长势强，分枝性强。中迟熟，播种至初收 50~75 天，延续采收 60~150 天。主侧蔓均可结果，坐果性好。肉质较紧实，品质较好。耐涝，较耐寒。每公顷产量 37.5~60 吨。

栽培要点 | 参照中度水瓜。

# 苦瓜

葫芦科苦瓜属一年生攀援性草本植物，学名 *Momordica charantia* L.，别名凉瓜等，古称锦荔枝、癞葡萄，染色体数 $2n=2x=22$。

苦瓜起源于古印度热带地区，中国在明代末朱橚撰的《救荒本草》中已有记载，现分布于全国，广州年栽培面积约 2 400 公顷。

**植物学性状**：根系较发达，侧根多。茎蔓生，5 棱，易发生侧蔓，侧蔓各节腋芽又能发生下一级侧蔓，形成比较繁茂的蔓叶系统。初生叶一对，对生，盾形，真叶互生，掌状裂叶，叶面绿色，叶背淡绿色，叶柄黄绿色。花单生，雌雄异花同株，雄花和雌花均为 5 片花瓣，黄色，异花授粉。果实为浆果，表面有瘤状突起，形状有纺锤形、短圆锥、长圆锥、长棒等，表皮深绿色、绿色、浅绿色、绿白色和白色等，成熟果橙黄色，果肉开裂。种子盾形，淡黄色、黄褐色或黑色，千粒重 150~180 克。

**类型**：按商品果实的特征可分为油瓜、大顶和珍珠 3 种类型。

1. **油瓜类型** 果实圆锥形、长圆锥形或棒形，瓜表面纵瘤明显，间有少量短条瘤或圆瘤，无刺状突起瘤。品种主要有绿宝石、碧绿三号、丰绿、长绿、碧丰 3 号、玉船 2 号、金秀 3 号、GL 924、越秀粗瘤、德宝 1 号、万绿早、新秀 128 等。

2. **大顶类型** 果实短圆锥形，果长在 16 厘米以下，蒂部直径在 8 厘米以上，条瘤与圆瘤相间，无刺状突起瘤。较耐寒，不耐热，适宜春季种植。品种主要有大顶苦瓜、翠绿 3 号等。

3. **珍珠类型** 果实圆锥形、长圆锥形或棒形，瓜表面布满圆瘤，部分品种间有少量短条瘤，瘤状突起尖、圆或钝。品种主要有碧珍 1 号、珠江苦瓜等。

**栽培环境与方法**：喜温，较耐热，不耐寒。种子发芽适温 30~35℃，温度在 20℃以下时发芽缓慢，13℃以下停止发芽。开花结果期适温为 20~25℃，15℃以下难以结果，30℃以上对生长和结果都不利。对光周期要求不严格，喜光不耐阴。直播和育苗移栽均可，在广州地区 3—8 月播种。

**收获**：花后 12~15 天为商品果的适宜采收期。

**病虫害**：病害主要有白粉病、枯萎病、炭疽病、疫病等，虫害主要有瓜实蝇、瓜绢螟、斜纹夜蛾、甜菜夜蛾等。

**营养及功效**：富含粗纤维、维生素 C、矿物质等。嫩果中糖苷含量高。味苦，性寒，具清热凉血、利尿、明目等功效。

# Bitter gourd

*Momordica charantia* L., an annual climbing herb. Family: Cucurbitaceae. Synonym: Bitter cucumber, Bitter melon, Balsam pear. The chromosome number: $2n=2x=22$.

Bitter gourd originated in the tropical region of ancient India, and was recorded in old literatures in 14th century in China. Bitter gourd, distributed throughout China. At present, the cultivation area is about 2400 hm$^2$ per annum in Guangzhou.

Botanical characters: Strong root system and many lateral root. Stems vine, 5 edges, with many laterals. A pair primary leaves, opposite, scutiform. Leaf alternate and palmate, green leaf, light green back and yellow-green leaf stalk. Flowers solitary, monoecious, male and female flowers with 5 petals, yellow in color, ovary, cross pollination. Fruit with protuberance in surface.

Fruit shape is spindle, cylindrical, short or long conicform, etc. Young fruit, edible, dark green, green, light green, green-white white, etc. Seeds sctiform, pale yellow, brown or black in color. The weight per 1000 seeds is 150~180 g.

Types: According to fruit characteristics, the bitter gourd cultivars can be divided into 3 types.

1. Oil bitter gourd. Fruit cylindrical, conicform or long conicform with longitudinal protuberance on the exterior, a few sliced tumor or round tumor, no stingless protuberance. The cultivars include Lübaoshi, Bilü No.3, Fenglü, Changlü, Bifeng No. 3, Yuchuan No. 2, Jinxiu No. 3, GL 924, Yuexiuculiu, Debao No. 1, Wanlüzao, Xinxiu 128, etc.

2. Broad-shoulder bitter gourd. Fruit short conicform, less than 16 cm in length and more than 8 cm in pedicle diameter, with sliced tumor and round tumor, no stingless protuberance. Broad-shoulder bitter gourd is suitable for the spring cultivation, due to with cold tolerance while no heat tolerance. The cultivars include Dading, Cuilü No.3, etc.

3. Pearl bitter gourd. Fruit cylindrical, conicform or long conicform with round protuberance on the exterior, few sliced tumor or tumor, round and blunt stingless protuberance. The cultivars include Zhujiang, Bizhen No.1, etc.

Cultivation environment and methods: Bitter gourd is heat tolerant while no cold tolerant. The suitable temperature for seed germination is 30~35℃, and germinate slowly under 20 ℃, and stop at 13℃. The suitable temperature for blossom and fruit is 20~25℃. Fruit-setting is diffcult below 15℃. The growth and fruit development is inhibited over 30 ℃. Bitter gourd require sunny weather. Direct seeding or seedling transplanting. Sowing dates: March to August.

Harvest: Bitter gourd fruit can harvest 12~15 days after anthesis.

Nutrition and efficacy: Bitter gourd is rich in crude fibers, vitamin C, minerals, etc. and has high glycoside content in young fruits.

## 绿宝石

Lübaoshi bitter gourd

**品种来源**

广东省农业科学院蔬菜研究所育成的杂种一代，1999年通过广东省农作物品种审定。

📍 **分布地区** | 全市各区。

**特　征** | 主蔓第1雌花节位第19~22节。果长圆锥形，长25~27厘米，横径6.0~7.0厘米，肉厚1.1~1.2厘米，条瘤状，瘤沟较浅，果皮绿色，单果重350~450克。

**特　性** | 生长势强。早熟，播种至初收春季67~72天、夏秋季45~50天，延续采收60~90天。田间表现中抗白粉病和枯萎病，较耐寒，耐热性和耐湿性较强。味微苦，品质中等。每公顷产量50~65吨。

**栽培要点** | 以早熟栽培为主，保护地1—2月育苗、3—4月定植，也可在3—8月育苗移栽或直播。每667米²用种量为100~150克，种植600~900株。搭人字架或平棚。生长期间及时追肥，注意病虫害防治。

## 丰绿

Fenglü bitter gourd

**品种来源**

广东省农业科学院蔬菜研究所育成的杂种一代，2010年通过广东省农作物品种审定。

📍 **分布地区** | 全市各区。

**特　征** | 主蔓第1雌花节位第23~25节。果长圆锥形，尾钝，长26~28厘米，横径6.5~8.0厘米，肉厚1.2~1.4厘米，条瘤状，瘤沟浅，果皮浅绿色，单果重450~550克。

**特　性** | 生长势强。中熟，播种至初收春季75~80天、夏秋季51~56天，延续采收60~120天。田间表现中抗白粉病和枯萎病，耐热性和耐湿性强，耐寒性中等。苦味较淡，品质中等。每公顷产量60~85吨。

**栽培要点** | 播种期3—8月，育苗移栽。每667米²用种量为50~100克，种植300~600株。搭人字架或平棚。因生长与结果期长，宜施足基肥并及时追肥，注意病虫害防治。

## 长绿 Changlü bitter gourd

### 品种来源

广东省农业科学院蔬菜研究所育成的杂种一代，2009年通过广东省农作物品种审定。

📍 分布地区｜全市各区。

特　　征｜主蔓第 1 雌花节位第 20~23 节。果长圆锥形，尾钝，长 26~28 厘米，横径 6.5~7.5 厘米，肉厚 1.1~1.2 厘米，条瘤状，瘤沟较深，果皮绿色，色泽较深，单果重 450~550 克。

特　　性｜生长势强。中熟，播种至初收春季 70~75 天、夏秋季 50~55 天，延续采收 60~120 天。耐热性和耐寒性较强，耐湿性强。味微苦，品质中等。每公顷产量 55~75 吨。

栽培要点｜参照丰绿苦瓜。

## 碧绿三号 Bilü No.3 bitter gourd

### 品种来源

广东省农业科学院蔬菜研究所育成的杂种一代，2005年通过广东省农作物品种审定。

📍 分布地区｜全市各区。

特　　征｜主蔓第 1 雌花节位第 17~21 节。果长圆锥形，长 26~29 厘米，横径 6.5~7.5 厘米，肉厚 1.1~1.2 厘米，条瘤状，瘤沟较深，果皮绿色，单果重 400~500 克。

特　　性｜生长势强。早熟，播种至初收春季 67~72 天、夏秋季 47~52 天，延续采收 60~120 天。味微苦，品质中等。每公顷产量 55~75 吨。

栽培要点｜以早熟栽培为主，保护地 1—2 月育苗，3—4 月定植，也可 3—8 月育苗移栽。每 667 米² 用种量为 50~100 克，种植 300~600 株。搭人字架或平棚。施足基肥，生长结果期间及时追肥，注意病虫害防治。

## 碧丰 3 号 Bifeng No.3 bitter gourd

### 品种来源

广州市农业科学研究院育成的杂种一代。

📍 分布地区｜全市各区。

特　　征｜主蔓第 1 雌花节位第 22~24 节。果长圆锥形，尾钝，长 26~28 厘米，横径 6.5~8.0 厘米，肉厚 1.2~1.4 厘米，条瘤状，瘤沟较浅，果皮绿色，单果重 450~550 克。

特　　性｜生长势强。中熟，播种至初收春季 75~80 天、夏秋季 51~56 天，延续采收 60~120 天。抗白粉病，中抗枯萎病，耐热、耐湿能力强，耐寒性中等。苦味较淡，品质中等。每公顷产量 60~85 吨。

栽培要点｜参照丰绿苦瓜。

## 金秀 3 号

Jinxiu No.3 bitter gourd

> 品种来源

汕头市金韩种业有限公司 2012 年育成的杂种一代。

📍 分布地区｜全市各区。

特　　征｜主蔓第 1 雌花节位第 21~23 节。果长圆锥形，长 28~32 厘米，横径 7.0~8.0 厘米，肉厚 1.1~1.3 厘米，条瘤状，瘤沟深，果皮绿色，色泽深，单果重 500~600 克。

特　　性｜生长势强。中熟，播种至初收春季 72~77 天、夏秋季 50~55 天，延续采收 60~120 天。田间表现抗白粉病、感枯萎病，耐热性和耐湿性强，耐寒性较强。味微苦，品质中等。每公顷产量 65~85 吨。

栽培要点｜参照丰绿苦瓜。

## 玉船 2 号

Yuchuan No.2 bitter gourd

> 品种来源

汕头市金韩种业有限公司 2009 年育成的杂种一代。

📍 分布地区｜全市各区。

特　　征｜主蔓第 1 雌花节位第 23~26 节，果长圆锥形，长 29~33 厘米，横径 7.0~8.0 厘米，肉厚 1.1~1.3 厘米，条瘤状，瘤沟深，果皮绿色，色泽较深，单果重 500~600 克。

特　　性｜生长势强。中迟熟，播种至初收春季 77~82 天、夏秋季 53~58 天，延续采收 60~120 天。味微苦，品质中等。每公顷产量 65~85 吨。

栽培要点｜参照丰绿苦瓜。

## GL 924

GL 924 bitter gourd

> 品种来源

广东省良种引进服务公司 2008 年自泰国引进推广的杂种一代。

📍 分布地区｜全市各区。

特　　征｜主蔓第 1 雌花节位第 20~23 节。果长圆锥形，长 25~28 厘米，横径 6.5~7.5 厘米，肉厚 1.1~1.2 厘米，条瘤状，瘤沟较浅，果绿色，单果重 400~500 克。

特　　性｜生长势强。中熟，播种至初收春季 70~75 天、夏秋季 48~53 天，延续采收 60~120 天。味微苦，品质中等。每公顷产量 60~80 吨。

栽培要点｜参照丰绿苦瓜。

## 万绿早

Wanlüzao bitter gourd

**品种来源**

广州市金苗种子有限公司 2008 年引进推广的杂种一代。

分布地区｜全市各区。

特　　征｜主蔓第 1 雌花节位第 18~22 节。果长圆锥形，长 28~30 厘米，横径 6.5~8.0 厘米，肉厚 1.1~1.2 厘米，条瘤状，果皮绿色，单果重 450~550 克。

特　　性｜生长势强。早熟，播种至初收春季 67~72 天、夏秋季 45~50 天，延续采收 60~120 天。味微苦，品质中等。每公顷产量 55~75 吨。

栽培要点｜参照碧绿三号苦瓜。

## 越秀粗瘤苦瓜

Yuexiuculiu bitter gourd

**品种来源**

广州市华艺种苗行有限公司 2013 年引进推广的杂种一代。

分布地区｜全市各区。

特　　征｜主蔓第 1 雌花节位第 19~24 节。果长圆锥形，长 27~32 厘米，横径 6.5~8.0 厘米，肉厚 1.1~1.3 厘米，条瘤状，瘤条粗大顺直，瘤沟深，果皮绿色，色泽深，单果重 500~600 克。

特　　性｜生长势强。中熟，播种至初收春季 72~77 天、夏秋季 50~55 天，延续采收 60~120 天。田间表现抗白粉病、较抗枯萎病，耐热性强，耐寒性较强，耐湿性强。味微苦，品质中等。每公顷产量 60~80 吨。

栽培要点｜参照丰绿苦瓜。

## 新秀 128

Xinxiu 128 bitter gourd

**品种来源**

广州市大农园艺种子有限公司 2012 年自泰国引进推广的杂种一代。

分布地区｜全市各区。

特　　征｜主蔓第 1 雌花节位第 21~25 节。果长圆锥形，长 28~33 厘米，横径 6.5~8.0 厘米，肉厚 1.1~1.2 厘米，条瘤状，瘤沟深，果皮绿色，色泽深，单果重 500~600 克。

特　　性｜生长势强。中熟，播种至初收春季 72~77 天、夏秋季 50~55 天，延续采收 60~120 天。味微苦，品质中等。每公顷产量 60~80 吨。

栽培要点｜参照丰绿苦瓜。

# 德宝 1 号

Debao No.1 bitter gourd

# 大顶苦瓜

Dading bitter gourd

**品种来源**

引自江门。

📍 分布地区 | 全市各区。

特　征 | 主蔓第 1 雌花节位第 15~20 节。果短圆锥形，果长 13~16 厘米，肩宽 10~12 厘米，肉厚 1.2~1.5 厘米，条瘤与圆瘤相间，果皮深绿色，单果重 350~500 克。

特　性 | 生长势中等。早熟，播种至初收春季 65~70 天、夏秋季 45~50 天，延续采收 45~75 天。田间表现中抗白粉病、较感枯萎病，耐寒性强，耐热性中等，耐湿性强。味甘，微苦，品质优。每公顷产量 30~45 吨。

栽培要点 | 以早熟栽培为主，保护地 1—2 月育苗，3—4 月定植，也可在 7—8 月播种育苗。每 667 米² 用种量为 200~300 克，种植 1 200~1 800 株。以搭竖排架为主。注意整蔓整枝，选留粗壮的母枝结果。施足基肥，结果期间勤追肥，注意病虫害防治。

**品种来源**

广州市华叶种苗科技有限公司 2011 年引进推广的杂种一代。

📍 分布地区 | 全市各区。

特　征 | 主蔓第 1 雌花节位第 20~24 节。果长圆锥形，长 27~31 厘米，横径 6.5~7.5 厘米，肉厚 1.1~1.2 厘米，条瘤状，瘤沟深，果皮绿色，色泽深，单果重 450~600 克。

特　性 | 生长势强。中熟，播种至初收春季 71~76 天、夏秋季 48~53 天，延续采收 60~120 天。田间表现抗白粉病、较感枯萎病，耐热性强，耐寒性较强，耐湿性强。味微苦，品质中等。每公顷产量 60~80 吨。

栽培要点 | 参照丰绿苦瓜。

## 翠绿 3 号 Cuilü No.3 bitter gourd

品种来源

广东省农业科学院蔬菜研究所育成的杂种一代，2005年通过广东省农作物品种审定。

分布地区 | 全市各区。

特　　征 | 主蔓第 1 雌花节位第 16~19 节，果短圆锥形，果长 14~17 厘米，肩宽 11~13 厘米，肉厚 1.2~1.5 厘米，条瘤与圆瘤相间，果皮深绿色，单果重 350~450 克。

特　　性 | 生长势中等。早熟，播种至初收春季 65~70 天、夏秋季 45~50 天，延续采收 45~75 天。田间表现中抗白粉病和枯萎病，耐寒性强，耐热性中等，耐湿性强。味甘，微苦，品质优。每公顷产量 30~45 吨。

栽培要点 | 参照大顶苦瓜。

## 碧珍 1 号 Bizhen No.1 bitter gourd

品种来源

广州市农业科学研究院 2013 年育成的杂种一代。

分布地区 | 全市各区。

特　　征 | 主蔓第 1 雌花节位第 15~18 节。果长圆锥形，果长 25~27 厘米，横径 5.5~6.5 厘米，肉厚 1.1~1.4 厘米，圆瘤，瘤状突起明显，果皮绿色，单果重 350~450 克。

特　　性 | 生长势强。早熟，播种至初收春季 65~70 天、夏秋季 45~50 天，延续采收 45~90 天。苦味较浓，质较爽脆，品质优。每公顷产量 45~60 吨。

栽培要点 | 以早熟栽培为主，保护地 1—2 月育苗，3—4 月定植，也可在 3—8 月播种育苗。适当密植，每 667 米² 用种量为 100~200 克，种植 600~1 200 株。施足基肥，结果期间要勤追肥，注意病虫害防治。

## 珠江苦瓜 Zhujiang bitter gourd

品种来源

广东省良种引进服务公司 2012 年自泰国引进推广的杂种一代。

分布地区 | 全市各区。

特　　征 | 主蔓第 1 雌花节位第 20~24 节，果长圆锥形，果长 27~32 厘米，横径 6.5~7.5 厘米，肉厚 1.1~1.3 厘米，圆瘤间少量条瘤，果皮绿色，单果重 450~550 克。

特　　性 | 生长势强。早中熟，播种至初收春季 68~73 天、夏秋季 50~55 天，延续采收 60~100 天。苦味较浓，质爽脆，品质优。每公顷产量 50~70 吨。

栽培要点 | 3—8 月播种育苗，保护地 1—2 月育苗，3—4 月定植。每 667 米² 用种量为 50~100 克，种植 300~600 株。搭人字架或平棚。施足基肥，生长与结果期间及时追肥，注意病虫害防治。

# 节瓜

葫芦科冬瓜属冬瓜种的变种，一年生攀援性草本植物，学名 *Benincasa hispida* Cogn. var. *chieh-qua* How.，别名毛瓜等，染色体数 $2n=2x=24$。

节瓜起源于中国南部和东印度，是广东特产蔬菜之一，栽培历史悠久，300多年前广州已有记载，目前年栽培面积约1500公顷。

**植物学性状**：根系较强大。茎蔓生，5棱，中空，绿色，被茸毛，易生侧蔓。蔓茎节生分枝卷须。叶为互生单叶，掌状，5~7裂，深绿色、绿色、浅绿色或黄绿色。雌雄同株异花，花单生，花冠黄色。异花授粉。果实短圆柱形到长圆柱形，深绿色、绿色到黄绿色，密被茸毛，成熟果被蜡粉或无；肉白色、浅绿色或绿色。嫩果与成熟果均可食用，以嫩果为主。种子近椭圆形，扁平，浅黄白色，种皮光滑、粗糙或具突起边缘；大籽千粒重40~60克，小籽千粒重12~15克。

**类型**：按果的肉质、表皮斑点等可分为普通和翡翠2种类型。

1. **普通类型** 分枝力较强或中等，前期以主蔓结瓜为主，中后期侧蔓结瓜，采收嫩瓜为主。瓜皮绿色、深绿色，有星点，肉白色，肉质致密，煮熟时较绵，味微甜。适宜春、秋种植。主栽品种有冠星2号、冠华4号、夏冠1号、粤秀、粤农、玲珑、丰乐、杂优连环节、丰冠等。

2. **翡翠类型** 是近年兴起的新类型。分枝力强，以侧蔓结瓜为主，采收嫩瓜和老瓜。瓜皮绿色、浅绿色，有星点，并有较多的浅黄色斑点，肉浅绿色或绿色，肉质致密，煮熟时较硬或爽脆，味微甜。抗逆性较强，适宜春、夏、秋种植。主栽品种有碧绿翡翠、冠玉1号、宝玉等。

**栽培环境与方法**：喜温暖，怕涝，耐强光。生长适温20~30℃，25℃和相对湿度85%以上较宜坐果和果实发育，高温干燥或温度低于20℃不利坐果和果实发育。广州地区春植播期1—3月，4—5月收获；夏植播期4—6月，6—8月收获；秋植播期7—8月，9—10月收获。春植育苗移栽，夏、秋植直播和育苗移栽均可。

**收获**：一般开花后8~10天、果重0.5千克左右即可采收嫩果；采收老熟果一般开花后30天采收。

**病虫害**：病害主要有猝倒病、疫病、枯萎病、白粉病、病毒病等，虫害主要有蓟马、瓜绢螟、美洲斑潜蝇、烟粉虱等。

**营养及功效**：含碳水化合物、蛋白质、维生素C等多种营养和较低的钠。味甘，性平，具清热、解暑、解毒、利尿等功效。

# Chieh-qua

*Benincasa hispida* Cogn. var. *chieh-qua* How., an annual climbing herb. Family: Cucurbitaceae. Synonym: Mao gourd, Chinese squash. The chromosome number: $2n = 2x = 24$.

Chieh-qua, originated in southern China and east India, is one of the special local vegetables in Guangdong. Chieh-qua has a long history of cultivation and is recorded in Guangzhou for 300 years. At present, the cultivation area is about 1500 hm$^2$ per annum in Guangzhou.

Botanical characters: Strong root system. Vine stems, 5 edges, hollow, green in color, coated with hairs, with many lateral. Leaf alternate, palmate with 5~7 lobes, margin serrate, leaf surface, back and petioles covered with stiff hairs, dark green, green, light green and yellow-green in color. Flowers solitary, monoecious, yellow in color, cross pollination. Fruit short or long cylindrical. Dark green, green, yellow-green in color, coated with hairs, with or without wax, powders at maturity, flesh white, light green or green in color. Both young and mature fruit can be cooked, and the former is better. Seed nearly oval, flat, light yellow-white in color, smooth or with protuberant edge. The weight per 1000 seeds is 40~60 g for large seeds, 12~15 g for small seeds.

Types: According to the fruit flesh, exterior spots, chieh-qua cultivars can be classified into 2 types.

1. Common type. The plant has medium or abound lateral vine. Fruit-setting occur on main vine at early stage while lateral vine at later stage. Fruit is green or dark green in color with spot. Flesh is white and compact, slightly sweet. The cultivars include Guanxin No.2, Guanhua No.4, Xiaguan No.1, Yuexiu, Yuenong, Linglong, Fengle, etc.

2. Emerald type. Emerald type chieh-qua has vigorous lateral vine. Fruit-setting mainly occur on lateral vine. Fruit green or light green with many light yellow spot. Flesh is white, light green and green in color and compact, crispy, slightly sweet. Emerald type chieh-qua has strong stress tolerance and the suitable cultivation seasons are spring, summer and autumn. The cultivars include Bilüfeicui, Guanyu No.1, Baoyu, Zayoulianhuanjie, Fengguan, etc.

Cultivation environment and methods: Warm weather is favour for chieh-qua growth. The suitable temperature for chieh-qua growth is 20~30 ℃. The suitable temperature and humidity for blossom and fruit setting is 25 ℃ and above 85%, Fruit-setting is difficult below 20 ℃ or in hot dry climate. Spring sowing dates: January to March and harvesting dates: April to May. Summer sowing dates: April to July and harvesting dates: June to August. Autumn sowing dates: July to August and harvesting dates: September to October. Seedling transplanting in spring, direct sowing or seedling transplanting in summer and autumn.

Harvest: The young fruit harvest 8~10 days after anthesis, about 0.5 kg. The mature fruit harvest 30 days after anthesis.

Nutrition and efficacy: Chieh-qua is rich in carbohydrates, proteins, vitamin C and lower sodium.

## 冠星 2 号

Guanxing No.2 chieh-qua

**品种来源**

广州市农业科学研究院育成的杂种一代，1999 年通过广东省农作物品种审定。

📍 分布地区 | 白云、南沙、番禺、增城、从化等区。

特　　征 | 主蔓长 400~450 厘米。叶长 20~24 厘米，宽 25~32 厘米，绿色。春播主蔓第 4~6 节、秋播主蔓第 13 节着生第 1 雌花，以后每隔 3~4 节着生 1 雌花或连续着生雌花。果实圆柱形，长 18~20 厘米，横径 7~8 厘米，深绿色，有光泽，星点稍多，无棱沟，肉厚 1.3~1.5 厘米，单果重 500~600 克。

特　　性 | 生长势强，分枝性较强。早中熟，播种至初收春播约 85 天、夏播 40 天、秋播 45 天，延续采收 35~50 天。适应性强，田间表现耐热、耐雨水。肉质致密，味微甜，品质优。每公顷产量春植 45~53 吨，夏、秋植 23~30 吨。

栽培要点 | 播种期 1—8 月，催芽直播或育苗移栽。每 667 米² 用种量 100~150 克，苗期 20~35 天，早播宜采用薄膜覆盖防寒。支架栽培，双行植，行距 60 厘米，株距春植 35~40 厘米、夏秋植 30~35 厘米。以有机肥和磷肥、钾肥作基肥，施足基肥，苗期适当控制肥水，以防徒长，开花结果期施重肥。摘除 1 米以下侧蔓。人工辅助授粉，及时采收。注意猝倒病、疫病、蓟马、瓜绢螟等病虫害防治。4—11 月收获。

## 冠华 4 号

Guanhua No.4 chieh-qua

**品种来源**

广州市农业科学研究院育成的杂种一代，2005 年通过广东省农作物品种审定。

📍 分布地区 | 从化、白云等区。

特　　征 | 主蔓长 450~500 厘米。叶长 19~24 厘米，宽 25~32 厘米，绿色。春播主蔓第 6~8 节、秋播主蔓第 10~13 节着生第 1 雌花，以后每隔 2~4 节连续着生 2 雌花。果实圆柱形，头尾均匀，长 15~17 厘米，横径 7~8 厘米，深绿色，有光泽，星点较少，无棱沟，肉厚 1.4~1.5 厘米，单果重 480~525 克。

特　　性 | 生长势强，分枝力强。早熟，播种至初收春播约 77 天、夏播 45 天、秋播 49 天，延续采收 35~50 天。高抗枯萎病，适应性强，田间表现耐涝性、耐寒性、耐热性较强。商品瓜率高，肉质致密、脆嫩，味微甜，品质优。每公顷产量春、秋植 38~60 吨，夏植 23~30 吨。

栽培要点 | 参照冠星 2 号。

## 粤农
Yuenong chieh-qua

**品种来源**

广东省农业科学院蔬菜研究所育成的杂种一代,1999年通过广东省农作物品种审定。

📍 **分布地区** | 白云、从化、增城等区。

**特　征** | 主蔓长450~500厘米。叶长22~25厘米,宽30~34厘米,深绿色。主蔓第5~9节着生第1雌花,以后每隔2~3节着生1雌花或连续出现雌花。果实短圆柱形,长15厘米,横径6厘米,深绿色,有光泽,无棱沟,星点较少,肉厚1.2~1.5厘米,单果重300~350克。

**特　性** | 生长势强,分枝性中等。早熟,播种至初收春播约80天,秋播40~45天,延续采收30~40天。适应性广,田间表现抗枯萎病,耐寒性较强。肉质嫩滑,味微甜,品质优。每公顷产量30~45吨。

**栽培要点** | 播种期春植1—3月、秋植7月至8月上旬。其余参照冠星2号。

## 夏冠1号
Xiaguan No.1 chieh-qua

**品种来源**

广东省农业科学院蔬菜研究所育成的杂种一代,2005年通过广东省农作物品种审定。

📍 **分布地区** | 白云、番禺、从化、增城等区。

**特　征** | 主蔓长450~500厘米。叶长22~25厘米,宽28~34厘米,绿色。春播主蔓第10~11节、秋播主蔓第12~13节着生第1雌花。果实圆柱形,长16~17厘米,横径5.5~6.0厘米,青绿色,有光泽,有星点,无棱沟,肉厚1.5厘米左右,单果重450~500克。

**特　性** | 生长势强,分枝力强。早熟,播种至初收春播约78天,夏、秋播45~51天,延续采收35~50天。高抗枯萎病,适应性强,田间表现耐涝性、耐热性较强。肉质致密,味微甜,品质优。每公顷产量60吨左右。

**栽培要点** | 参照冠星2号。

## 丰乐 Fengle chieh-qua

**品种来源** 广东省农业科学院蔬菜研究所育成的杂种一代，2003年通过广东省农作物品种审定。

**分布地区** | 增城、从化等区。

**特　征** | 主蔓长450~500厘米。叶长23~26厘米，宽30~34厘米，深绿色。主蔓第10~15节着生第1雌花。果实长圆柱形，长17~18厘米，横径5.5厘米左右，深绿色，有光泽，有星点，无棱沟，肉厚1.2厘米左右，单果重400克左右。

**特　性** | 早中熟，播种至初收春播70~85天、秋播48~55天，延续采收35~50天。耐贮运。肉质致密，味微甜，品质优。每公顷产量25~35吨。

**栽培要点** | 播期春植1—3月、秋植7—8月。其余参照冠星2号。

## 玲珑 Linglong chieh-qua

**品种来源** 广东省农业科学院蔬菜研究所育成的杂种一代，2012年通过广东省农作物品种审定。

**分布地区** | 番禺、从化、增城等区。

**特　征** | 主蔓长450~500厘米。叶长18~23厘米，宽25~31厘米，绿色。主蔓第6~10节着生第1雌花。果实圆柱形，长16~18厘米，横径7~8厘米，深绿色，有光泽，星点小而多，无棱沟，肉厚1.5~1.6厘米，单果重500克左右。

**特　性** | 早中熟，播种至初收春播约84天、秋播49天，延续采收35~43天。适应性强。肉质致密，味微甜，品质优。每公顷产量35~45吨。

**栽培要点** | 播种期春植1月下旬至3月、秋植7—8月。其余参照冠星2号。

## 粤秀 Yuexiu chieh-qua

**品种来源** 广州卓艺种业有限公司2008年育成的杂种一代。

**分布地区** | 白云、从化等区。

**特　征** | 主蔓长450~500厘米。叶长21~27厘米，宽29~36厘米，绿色。春播第7~8节、秋播第10~13节着生第1雌花。果实圆柱形，长15~18厘米，横径6~8厘米，深绿色，有光泽，星点较少，无棱沟，肉厚1.3~1.4厘米，单果重400~500克。

**特　性** | 生长势强，分枝力强。早熟，播种至初收春播70~75天、秋播40~45天，延续采收35~50天。肉质致密，味微甜，品质优。每公顷产量33~60吨。

**栽培要点** | 参照冠星2号。

## 七星仔

Qixingzai chieh-qua

**品种来源** 

地方品种。

**分布地区** | 白云、从化、黄埔等区少量种植。

特　　征 | 主蔓长350~450厘米。叶长18厘米，宽18厘米，深绿色，叶缘具细齿。主蔓第3~7节着生第1雌花，以后每隔4~5节着生1雌花。果实长圆柱形，长16~20厘米，横径6.0~6.5厘米，青绿色，有星点，有光泽，被白色茸毛，肉厚1.5~2.0厘米，白色，单果重300~400克，成熟果被白色蜡粉。

特　　性 | 生长势强，分枝性中等偏强。中熟，播种至初收春播约85天、秋播约45天，延续采收35~45天。以主蔓结果为主。适应性广，味微甜，品质优。每公顷产量23~30吨。

栽培要点 | 播种期1—8月。其余参照冠星2号。

## 碧绿翡翠

Bilüfeicui chieh-qua

**品种来源** 

广东省良种引进服务公司2006年从泰国引进的杂种一代。

**分布地区** | 从化、白云等区。

特　　征 | 主蔓长600厘米以上。叶长23厘米，宽28厘米，深绿色。春播第18~25节、秋播第15~20节着生第1雌花，以后每隔4~6节着生1雌花或连续出现雌花。果实棒形，嫩瓜长20厘米以上，横径5~7厘米，浅绿色，较多浅黄色斑点，无棱沟；肉厚1.1~1.5厘米，浅绿色，瓜腔小，单果重350~550克。老瓜长30厘米以上，单果重3.5千克左右，表面被蜡粉。

特　　性 | 生长势强，分枝性强。中熟，播种至初收春播约98天，夏、秋播46天，延续采收50~90天。主侧蔓结果，坐果性强。肉质碧绿致密、爽甜，品质优。适应性强，抗病性强，耐寒性、耐热性较强；耐贮运，货架期长。每公顷产量60~90吨。

栽培要点 | 播期春植为1—3月、秋植为7—8月。催芽直播或育苗移栽。单行疏植，株距1米，行距4米，每667米$^2$种植300株以下，拱棚或平棚架栽培。以有机肥和磷肥、钾肥作基肥，施足基肥，苗期适当控制肥水，以防徒长，开花结果期施重肥。适当压蔓后再引蔓上架，结果前摘除基部侧蔓。人工辅助授粉，及时采收。4—11月收获。

## 冠玉 1 号

Guanyu No.1 chieh-qua

> 品种来源

广州市农业科学研究院 2013 年育成的杂种一代。

分布地区｜白云、番禺、增城等区。

特　　征｜主蔓长 600 厘米以上。叶长 20~25 厘米，宽 25~30 厘米，深绿色。春播第 10~15 节、秋播第 13~17 节着生第 1 雌花，以后每隔 3~6 节着生 1 雌花或连续出现雌花。果实长圆筒形，头尾均匀，嫩瓜长 20~25 厘米，横径 6~8 厘米，深绿色，较多浅黄色斑点，无棱沟；肉厚 1.2~1.6 厘米，瓜腔小，浅绿色，单果重 500~660 克。老瓜长 35 厘米左右，单果重 3 千克左右，表面无粉。

特　　性｜生长势强，分枝性强。早中熟，播种至初收春播约 84 天，夏、秋播约 50 天，延续采收 50~90 天。主侧蔓结果，坐果性强。肉质翠绿致密、爽甜，品质优。适应性强，田间表现较抗枯萎病，耐寒性、耐热性、耐旱性较强，耐涝性稍强；耐贮运，货架期长。每公顷产量 60~90 吨。

栽培要点｜播种期 1—8 月，每 667 米² 人字架栽培 1 200~1 500 株、平棚栽培 600~800 株。其余参照碧绿翡翠。

## 宝玉

Baoyu chieh-qua

> 品种来源

广州亚蔬园艺种苗有限公司 2010 年育成的杂种一代。

分布地区｜白云、番禺、从化等区。

特　　征｜主蔓长 600 厘米以上。叶长 25~31 厘米，宽 28~33 厘米，绿色。春播第 9~12 节、秋播第 18 节左右着生第 1 雌花，以后每隔 3~5 节着生 1 雌花或连续出现雌花。果实长圆筒形，头尾均匀，嫩瓜长 18~25 厘米，横径 6~8 厘米，翠绿色带绿白斑点，无棱沟；肉厚 1.2~1.6 厘米，浅绿色，瓜腔小，单果重 550~750 克。老瓜长 35 厘米以上，肉厚 4 厘米左右，单果重 2.5 千克左右，表面无粉。

特　　性｜生长势强，分枝性强。早熟，播种至初收春播约 84 天，夏、秋播 50~55 天，延续采收 40~50 天。主侧蔓结果，坐果性强。肉质翠绿致密、爽甜，品质优。适应性强，田间表现耐热性、耐旱性较强，耐涝性稍强，耐寒性中等；耐枯萎病，较抗疫病。耐贮运，货架期长。每公顷产量 55~90 吨。

栽培要点｜播种期 1—8 月，每 667 米² 人字架栽培 1 200~1 500 株、平棚栽培 600~800 株。其余参照碧绿翡翠。

## 杂优连环节

Zayoulianhuanjie chieh-qua

### 品种来源

广东省农业科学院蔬菜研究所 2006 年育成的杂种一代。

分布地区 | 增城、从化等区。

特　　征 | 主蔓长 500 厘米以上。叶长 23 厘米，宽 28 厘米，绿色。主蔓第 5~9 节着生第 1 雌花。果实长圆柱形，长 28~35 厘米，横径约 5 厘米，浅绿色，有光泽，无棱沟，肉厚 1.5 厘米以上，单果重 500 克左右。

特　　性 | 生长势中等，分枝性强。早熟，播种至初收春播 65~75 天、秋播 45 天左右，延续采收 30~40 天。田间表现抗逆性中等。瓜肉硬实，味微甜，品质优。每公顷产量 60 吨以上。

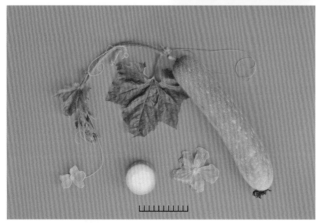

栽培要点 | 播种期春植 1—3 月、秋植 7—8 月。每 667 米$^2$ 种植 2 400 株左右。其余参照碧绿翡翠。

## 丰冠

Fengguan chieh-qua

### 品种来源

广东省农业科学院蔬菜研究所育成的杂种一代，2009 年通过广东省农作物品种审定。

分布地区 | 增城、从化等区。

特　　征 | 主蔓长 450~500 厘米。叶长 23~27 厘米，宽 30~35 厘米，深绿色。第 6~10 节着生第 1 雌花；第 8~11 节着生第 1 个瓜。果实长圆筒形，长 24~27 厘米，横径 6.5~7.0 厘米，绿色，无棱沟，浅黄色斑点较小，肉厚 1.4~1.6 厘米，单果重 650 克左右。

特　　性 | 生长势强，分枝力中等。中熟，播种至初收春播 85 天左右、秋播 50 天左右，延续采收 38~43 天。肉质致密，品质较好。高抗枯萎病，田间表现耐热性、耐寒性、耐涝性和耐旱性强。每公顷产量 50~60 吨。

栽培要点 | 播种期 1—8 月。每 667 米$^2$ 种植 2 400 株左右。其余参照碧绿翡翠。

# 冬瓜

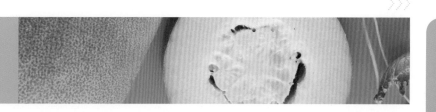

葫芦科冬瓜属一年生攀援性草本植物，学名 Benincasa hispida Cogn.，别名东瓜，染色体数 $2n=2x=24$。

冬瓜起源于中国南部和东印度，最早文字记载见于秦汉时期《神农本草经》，是广东特产蔬菜之一，300多年前的文献就把冬瓜列为"广瓜"之一，目前在广州栽培普遍，年栽培面积约 1 500 公顷。

**植物学性状**：根系发达。茎蔓生，5 棱，中空，绿色，被茸毛，易生侧蔓。蔓茎节生分枝卷须。单叶互生，掌状，5~7 裂，叶深绿色、绿色。雌雄异花同株，花单生，花冠黄色，异花授粉。果实有扁圆形、短圆柱形到长圆柱形等，皮色浅绿色至墨绿色，密被茸毛，成熟果茸毛渐少，被蜡粉或无；肉白色。嫩果与成熟果均可食用，以成熟果为主。种子近椭圆形，扁平，浅黄白色，种皮光滑或具突起边缘；千粒重 40~60 克。

**类型**：按果实大小可分为大型、中型和小型 3 种类型。按果实表皮颜色和被蜡粉与否，又可分为青皮冬瓜和灰（粉）皮冬瓜。

1. **大型冬瓜** 中晚熟或晚熟。果实短圆柱形至长圆柱形，被白蜡粉或无。单果重在 10 千克以上，每株采收 1 个，采收老熟瓜上市。主栽品种有黑皮冬、黑优 2 号、铁柱、灰皮冬瓜等。

2. **中型冬瓜** 较早熟或中熟。果实短圆柱形至长圆柱形，单果重 5~10 千克，每株可采收 1~2 个，采收老熟瓜上市。主栽品种有东莞黑皮冬等，广州较少种植。

3. **小型冬瓜** 早熟或较早熟。果较小，扁圆形、短圆柱形或长圆柱形，单果重 0.7~2 千克，每株可采收十至二十多个，主侧蔓均可结果，采收嫩瓜或老熟瓜上市。主栽品种有迷你小冬瓜等。

**栽培环境与方法**：喜温耐热，怕涝。生长适温 20~30℃；25℃左右较宜蔓叶生长和开花结果，15℃以下不利于开花结果。广州地区春植播期 12 月至翌年 3 月，5—7 月收获；秋植播期 6—7 月，10 月收获。春植育苗移栽，秋植直播和育苗移栽均可。

**收获**：小型冬瓜采收嫩果一般开花后 10 天左右采收；采收老熟果一般开花后 40 天以上、皮色转色后采收。

**病虫害**：病害主要有猝倒病、疫病、枯萎病、白粉病、病毒病等，虫害主要有蓟马、瓜绢螟、美洲斑潜蝇、烟粉虱等。

**营养及功效**：富含多种维生素和钾、钙、磷、铁等人体所需元素。味甘，性凉，具清热解毒、利尿消痰、祛湿解暑、除烦止渴、消肿减肥等功效。

Benincasa hispida Cogn., an annual climbing herb. Family: Cucurbitaceae. Synonym: Tallow gourd, Ash gourd. The chromosome number: $2n=2x=24$.

Wax gourd originated in southern China and east India, is one of the special local vegetables in Guangdong which is listed as one of "Guangzhou gourd" in the literature of more than 300 years ago. At present, the cultivation area is about 1500 ha per annum in Guangzhou.

# Wax gourd

Botanical characters: Strong root system. Vine stems, 5 edges, hollow, green in color, coated with hairs, with many lateral. Leaf alternate and palmate with 5~7 lobes, margin serrate, dark green or green in color. Flowers solitary, monoecious, yellow in color, cross pollination. Fruit oval, short or long cylindrica, light green to dark green in color, coated with hairs. Mature fruit with or without wax powders, white in color. Young gourd and mature gourd are edible, mainly the later. Seed nearly oval, flat, light yellow-white in color, seed coat smooth or with protuberant edge. The weight per 1000 seeds is 40~60 g.

Types: According to the size of the fruit, the cultivars can be divided into 3 types: large, medium and small wax gourd. According to the skin color coated with wax or not, the cultivars can be divided into 2 types: Green skin and Grey skin type.

1. Large wax gourd: Middle-late or late maturing cultivars. Fruit short or long cylindrical in shape, light green to dark green in color, The weight per gourd is above 10 kg. One fruit per plant is set. The cultivars include Dark skin wax gourd, Heiyou No.2, Tiezhu, Grey skin wax gourd, etc.

2. Medium wax gourd: Early or medium maturing cultivars. Fruit short or long cylindrical in shape, light green to dark green in color. The weight per gourd is 5~10 kg. One to two fruit per plant is set. The cultivars include Dongguan dark skin wax gourd.

3. Small wax gourd: Early maturing cultivars. Fruit oval, short or long cylindrical in shape, light green to dark green in color. The weight per gourd is 1~2 kg. A dozen fruits per plant are set. The cultivars include Holland green skin wax gourd, Japanese green skin wax gourd, Taiwan mini 3201.

Cultivation environment and methods: Warm weather is favour for wax gourd growth. The suitable temperature for wax gourd growth is 20~30℃ for vine and leaf growth, 25℃ is for blossom and fruit setting. Blossom and fruit setting is difficult below 15℃. Spring sowing dates: December to March the following year and harvesting dates: May to July. Autumn sowing dates: June to July and harvesting dates: October. Seedling transplant in spring, direct sowing or seedling transplanting in autumn.

Harvest: The young fruit harvest 10 days after anthesis and the mature fruit harvest 40 days after anthesis.

Nutrition and efficacy: Wax gourd is rich in vitamins, potassium, calcium, phosphorus, iron, etc.

## 黑优 2 号

Heiyou No.2 wax gourd

**品种来源**

广东省农业科学院蔬菜研究所育成的杂种一代，2010年通过广东省农作物品种审定。

分布地区｜增城、南沙、番禺等区。

特　　征｜主蔓长 350 厘米以上。叶长 25~30 厘米，宽 26~32 厘米，绿色。主蔓第 14~16 节着生第 1 雌花，以后每隔 4 节着生 1 雌花或连续 2 节着生雌花。果实长圆柱形，长 64.5 厘米，横径 22.0 厘米，墨绿色，有光泽，头尾均匀，浅棱沟，肉厚 6.0 厘米，白色，单果重 14 千克左右。

特　　性｜早熟，生育期春播 120 天、秋播 95 天。生长势强，侧蔓较多。瓜皮转墨绿色早，肉质致密，品质优。抗枯萎病，中抗疫病，田间表现抗病毒病。每公顷产量 75~92 吨。

栽培要点｜播种期春季 2 月初、夏秋季 6 月底至 7 月中旬。支架栽培，单行植，每 667 米$^2$ 种植 500~550 株。多施基肥，开花期及时补充磷肥、钾肥。坐果前摘除全部侧蔓，坐果后留 2~3 条侧蔓。人工辅助授粉，每株预留 2 个幼瓜，待幼果到 0.5 千克时再择优去劣定 1 个瓜，定瓜位置控制在 25~30 节。定瓜后，留 10~12 片健全叶打顶。果重 1~1.5 千克时吊果。注意防治病虫害。6—7 月或 10—11 月采收。

## 铁柱

Tiezhu wax gourd

> 品种来源

广东省农业科学院蔬菜研究所育成的杂种一代，2013年通过广东省农作物品种审定。

📍 分布地区 | 增城、南沙、番禺等区。

特　　征 | 主蔓长 350 厘米以上。叶长 25~30 厘米，宽 26~32 厘米，绿色。主蔓第 15~17 节着生第 1 雌花，以后每隔 4 节着生 1 雌花或连续多节着生雌花。果实长圆柱形，瓜形瘦长，尾部钝尖，长 80~100 厘米，横径 17~20 厘米，墨绿色，浅棱沟，肉厚 6.6~6.8 厘米，白色，单果重 16 千克左右。

特　　性 | 中晚熟，生育期春播 120 天、秋播 95 天。生长势强，分枝性中等。肉质致密，囊腔小，品质优。耐贮运，抗枯萎病，中抗疫病。每公顷产量 93~97 吨。

栽培要点 | 参照黑优 2 号。

## 灰皮冬瓜（灰斗）

Grey skin wax gourd

> 品种来源

地方品种。

📍 分布地区 | 从化、增城等区。

特　　征 | 主蔓长 450 厘米以上。叶长 20~22 厘米，宽 30~32 厘米，深绿色。主蔓第 12~15 节着生第 1 雌花，以后每隔 4~5 节着生 1 雌花或连续多节着生雌花。果实长圆柱形，长 40~60 厘米，横径 20~25 厘米，深绿色，被白色蜡粉，肉厚 4~5 厘米，白色。单果重 15~20 千克。种子边缘有棱，白色。

特　　性 | 中晚熟，生长期 140~150 天，侧蔓多。肉质致密，耐贮运。纤维较多。田间表现抗性强，耐旱，耐热。每公顷产量 80~90 吨。

栽培要点 | 播种期春季 2—3 月。每 667 米$^2$ 种植 550~600 株。6—7 月采收。其余参照黑优 2 号。

## 黑皮冬

Dark skin wax gourd

> 品种来源

地方品种。

📍 分布地区｜全市各区。

特　　征｜主蔓长 450~500 厘米。叶长 19 厘米，宽 28 厘米，深绿色。主蔓第 16~22 节着生第 1 雌花，以后每隔 4~5 节着生 1 雌花或连续多节着生雌花。果实长圆柱形，长 41 厘米，横径 25 厘米，墨绿色，肉厚 6.5 厘米，白色。单果重 13~15 千克。

特　　性｜晚熟，春播生长期 170~180 天，侧蔓多。肉质致密，耐贮运。味清淡，品质优。每公顷产量 45~53 吨。

栽培要点｜播种期春季 1—3 月、秋季 7 月。每 667 米² 种植 550~600 株。5—6 月或 10 月采收。其余参照黑优 2 号。

## 迷你小冬瓜

Mini wax gourd

> 品种来源

广州农达种子科技有限公司引进。

📍 分布地区｜番禺、南沙、增城等区。

特　　征｜主蔓长 10 米以上。叶长 24 厘米，宽 31 厘米，绿色。主蔓第 5~8 节着生第 1 雌花，以后每隔 3~6 节着生 1 雌花或连续多节着生雌花。果实短圆筒形，嫩果长 14 厘米，横径 11 厘米，青绿色，肉厚 3.5 厘米，单瓜重 0.7~1 千克；老熟果青白色，表面无蜡粉，单果重 2.5 千克左右。每公顷可采收嫩瓜 45 000~60 000 个，产量 45~60 吨。

特　　性｜早熟，春播生长期 170~180 天。生长势强，侧蔓多。瓜形整齐，肉质致密、爽脆细腻，味清淡，品质优。耐贮运。田间表现耐热性强。

栽培要点｜播种期春季 1—3 月、秋季 7—8 月。春季催芽后育苗移栽，秋季催芽后直播或育苗移栽，棚架栽培，单行植，每 667 米² 种植 200 株。多施基肥，当蔓长到 1 米左右应适当追肥，以后根据生长情况，一般采收 1 次追肥 1 次。当蔓长超过 1 米时向同一方向均匀牵引，及时去除没有到达棚顶的侧枝。开花期如遇阴雨天需人工辅助授粉，注意防治病虫害。5—6 月或 10—11 月采收，嫩果开花后 10~15 天采收，老熟果开花后 45 天左右采收。

# 瓠瓜

葫芦科葫芦属一年生攀援、短日性草本植物，学名 Lagenaria siceraria (Molina) Standl.，别名蒲瓜、扁蒲、葫芦、夜开花等，染色体数 $2n=2x=22$。

中国南北各地均有栽培，南方较普遍。广州栽培历史悠久，公元9世纪已有记载，目前年栽培面积约300公顷。播种期12月至翌年8月，采收期4—11月。

**植物学性状**：浅根系，侧根发达，但再生力弱。茎蔓生，5棱，侧枝多。叶心脏形，浅裂，茎叶均密被白色茸毛。雌雄同株异花，花单生，花冠白色，晚上或弱光下开花，故俗称"夜开花"；以侧蔓结果为主。瓠瓜形状多样，以嫩果为食，皮浅绿色至绿色，果肉白色。种子近楔形，扁，灰黄色，千粒重125~170克。

**类型**：广州栽培的瓠瓜根据瓜形可分为：

1. **短蒲类型**　短圆柱形，皮浅绿色，较耐寒，是春季栽培的主要类型，主栽品种有绿富短蒲瓜、青秀蒲瓜、油青早1号蒲瓜等。

2. **葫芦类型**　近葫芦形，皮浅绿色至绿色，下部较大，上部较小，略窄腰，较耐热，是夏秋栽培主要类型。

**栽培环境与方法**：喜温暖、湿润的气候和有机质丰富的土壤。其适宜温度，种子发芽为30~35℃，生长发育为20~25℃，5℃以下受害。一般葫芦型蒲瓜比短蒲型耐热，短蒲型比葫芦型蒲瓜耐寒。苗期短日照有利于雌花形成。

**收获**：开花后11~20天，皮色变淡而略带白色，肉质坚实且富有弹性时采收。

**病虫害**：病害主要有疫病、白粉病、病毒病等；虫害主要有美洲斑潜蝇、瓜实蝇等。

**营养及功效**：含有较丰富的蛋白质、糖分、有机酸和各种维生素。味甘，性平，具清热利水、止渴、解毒等功效。

Lagenaria siceraria (Molina) Standl., an annual climbing herb. Family: Cucurbitaceae. Synonym: White-flower gourd, Calabash gourd, Trumpet gourd. The chromosome number: $2n=2x=22$.

Bottle gourd is cultivated widely in China, and more common in the South. It has long been cultivated in Guangzhou. It was mentioned in old literatures during the ninth century. At present, the cultivation area is about 300 $hm^2$ per annum in Guangzhou. Sowing dates: December to August the following year, harvesting dates: April to November.

Botanical characters: Strong lateral root system. Stem vine, 5 edges, with a lot of laterals. Leaf cordate, lobed, coated with dense white stiff hairs. Flowers solitary, monoecious, white in color. Fruit-setting mainly occur on lateral vine. Pepo variable in shape, young fruits (edible) with skin light green and flesh white in color, Seeds nearly cuneate, flat, yellow-grey in color. The weight per 1000 seeds is 125~170 g.

# Bottle gourd

# 葫芦瓜

Gourd calabash

**品种来源**

地方品种。

📍 **分布地区** | 南沙、番禺等区。

Types: According to fruit shape, bottle gourd cultivars in Guangzhou is divided into:

1. Short calabash: Fruit short cylindrical, light green skin in color. Due to the cold-tolerance, they are cultivated in spring. The cultivars include Lüfuduan, Qingxiu, Youqingzao No.1, etc.

2. Calabash: Nearly pear-shaped, light green to green in color. Larger at the base, while smaller near fruit stalk with a slender middle part. Due to the heat-tolerance, they are cultivated in summer and autumn.

Cultivation environment and methods: The rich organic soil, warm and humid climate are favour for bottle gourd growth. The suitable temperature: seed germination, 30~35℃, plant growth and development, 20~25℃. It was injured below 5℃. The formation of female flowers requried short-day at seedling stage.

Harvest: Bottle gourd young fruit harvest 11~20 days after anthesis.

Nutrition and efficacy: Bottle gourd is rich in proteins, sugars, organic acids and vitamins.

**特　　征** | 叶片绿色。第1雌花着生节位从主蔓到侧蔓共12~22节。果实近葫芦形，瓜长22~30厘米，横径13~16厘米，浅绿色，肉白色，单果重750~1 000克。

**特　　性** | 生长势强，侧蔓多。中早熟，播种至初收60~70天，延续采收40~50天。侧蔓结果为主，坐果性强。肉质嫩滑，品质优。较耐热。每公顷产量可达52.5吨以上。

**栽培要点** | 采用平棚栽培，播种期春季2—3月，秋季7—8月，夏、秋季栽培优势较明显，长至6~7节时打顶，及时引蔓上架。注意防治病虫害。

## 青秀蒲瓜 Qingxiu calabash

> 品种来源

广州市农业科学研究院育成的杂种一代。

📍 分布地区 | 南沙、番禺、增城、从化等区。

特　　征 | 叶片绿色。第1雌花着生节位从主蔓到侧蔓共 10~15 节。果实圆柱形，长 22~25 厘米，横径 7~8 厘米，皮绿色，有光泽，肉白色，单瓜重 500~750 克。

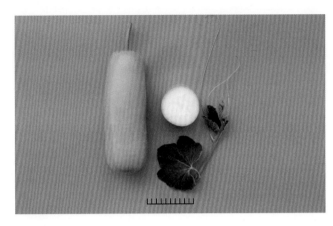

顷产量可达 60 吨以上。

栽培要点 | 播种期春植 12 月末至翌年 3 月初，秋植 7 月至 8 月中旬。每 667 米$^2$ 种植 600 株。其余参照葫芦瓜。

## 绿富短蒲瓜 Lüfuduan calabash

> 品种来源

广州市金苗种子有限公司育成。

📍 分布地区 | 南沙、番禺、增城、从化、白云等区。

特　　征 | 叶片绿色。第1雌花着生节位从主蔓到侧蔓共 13~20 节。果实圆柱形，长 28~32 厘米，横径 6.5~7 厘米，浅绿色，肉白色，单果重 750~1 000 克。
特　　性 | 生长势强，侧蔓较多。早熟，播种至初收 50~70 天，延续采收 40~50 天。侧蔓结果为主，坐果性强，耐寒性较强。品质优。每公顷产量 45~67.5 吨。
栽培要点 | 平棚栽培。春播为 12 月至翌年 2 月，秋播 7—8 月。每 667 米$^2$ 种植 400~450 株。

特　　性 | 生长势强，侧蔓多。早熟，播种至初收 50~60 天，延续采收 40~50 天。侧蔓结果为主，雌花多，坐果性强，较耐寒。果肉味甜、嫩滑、品质优。每公顷产量可达 60 吨以上。
栽培要点 | 平棚栽培为佳，春季 2—3 月保护地播种育苗，气温稳定在 15℃以上时定植；秋季 7—8 月播种，每 667 米$^2$ 种植 400~450 株。春季表现优于秋季。主蔓不必打顶。加强肥水管理，注意防治病虫害。

## 油青早 1 号蒲瓜 Youqingzao No.1 calabash

> 品种来源

张水江菜种店育成。

📍 分布地区 | 南沙、番禺、增城、从化等区。

特　　征 | 叶片绿色。第1雌花着生节位从主蔓到侧蔓共 11~16 节。果实短圆柱形，果长约 30 厘米，皮色油绿，有光泽，果肉白色，质细嫩，单果重约 1 500 克。
特　　性 | 生长势强，坐果力强，采收期长。每公

# 黄瓜

葫芦科甜瓜属中幼果具刺的栽培种，一年生攀援性草本植物，学名 Cucumis sativus L.，别名青瓜、胡瓜，染色体数 $2n=2x=14$。

黄瓜原产于印度北部地区，是中国古代栽培的主要瓜类蔬菜之一，文字记载最早见于南北朝北魏时期贾思勰《齐民要术》，当时名胡瓜。黄瓜一名是唐代陈藏器《本草拾遗》首次著录的。现南北各地普遍栽培，可以周年生产。广州年栽培面积约5 200公顷。

**植物学性状**：根系浅。茎无限生长，分枝多，断面具4~5棱，表皮有刺毛。叶对生，长椭圆形，真叶互生，五角掌状或心形，深绿色，被茸毛。花为退化型单性花，雌雄花交替发生。异花授粉。果实为假果，幼果表皮白色至绿色；果面平滑或有瘤状突起，瘤的顶端着生黑刺或白刺，刺黑色或白色；果筒形至长棒形。种子披针形，扁平，种皮黄白色，千粒重22~42克。

**类型**：按品种的分布区域、形态特征及生物学性状，可分为华北型、华南型和小型黄瓜。

1. **华北型黄瓜** 长势中等，较耐低温，对日照长短反应不敏感。嫩果棍棒状，绿色，瘤密，多白刺。主栽品种有园丰园6号、沃林9号、津优黄瓜等。

2. **华南型黄瓜** 较繁茂，较耐湿热，属短日性植物。果形多筒状，中小型，刺瘤稀小，多黑刺。嫩果绿色、绿白色或黄白色；老熟果黄褐色，有网纹。主栽品种有早青4号、先锋大吊瓜、金山黄瓜等。

3. **小型黄瓜** 长势中等，分枝性强，主侧蔓均可结瓜。雌性强，多花多果，果实短小。主栽品种有夏美伦等。

**栽培环境与方法**：喜温，不耐寒，不耐高温，生长适温18~28℃。喜光，喜湿，不耐旱。可分为露地栽培和保护地栽培两大类型。

1. **露地栽培** 露地栽培可分春、夏、秋茬黄瓜。春茬在2月播种育苗或直播。夏茬在6月初播种。秋茬在8月中下旬播种。每667米$^2$用种量150~200克。

2. **保护地栽培** 保护地栽培有大、中、小棚以及大型连栋温室栽培，近年来发展迅速。每667米$^2$用种量100~150克。

**收获**：初期宜收嫩瓜，盛收期可以长至商品成熟期采收。

**病虫害**：病害主要有霜霉病、白粉病、细菌性角斑病、花叶病、蔓枯病、疫病、枯萎病、炭疽病、黑星病、猝倒病等，虫害主要有瓜蚜、黄守瓜、白粉虱、烟粉虱、美洲斑潜蝇、瓜实蝇、瓜绢螟等。

**营养及功效**：含有丰富的蛋白质、脂肪、糖类、多种维生素以及钙、磷、铁、钾、钠、镁等。味甘，性寒，具清热、利尿、解毒等功效。

# Cucumber

*Cucumis sativus* L., an annual climbing herb. Family: Cucurbitaceae. Synonym: Cherkin, Courgette. The chromosome number: $2n=2x=14$.

Cucumber is native to north India, and one of main gourd vegetable in ancient China. Cucumber is widely cultivated in China, and cultivated year round. At present, the cultivation area is about 5200 $hm^2$ per annum in Guangzhou.

Botanical characters: Shallow root system. Stem is indeterminate growth and branched, 4~5 edges. The rind with bristle. Cotyledon opposite, oblong or long obovate. True leaves alternate, palmate or heart-shaped, deep green, covered with fluff. Unisexual flower, cross-pollinated. Spurious fruit, short to long cylindrical. Young fruit white to green rind smooth or with protuberance, black or white prickle on protuberance. Seeds lanceolate, flat, yellow white. The weight per 1000 seeds is 22~42 g.

Types: According to cultivation area and biological character, cucumber can be classified into 3 types:

1. North China cucumber: Plant medium size, with tolerance to low temperature, not sensitive to the day-length. Young fruit long cylindrical, green, dense protuberance, many white thorn. The cultivars include Yuanfengyuan No.6, Wolin No.9, Jinyou, etc.

2. Southern China cucumber: Plant luxuriant with heat-tolerance, short-day plant. Fruit cylindrical, small or midium size, few protuberance, with black prickle. Young fruit green, green white or yellow white. Ripe fruit, brown, netted. The cultivars include Zaoqing No.4, Xianfeng, Jinshan, etc.

3. Mini cucumber: Plant medium size, strong branching, fruit-setting occur on main and lateral vine. The female is strong, more flower, more fruit, small fruit. The cultivars include Xiameilun, etc.

Cultivation environment and method: The warm, sunny and moist weather is favour for cucumber growth. The suitable temperature is 18~28℃. Cucumber cultivation can be divided into field cultivation and protected cultivation.

1. Field cultivation: Cucumber can be cultivated in spring, summer and autumn. Sowing dates: spring, February; summer, early June; autumn, late August. It needs 2250~3000 g seed per ha.

2. Protected cultivation: The area of protected cultivation is increasing rapidly. It needs 1500~2250 g seed per ha.

Harvest: Young cucumber fruits are harvested. Also mature fruits are harvested at midium and late fruit-setting stage.

Nutrition and efficacy: Cucumber is rich in proteins, fats, carbohydrates, vitamins, cellulose, calcium, phosphorus, iron, potassium, sodium, magnesium, etc.

## 粤秀 8 号

Yuexiu No.8 cucumber

**品种来源**　广东省农业科学院蔬菜研究所育成的杂种一代，2008年通过广东省农作物品种审定。

**分布地区** | 全市各区。

**特　　征** | 叶长 22.5 厘米，宽 20.8 厘米，深绿色。主蔓第 6 节左右着生第 1 雌花，以后每隔 4~5 节着生 1 雌花。瓜长 34.2~39.2 厘米，横径 3.3~3.6 厘米，瓜条顺直棒形，瓜色深绿，有光泽，顶部黄色纵纹不明显，刺瘤密，白色，瓜把短，肉厚 1.2~1.3 厘米，单果重 360 克左右。

**特　　性** | 生长势强。中早熟，播种至初收春季 57 天、秋季 46 天。中抗枯萎病，高感疫病和炭疽病；耐寒性与耐涝性中等，耐热性强。品质优。每公顷产量 30~35 吨。

**栽培要点** | 适合春季大棚及春、夏秋露地栽培。每 667 米² 种植 3 000~3 500 株。施足基肥，勤追肥，有机肥、化肥、生物肥交替使用。生长中后期可结合防病喷叶面肥，每周 1 次，以提高中后期产量。生长期间注意防治病害，及时清理底部老叶。中部侧枝见瓜后留 2 叶掐尖。及时采收商品瓜。

## 中农 8 号

Zhongnong No.8 cucumber

**品种来源**　中国农业科学院蔬菜花卉研究所 1993 年育成的杂种一代。

**分布地区** | 全市各区。

**特　　征** | 叶长 22 厘米，宽 20 厘米，深绿色。主蔓第 4~7 节着生第 1 雌花，以后每隔 3~5 节着生 1 雌花。瓜长 35~40 厘米，横径 3.0~3.5 厘米，瓜条棒形，瓜把短，瓜皮色深绿、有光泽，顶部黄色纵纹不明显，瘤小，刺密、白色，单果重 200~300 克。

**特　　性** | 生长势强，分枝较多。迟熟，播种至初收约 70 天。耐贮运。耐低温、弱光能力强。田间表现对霜霉病、白粉病、枯萎病、炭疽病、黄瓜花叶病毒病等多种病害抗性较强。肉质脆，味甜，品质优。每公顷产量 30~35 吨。

**栽培要点** | 参照粤秀 8 号。

## 沃林 9 号

Wolin No.9 cucumber

**品种来源** 天津德瑞特种业有限公司引进的杂种一代。

**分布地区** | 全市各区。

**特　　征** | 叶长 23 厘米，宽 21 厘米，深绿色。主蔓第 4~6 节着生第 1 雌花，以后每隔 5~6 节着生 1 雌花。瓜长 34~36 厘米，横径 3.5~3.8 厘米，肉厚 1.2~1.4 厘米，瓜条顺直，瓜色深绿，有光泽，果顶黄条纹不明显，尖，瘤较粗，刺密、白色，单果重 300~350 克。

**特　　性** | 生长势强。早熟，播种至初收春季 55 天、秋季 43 天。田间表现抗枯萎病、炭疽病、霜霉病、白粉病，耐寒性与耐涝性中等。肉质脆，淡绿色，味微甜，品质优。每公顷产量 30~35 吨。

**栽培要点** | 参照粤秀 8 号。

## 早青 4 号

Zaoqing No.4 cucumber

**品种来源** 广东省农业科学院蔬菜研究所育成的杂种一代，2011 年通过广东省农作物品种审定。

**分布地区** | 全市各区。

**特　　征** | 叶长 22 厘米，宽 20 厘米，深绿色。主蔓第 5~7 节着生第 1 雌花，以后每隔 4~6 节着生 1 雌花。瓜长 22.2~24.1 厘米，横径 4.6~4.9 厘米，短圆筒形，青绿色，瘤小，刺疏、白色，肉厚 1.2~1.4 厘米，单果重 350~420 克。

**特　　性** | 生长势强。早熟，播种至初收春季 59 天、秋季 41 天。中抗枯萎病，高感炭疽病、疫病。田间表现耐热性、耐寒性强，耐涝性、耐旱性中等。品质优。每公顷产量 33~38 吨。

**栽培要点** | 春季播种期为 1—3 月，采用浸种催芽后育苗或地膜覆盖直播；夏、秋季 7—9 月播种，直播，双行植，株距 25~30 厘米，结合整地重施基肥。当植株具 2 片真叶时开始追肥。结合中耕除草培土培肥。每采收 2 次追肥 1 次，每次每公顷施复合肥 80 千克。注意及时采收根瓜，加强病虫害综合防治。

## 金山黄瓜

Jinshan cucumber

> 品种来源

地方品种。

📍 分布地区 | 全市各区。

特　　征 | 叶长 23 厘米，宽 21.5 厘米，深绿色。主蔓第 5~8 节着生第 1 雌花，以后每隔 6~8 节着生 1 雌花。瓜长 28~32 厘米，横径 4.3~4.6 厘米，瓜圆筒形，黄褐色，无瘤无刺，无黄条斑纹，肉厚 1.9~2.1 厘米，单果重 480~520 克。

特　　性 | 生长势强。中迟熟，播种至初收春季 60 天、秋季 50 天。田间表现抗枯萎病、炭疽病、霜霉病、白粉病，耐寒性与耐涝性中等。品质优。每公顷产量 35~40 吨。

栽培要点 | 参照早青 4 号。

## 夏丰 606 大吊瓜

Xiafeng 606 cucumber

> 品种来源

广州市华艺种苗行有限公司的杂种一代。

📍 分布地区 | 全市各区。

特　　征 | 叶长 23 厘米，宽 22 厘米，深绿色。主蔓第 7~9 节着生第 1 雌花，以后每隔 5~7 节着生 1 雌花。瓜长 23~28 厘米，横径 5.5~6.5 厘米，瓜圆筒形，绿白色，上部青绿布白纹点，下部白条斑有少许白瓜刺，肉厚 2.3~2.5 厘米，单果重 350~500 克。

特　　性 | 生长势强。中早熟，播种至初收春季 55 天、秋季 45 天。田间表现抗性较强，耐热，耐湿，耐肥水，对日照要求不严格，采收期长。瓜条耐老化，耐贮运。品质优。每公顷产量 30~35 吨。

栽培要点 | 参照早青 4 号。

## 先锋大吊瓜

Xianfeng cucumber

**品种来源**

广州市大农园艺种子有限公司的杂种一代。

📍 **分布地区** | 全市各区。

**特　　征** | 叶长22厘米，宽21厘米，深绿色。主蔓第6~8节着生第1雌花，以后每隔6~9节着生1雌花。瓜长28~30厘米，横径5.8~6.2厘米，瓜圆筒形，绿白色，刺瘤小、少，白色，黄条斑纹明显，肉厚2.0~2.2厘米，单果重550~600克。

**特　　性** | 生长势强。早熟，播种至初收春季55天、秋季43天。田间表现较抗枯萎病、炭疽病、霜霉病、白粉病，耐寒性与耐热性较强。品质优。每公顷产量30~35吨。

**栽培要点** | 参照早青4号。

## 夏美伦小黄瓜

Xiameilun mini cucumber

**品种来源**

北京天地园种苗有限公司引进。

📍 **分布地区** | 全市各区。

**特　　征** | 小果型黄瓜，强雌型。叶长21厘米，宽20厘米，深绿色。主蔓第5~6节着生第1雌花，以后每节着生1雌花。瓜长13~15厘米，横径2.5~3.0厘米，肉厚约0.8厘米，瓜把无，瓜身刺少而小，无黄条纹，瓜色黄绿，商品瓜率高，肉厚0.8~1.1厘米，单果重100~120克。

**特　　性** | 生长势强。早熟，播种至初收50天。田间表现抗霜霉病、枯萎病和白粉病。质脆，味微甜，口感好，品质优。每公顷产量15~18吨。

**栽培要点** | 保护性设施条件下栽培。苗龄25~30天，每667米²种植2 000~2 200株，要及时去除6节以下侧枝与雌花。对肥水需求量大，定植前施足基肥，中后期增施磷钾肥。全期防治刺吸式口器害虫，如白粉虱、蚜虫、蓟马等，以免传播病毒病。

# 南瓜 Pumpkin and squashes

葫芦科南瓜属中的一年生草本植物，起源于美洲大陆，栽培种有5个：中国南瓜（*Cucurbita moschata* Duch. ex Poir），又名倭瓜、饭瓜；印度南瓜（*C. maxima*），又名笋瓜、西洋南瓜、栗南瓜；美洲南瓜（*C. pepo*），又名西葫芦、美国南瓜；墨西哥南瓜（*C. ficifolia*）和灰籽南瓜（*C. mixta*），又名墨西哥南瓜；黑籽南瓜（*C. ficifolia*）。染色体数为$2n=2x=40$。

南瓜种类、类型及品种繁多，广州地区栽培的南瓜种类主要有中国南瓜、印度南瓜，以鲜食为主。中国南瓜代表性品种有蜜本南瓜等；印度南瓜以红皮品种为主，品种有东升南瓜等。

**栽培环境与方法**：春、秋季均可种植。种子发芽适温25~30℃，开花结果需要稍高温度，印度南瓜在平均气温超过23℃时，淀粉积累能力减弱。对日照长短要求不严格。有较强的耐旱能力，在生长旺期，需要充足水分，但忌积水。在高温干旱环境下易得病毒病。

**病虫害**：病害主要有病毒病、疫病、白粉病等，虫害主要有夜蛾、黄守瓜、蚜虫、粉虱等。

**收获**：鲜食型老南瓜充分成熟后采收，鲜食型嫩南瓜因品种而异。

**营养及功效**：含有丰富的淀粉、蛋白质、胡萝卜素、维生素B、维生素C和钙、磷等成分。味甘，性温，具补中益气、消炎止痛、解毒、杀虫等功效。

*Cucurbita* spp., an annual climbing herb. Family: Cucurbitaceae. The chromosome number: $2n=2x=40$. There are 5 species: *C. moschata* (Duch.) Poir.; *C. maxima*; *C. pepo* L.; *C. ficifolia* and *C. mixta*.

There are many species and types in pumpkin and squashes. Pumpkin and *C. maxima* are widely cultivated in Guangzhou. The cultivars include Miben, Dongsheng, etc.

Cultivation environment and methods: Pumpkin and squashes can be cultivated in spring and autumn in Guangzhou. The suitable temperature for seed germination is 25~30℃. The starch accumulation decreased above 23℃ (average). Pumpkin and squashes have drought-tolerance and no waterlog-tolerance, they requrie sufficient irrigation. Virus diseases increase under high temperature and dry condition.

Harvest: Pumpkins and squashes are harvested while the fruit are mature, and also harvested while the fruit are very immature.

Nutrition and efficacy: Pumpkin and squashes are rich in starchs, proteins, carotene, vitamin B, vitamin C, calcium, phosphorus, etc.

## 蜜本南瓜 Miben pumpkin

**品种来源** 汕头市白沙蔬菜原种研究所育成。

**分布地区** | 全市各区。

**特　　征** | 蔓生。叶片钝角掌状，绿色，叶脉交界处有不规则斑纹。主侧蔓均可结果，主蔓第 1 雌花着生在第 15~18 节，花色金黄。瓜为棒槌形，顶端膨大，长 30~40 厘米，横径约 14.5 厘米，种子集中在顶端上，嫩果表皮浅绿色，熟果橙黄色，具网状花纹及白斑点，表皮被薄蜡粉，果肉橙红色，单果重约 3 千克。

**特　　性** | 中国南瓜类型。中迟熟，播种至初收 100~120 天。淀粉细腻，味甜，品质优。生长势强，分枝性强，耐贮运。田间表现高抗白粉病，耐病毒病和疫病。每公顷产量 15~22.5 吨。

**栽培要点** | 忌积水，选择排灌水方便的地块种植。春季 2 月上中旬播种育苗，每 667 米$^2$ 种植 300~350 株，预防苗期生长过旺，开花后重施肥；秋季 7 月播种，每 667 米$^2$ 种植 500~550 株，前期水肥管理要及时。充分成熟后采收。

## 金铃南瓜 Jinling pumpkin

**品种来源** 广州市农业科学研究院 2012 年育成。

**分布地区** | 全市各区。

**特　　征** | 蔓生。叶片钝角掌状，绿色，叶脉交界处有不规则斑纹。主侧蔓均可结果，主蔓第 1 雌花着生在第 15~21 节，第 1 雌花比第 1 雄花早开 5 天左右，花色金黄。瓜为梨形，棱沟不明显，纵径约 11 厘米，横径 15 厘米，肉厚 2.6 厘米，单果重约 1.3 千克。嫩果表皮浅绿色，熟果橙黄色，具网状花纹及白斑点，表皮被薄蜡粉，果肉橙红色。

**特　　性** | 中国南瓜类型。中迟熟，播种至初收 100~120 天。充分成熟后果肉具有怡人、浓郁的芳香味，淀粉细腻，味甜，品质优。生长势强，分枝性强，耐贮运。田间表现高抗白粉病，耐病毒病和疫病。每公顷产量 15~22.5 吨。

**栽培要点** | 参照蜜本南瓜。

## 盒瓜 He pumpkin

**品种来源** 地方品种。

**分布地区** | 全市各区。

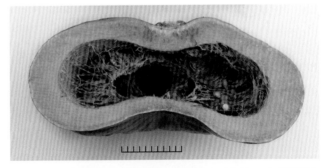

**特　　征** | 蔓生。叶长约 36 厘米，绿色。主侧蔓均可结果，主蔓第 1 雌花着生在第 18 节。果实扁圆形，纵径约 16 厘米，横径 24 厘米，果顶及果柄附近凹陷，表面有瘤状突出，嫩果绿色，熟果黄色，具网状花纹及白斑点，表皮被薄蜡粉，单果重约 2.5 千克。

**特　　性** | 中国南瓜类型。中迟熟，播种至初收约 100 天，延续采收 30~40 天。生长势强，侧蔓多，主侧蔓结果。耐寒力中等，较耐热，耐贮运。味甜，品质一般。田间表现高抗白粉病，耐病毒病和疫病。每公顷产量约 23 吨。

**栽培要点** | 参照蜜本南瓜。

## 一串铃

Yichuanling pumpkin

**品种来源**

湖南省衡阳市蔬菜研究所育成，1997 年通过衡阳市品种审定。

◎ 分布地区丨全市各区。

特　征丨蔓生。主蔓第 1 雌花节位约第 7 节。主蔓可连续坐瓜 2~4 个。嫩瓜圆球形，瓜皮具光泽，深绿色间有绿白色点条状花纹，果肉绿白色，单果重约 0.35 千克。老熟瓜扁圆形，单果重约 1.9 千克，瓜皮黄棕色，果肉黄色。

特　性丨中国南瓜类型。嫩食型为主，品质优。播种至初收约 58 天，第 1 雌花比第 1 雄花早开 7~10 天。田间表现抗病性强，春季不易化瓜。每公顷产量 22.5~30 吨。

栽培要点丨春季 2 月上中旬播种育苗，秋季 8 月中下旬播种，每 667 米$^2$ 种植 800~1 200 株。水肥管理要及时，注意防治白粉病等病害。

## 东升红皮南瓜

Dongsheng red skin pumpkin

**品种来源** 引自农友种苗（中国）有限公司。

◎ 分布地区丨全市各区。

特　征丨蔓生。叶缘圆，基本无缺刻，节间短，分枝性中等。主蔓第 1 雌花节位约第 8 节，以后局部节位可连续产生雌花，主侧蔓均可结果，花色金黄。幼果圆球形，充分膨大后扁圆形，果脐小。果柄圆形，无棱角，粗短，蒂部膨大。幼果表皮黄色，成熟后成橘红色。单果重约 1.5 千克，肉厚 3.2 厘米。

特　性丨印度南瓜类型。早熟，播种至初收春季约 90 天、秋季 70 天。适宜冷凉气候，田间表现不耐高温、高湿。淀粉细腻，味甜，水分少，品质优，耐贮运。每公顷产量 15~23 吨。

栽培要点丨忌积水，选择排灌方便的地块种植。适施基肥，及时管理。春植 2 月上中旬至 3 月中旬保温育苗，宜搭架栽培，畦宽约 2.0 米，株距 0.6 米，双行植；秋植 8 月下旬至 9 月中旬可直播，可爬地栽培，畦宽约 5.0 米，株距 0.6 米，相向双行植。注意防治病毒病、疫病、白粉病、蚜虫、美洲斑潜蝇、烟粉虱等。避免在高温、缺水和土壤贫瘠的环境种植。充分成熟后采收。

# 越瓜 Oriental pickling melon

葫芦科甜瓜属一年生蔓生草本植物,学名 Cucumis melo L. var. conomon Makino,别名白瓜、梢瓜等,染色体数 $2n=2x=24$。

越瓜起源于非洲中部,也有认为起源于中国及热带亚洲。在中国最早文字记载见于南北朝后魏贾思勰《齐民要术》,宋代已有"梢瓜"之称。中国以华南及台湾地区栽培较普遍,为广东特产蔬菜之一,广州栽培历史悠久,300 年前著述为常种"广瓜"之一,现年栽培面积约 500 公顷。

**植物学性状**:根系浅,侧根多。茎蔓生,有棱,主蔓基部节间较短,以后节间较长,茎节触地易生不定根,侧蔓结果。叶为互生单叶,绿色,心脏形,浅裂,叶柄被茸毛。雌雄同株异花,雌花单生,雄花单生或簇生,花冠黄色。异花授粉。果实圆筒形或棍棒形,有浅纵沟纹,嫩瓜浅绿色、绿色,或带深色条纹,果肉汁多、爽脆,白色或淡绿色;熟瓜黄色或黄白色,有香味,微甜。种子近披针形,稍扁,淡黄白色,千粒重 15~19 克。

**类型**:根据果实形状和有无深色纵条纹,可分为长度、中度、短度和青筋 4 种类型。目前主栽品种有长度白瓜(如长丰白瓜)和青筋白瓜等。

**栽培环境与方法**:喜光,耐热,耐旱,耐阴,耐湿。种子发芽适温 28~32℃,植株生长发育适温 20~32℃,15℃以下生长受抑制或停止生长,气温达 40℃时仍可正常生长,40℃以上易引起落花落果。对土壤质地要求不严。广州地区播种期 4—8 月,5—10 月收获。

**收获**:采收嫩果一般开花后 7~10 天,果实饱满,瓜皮有光泽时进行。

**病虫害**:病害主要有疫病、白粉病、霜霉病、病毒病等,虫害主要有蓟马、美洲斑潜蝇、烟粉虱、瓜实蝇、蚜虫等。

**营养及功效**:富含碳水化合物、多种维生素和矿物质。味甘,性寒,具清热解渴、利尿除烦、消暑益气等功效。

*Cucumis melo* L. var. *conomon* Makino, an annual climbing herb. Family: Cucurbitaceae. The chromosome number: $2n=2x=24$.

Oriental pickling melon has long been cultivated in Guangzhou, and was recorded as one of "Guangzhou gourd" 300 years ago. It is widely cultivated in southern China and Taiwan province. At present, the cultivation area is about 500 hm$^2$ per annum in Guangzhou.

Botanical characters: Shallow root system, many lateral root. Vine stem, with rid, short internode on the base of main vine, adventitious root occur on stem nodes, fruiting on the lateral vine. Leaf alternate, green, heart-shaped, lobed, stiff hairs cover the petiole. Flower monoecious, female flowers solitary, male flowers solitary or clusters, cross-pollinated. Yellow corolla. Fruits are cylinders or long cylinders, with furrow; fresh fruits are light green, green or dark-striped, fruit flesh juicy, crispness, white or light green; mature fruits are yellow or yellow white, fragrance, some sweet. Seed is nearly lanceolate, flat, light yellow white. The weight per 1000 seeds is 15~19 g.

Types:According to fruit shape and stripes, oriental pickling melon can be classified into long-body, middle-body, short-body and green-striped. Long-body and green-striped are widely cultivated.

Cultivation environment and methods: Oriental pickling melon has tolerance to heat, drought, shade and waterlog. The suitable temperature: seed germination, 28~32℃; plant growth, 20~32 ℃. The growth is inhibited below 15 ℃, and blossom and fruiting are difficult over 40 ℃. Sowing dates: April to August, harvesting dates: May to October in Guangzhou.

Harvest: Young fruits are harvested 7~10 days after anthesis.

Nutrition and efficacy: Oriental pickling melon is rich in carbohydrates, vitamins and minerals.

## 长丰白瓜

Changfeng oriental pickling melon

**品种来源**

广州市农业科学研究院 1990 年从地方品种长度白瓜中选育而成。

📍 **分布地区** | 白云、花都、从化、增城等区。

**特　征** | 蔓生。叶长 19 厘米，宽 18 厘米，绿色。主蔓第 1~2 节抽生侧蔓，侧蔓第 1~2 节着生雌花。果实长圆筒形，头尾较匀，长 35~40 厘米，横径 4~6 厘米，绿白色，浅纵沟纹，肉厚 1.5 厘米左右，白色。单果重 350~400 克。

**特　性** | 早熟，播种至初收 35~40 天，延续采收 20~30 天。生长势强，侧蔓多，靠侧蔓结果。田间表现耐热性、抗逆性较强。瓜质脆，味微甜，品质好。每公顷产量 20~30 吨。

**栽培要点** | 播种期 4—7 月，每 667 米² 用种量 75~100 克，催芽后直播或育苗移栽，苗期 10~12 天。爬地栽培，单行植，株距 15~20 厘米。施足腐熟有机质基肥，主蔓及子蔓、孙蔓 4~5 片真叶时打顶，开花结果期勤追肥，封行前重施追肥。注意防治病虫害。5—10 月采收。

## 青筋白瓜

Green rid oriental pickling melon

**品种来源**

地方品种。

📍 **分布地区** | 增城等区。

**特　征** | 蔓生。叶长 20 厘米，宽 19 厘米，绿色。主蔓第 2~3 节抽生侧蔓，侧蔓第 1~2 节着生雌花。果实长圆筒形，瓜匀直，长 35~45 厘米，横径 4~6 厘米，淡绿色，有明显绿色浅纵沟纹，肉厚 1.4 厘米左右，白色。单果重 450 克左右。

**特　性** | 早熟，播种至初收 35~40 天，延续采收 25~35 天。生长势强，侧蔓多，靠侧蔓结果。田间表现耐热性、抗逆性强。肉爽脆，味甜，品质优。每公顷产量 25 吨左右。

**栽培要点** | 播种期 3—8 月。其余参照长丰白瓜。

# 西葫芦

葫芦科南瓜属一年生藤本植物，学名 *Cucurbita pepo* L.，别名美洲南瓜、云南小瓜、荨瓜、茭瓜、白瓜、番瓜等，染色体数 $2n=2x=40$。

西葫芦原产于北美洲南部，现今世界各地广泛栽培。广州从20世纪60年代开始有栽培，现番禺、白云、增城等区有小面积栽培。

**植物学性状**：单叶，大型，掌状深裂，互生（矮生品种密集互生），叶面粗糙多刺。叶柄长而中空。花单性，雌雄同株。雄花花冠钟形，花萼基部形成花被筒，花粉粒大而重，具黏性。雌花子房下位，有一环状蜜腺。单性结实率低，冬季和早春昆虫少时需人工授粉。瓠果，形状有圆筒形、椭圆形和长圆柱形等多种。嫩瓜皮色有白色、白绿色、金黄色、深绿色、墨绿色或白绿色相间；老熟瓜的皮色有白色、乳白色、黄色、橘红色或黄绿相间。每果有种子300~400粒，种子白色或淡黄色，长卵形，种皮光滑，千粒重130~200克。

**类型**：西葫芦分矮生、半蔓生、蔓生三大类型，多数品种主蔓优势明显，侧蔓少而弱。主蔓长度：矮生品种节间短，蔓长通常在50厘米以下；半蔓生品种一般约80厘米；蔓生品种一般长达数米。具卷须，攀援或匍匐生长。

**栽培环境与方法**：一般育苗移植，2—3月播种，4—5月收获。

**营养及功效**：西葫芦富含维生素C和葡萄糖，钙含量极高。味甘，性温，具清热利尿、除烦止渴、润肺止咳、消肿散结等功效。

# Summer squash

*Cucurbita pepo* L., an annual herb. Family: Cucurbitaceae. Synonym: Marrow, Vegetable Marrow. The chromosome number: $2n=2x=40$.

Summer squash is native to southern America, and widely cultivated around the world. Summer squash has been cultivated since the 1960s in Guangzhou. A limited acreage is found in Panyu, Baiyun, Zengcheng.

Botanical characters: Simple leaf, large, palmate and parted, alternate, leaves surface rough with thorn. Petiole long and hollow. Flower monoecious. Male flower, bell-shaped corolla, pollen grains large. Female flower ovary inferior, with a cycling nectary. Poor parthenocarpy. Hand pollination is required in winter and early spring. Fruit cylinders, elliptical, long cylindrical. Young fruit white, white-green, gold, dark-green and white-green; mature fruit white, milky, yellow, orange-red and yellow-green. There are 300~400 seeds per fruit, seed white or light yellow, long oval, smooth. The weight per 1000 seeds is 130~200 g.

Types: There are dwarf, semi-vine, vine 3 types. The main vine is superiority in most cultivars. Length of main vine: dwarf, less than 50 cm usually; semi-vine about 80 cm; vine, may be meters in length.

Cultivation environment and methods: Seedling transplanting. Sowing dates: February to March, harvesting dates: April to May.

Nutrition and efficacy: Summer squash is rich in vitamin C and glucose, its calcium content is very high.

## 欧冠皇

Ouguanhuang summer squash

**品种来源**

广州市兴田种子有限公司。

**分布地区**｜番禺、南沙、白云、增城等区。

特　　征｜矮生，株高35~45厘米，开展度120~140厘米。有卷须。叶片长35厘米，宽35厘米，呈掌状，绿色，裂缺不深，有银灰色斑点；叶柄中空无刺，长42厘米。第1雌花着生于第7节，黄色。雄花柄带小刺。果实长圆柱形，长22厘米，横径6.0厘米，翠绿色，有大白星点，肉厚1.1厘米，绿白色。单果重350克。

特　　性｜早熟，早春播种至初收55~60天，延续采收20天左右。生长势强，主蔓结果。采收嫩瓜为主。较耐寒，易感染病毒病，较耐白粉病，抗疫病和枯萎病。皮质光滑，肉质细嫩，品质优。每公顷产量50~80吨。

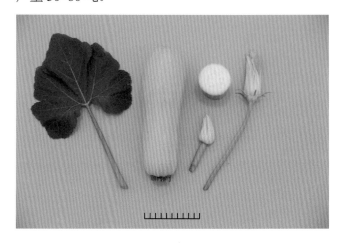

栽培要点｜播种期9月至翌年2月。使用营养杯薄膜棚保温育苗，于苗龄2~3片真叶时定植。畦宽1.2米（包沟），品字形种植，株距50厘米。用银灰色反光地膜覆盖，有利于驱避蚜虫，预防病毒病发生。每667米$^2$沟施腐熟基肥3 000千克、硫酸钾50千克，花前薄施追肥，花后重施肥。前期需要人工辅助授粉。当单瓜生长重达300克左右时应及时采摘。生长中后期要注意摘除老叶、病叶，保持通风透光，及早防治疫病、白粉病。全期重防蚜虫、瓜实蝇为害。

## 华丽

Huali summer squash

> 品种来源

广州南蔬农业科技有限公司。

📍 分布地区｜番禺、南沙、白云、增城等区。

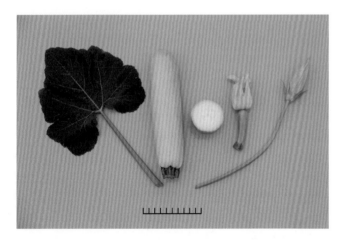

特　　征｜矮生，株高45~50厘米，开展度140~145厘米。叶片长38厘米，宽40厘米，呈掌状，绿色，有银灰色大斑点；叶柄中空带小刺，长43厘米。第1雌花着生于第6节，黄色。果实长圆柱形，长20厘米，横径4.8厘米，翠绿色，有白星点，肉厚1.0厘米，绿白色。单果重280克。

特　　性｜早熟，早春播种至初收55~60天，延续采收20天左右。雌性强，主蔓结果。采收嫩瓜为主。较耐寒，易感染病毒病，较耐白粉病。皮质光滑，肉质细嫩，品质优。每公顷产量50~75吨。

栽培要点｜参照欧冠皇。

## 百盛

Baisheng summer squash

> 品种来源

广州市大农园艺种子有限公司。

📍 分布地区｜番禺、南沙、白云、增城等区。

特　　征｜矮生，株高45~57厘米，开展度110~130厘米。茎深绿色，有卷须。叶片长33厘米，宽43厘米，呈掌状，绿色，裂缺不深，有银灰色大斑点；叶柄中空有小刺，长37厘米。第1雌花着生于第6节，黄色。雄花柄上无刺。果实长圆柱形，长21厘米，横径5.5厘米，翠绿色，有白星点，肉厚1.1厘米，绿白色。单果重320克。

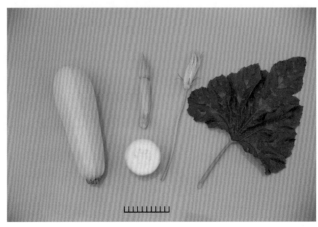

特　　性｜早熟，早春播种至初收50~55天，延续采收20天左右。雌性强，主蔓结果。采收嫩瓜为主。耐寒，易感染病毒病、白粉病，抗疫病和枯萎病。皮质光滑，肉质细嫩，品质优。每公顷产量50~80吨。

栽培要点｜参照欧冠皇。

# 佛手瓜 Chayote

葫芦科佛手瓜属多年生攀援性宿根草本植物，学名 *Sechium edule* (Jacq.) Swartz，别名梨瓜、佛掌瓜、万年瓜、隼人瓜、安南瓜、福寿瓜等。

佛手瓜原产于墨西哥、中美洲和西印度群岛，19世纪初由日本传入中国，广州有上百年的栽培历史，在郊区的农庄观赏长廊、农村的房前屋后有零星栽培。

**植物学性状**：根为弦线状须根，侧根粗长，第2年后可形成肥大块根。茎蔓性，长10米以上，分枝性强，节上着生叶片和卷须。叶互生，叶片与卷须对生；叶片呈掌状五角形，全缘，绿色或深绿色，被茸毛。雌雄同株异花，雌花多生于孙蔓上，开花迟于雄花。总状花序。异花授粉。虫媒花。果实梨形，如两掌合十，有佛教祝福之意，因此称之为"佛手""福寿"。果绿色至乳白色，果肉白色。种子扁平，纺锤形，无休眠期，成熟后如不及时采收，种子在果中就会很快萌发出，出现"胎萌"现象。

**栽培环境与方法**：根据果实的颜色分为白皮和绿皮两个品种，均以果实繁殖。生长适温为12~25℃，超过30℃生长受到抑制，低于5℃会产生冻害。一般12月至翌年2月播种，3—4月采收脆嫩枝梢，5—6月和9—12月采收果实。

**病虫害**：注意防治霜霉病、白粉病、蔓枯病、白粉虱及红蜘蛛等。

**营养及功效**：佛手瓜含锌较高，每100克含硒量30~50微克。味甘，性凉，具有理气和中、疏肝止咳的功效。

*Sechium edule* Swartz, a perennial herb. Family: Cucurbitaceae. Synonym: Chirstophine. The chromosome number: $2n=2x=28$.

Chayote is native to Mexico, Central America and western West Indies. It was introduced into China from Japan in early 19th century. It has been cultivated about 100 years in Guangzhou. It is distributed sparsely.

Botanical characters: Strong fibrous roots may develop into fleshy tuber. Stem vine, the length is more than 10 cm, branching vigorously, leaves and tendril on the segments. Leaf alternate, palmate, green or dark green. Flower, monoecious, male flower in racemes, female flower solitary. Entomophilous flowers and cross-pollinated. Fruits pear-shaped, 5 longitudinal grooves green or white green, flesh white. Seed is flat, spindle, it may germinate in the mature fruit.

Cultivation environment and methods: Chayote is classified into white and green 2 types. Propagated with fruit. The suitable growth temperature is 12~25℃, the growth is inhibited over 30 ℃, and hurt below 5 ℃. Sowing dates: December to February of the following year. Harvesting dates: tender shoot, March to April; fruit, May to June and September to December.

Nutrition and efficacy: Chayote fruit is rich in zinc. There are 30~50 μg selenium per 100 g tendril of chayote.

# 六、豆类
# VEGETABLE LEGUMES

- 长豇豆
- 菜豆
- 豌豆
- 四棱豆
- 菜用大豆

# 长豇豆

豆科豇豆属豇豆种中能形成长形豆荚的亚种，一年生缠绕草本植物，学名 Vigna unguiculata (L.) Walp. subsp. sesquipedalis (L.) Verdc.，别名豆角、长豆角、线豆角、带豆、裙带豆、腰豆、筷豆等，染色体数 $2n=2x=22$。

菜用长豇豆的起源中心在亚洲中南部，而我国是栽培豇豆的第二起源中心。据古书记载，长豇豆在我国至少有 1 500 年的栽培历史，以南方栽培面积最大，广州年栽培面积约 5 000 公顷。

植物学性状：根系发达，主根明显，再生能力弱。茎表面光滑，绿色或带紫红色。基生叶为对生单叶，以后均为三出复叶，互生。总状花序，花为蝶形花，花冠多淡紫色或紫色，少数白色或黄白色。果实为长荚果，细长，多呈条状或筷状，少数旋曲状，荚果常成对。嫩荚的果皮肥厚，荚色以淡绿色为主，其次为深绿色、白色、紫红色、花斑彩纹等。自花授粉。荚内含种子数一般为 16~24 粒，种子肾形，稍扁平或弯月形，种皮颜色有红、黑、白、褐、紫、花斑等多种，有些品种有不同颜色的条纹或花纹，无休眠期。千粒重为 130~180 克。

类型：依其生长习性可分为蔓生种、半蔓生种和矮生种。蔓生种无限生长，沿支柱左旋（逆时针方向）缠绕向上生长。矮生种株高 40~70 厘米，茎蔓直立或半开放，茎长至 4~8 节后，顶端即形成花芽，并发生侧枝，成为分枝较多的株丛。半蔓生种的茎蔓生长中等，一般高 100~200 厘米。广州地区栽培的主要是蔓生种中的长豇豆，主栽品种有丰产 2 号、穗丰 8 号、穗青豆角、丰产 6 号等。

栽培环境与方法：喜温耐热，生长适温 20~30℃，20℃以下豆荚发育受阻。喜充足光照，栽培用种多对光周期不明显。选择 2~3 年未种过豆科作物、土层深厚、肥沃、地势高、排水良好、通气性好的沙壤土或壤土种植。多采用直播，每 667 米$^2$ 用种量 1.5~2 千克。施足基肥，适当淋水。前期预防徒长，后期控制早衰。及时插架引蔓。

收获：一般开花后 10~12 天，豆荚充分长粗，种子刚刚膨大时及时采收。

病虫害：病害主要有白粉病、花叶病、枯萎病、煤霉病、锈病等，虫害主要有豆荚螟、小地老虎、叶蝉、美洲斑潜蝇、蓟马、豆蚜等。

营养及功效：富含蛋白质、碳水化合物、纤维素及各种维生素和矿物元素等。味甘、咸，性平，具理中益气、补肾健脾胃、和五脏等功效。

# Asparagus bean

## 丰产 2 号

Fengchan No. 2 asparagus bean

*Vigna unguiculata* Walp. supsp. *sesquipedalis* (L.) Verdc., an annual twining herb. Family: Leguminosae. Synonym: Yardlong bean, Vegetable cowpea. The chromosome number: $2n=2x=22$.

Asparagus bean is originated in central and southern Asia, It has been cultivated more than 1500 years in China. At present, the cultivation area is about 5000 $hm^2$ per annum in Guangzhou.

Botanical characters: Asparagus bean with obvious taproot system. Stems smooth, green or purple. The first pair of true leaves is simple, and the followings are alternating trifoliolate. Inflorescence: axillary racemes, papilionaceous, light-violet or purple, a few white or yellowish white in color. Linear pods, usually in pairs. Young pod succulent, mainly light green, few dark green, white, purple or variegated, etc. Self-pollination. 16~24 seeds in one pod. Seeds reniform, slightly flattened or meniscus. Red, black, white, brown, purple and variegated, etc. The weight per 1000 seeds is 130~180 g.

Types: According to the growth character, asparagus bean can be divided into three types: vine, semi-vine and dwarfed. The vine type twining anticlockwise, indeterminate. The dwarf type 40~70 cm in height, stem erect. The apical flower bud forms when the plant has 4~8 nodes, and the lateral branch develops. Semi-vine type 100~200 cm in height. The vine type is the main type asparagus bean cultivated in Guangzhou. The cultivars include Fengchan No.2, Suifeng No.8, Suiqing and Fengchan No. 6, etc.

Cultivation environment and methods: The suitable temperature for asparagus bean growth is 20~30℃, the pods growth retards below 20 ℃. Sunny, deep, fertile, good drainage sandy loam or loam are favourable to the growth of asparagus beans. Usually direct seedling, 22.5~30 kg seeds per ha. It should be vine supported.

Harvest: The young pods are harvested 10~12 days after anthesis.

Nutrition and efficacy: Asparagus bean is rich in proteins, carbohydrates, dietary fibers, vitamins, minerals, etc.

## 品种来源

广东省农业科学院蔬菜研究所育成，2001 年通过广东省农作物品种审定。

分布地区｜全市各区。

特　　征｜蔓生。主茎第 6~8 节开始着生花序，花蓝紫色。荚浅油绿色，表面平滑，荚长 58 厘米，横径 0.8 厘米，单荚重 24 克，双荚率高。种子肾形，黑色。

特　　性｜生长势强。中熟，播种至初收 65 天左右，可延续采收 25~30 天。耐热性、耐湿性较好。纤维少，品质好。每公顷产量 24~30 吨。

栽培要点｜播种期 3—8 月，直播或育苗移栽，株距 12~15 厘米，行距 40 厘米。施足基肥，适当追肥和淋水，前期预防徒长，后期控制早衰。及时搭架引蔓。注意防治豆荚螟、蓟马、叶蝉及豆蚜。

## 丰产 6 号

Fengchan No.6 asparagus bean

**品种来源** | 广东省农业科学院蔬菜研究所育成，2011 年通过广东省农作物品种审定。

📍 **分布地区** | 全市各区。

**特　　征** | 蔓生。主茎第 4.5~5.5 节开始着生花序，花浅蓝紫色。荚浅绿白色，荚面微凸，荚长 58 厘米，横径 0.9 厘米，单荚重 28.5 克，双荚率高。种子肾形，红白两色。

**特　　性** | 生长势强。中熟，播种至初收春植约 64 天、秋植约 43 天，可延续采收 30~35 天。中抗枯萎病，耐热性、耐涝性和耐旱性强，耐寒性中等。纤维少，品质好。每公顷产量 23~30 吨。

**栽培要点** | 参照丰产 2 号。

## 宝丰油豆角

Baofeng asparagus bean

**品种来源** | 广东省农业科学院蔬菜研究所育成。

📍 **分布地区** | 全市各区。

**特　　征** | 蔓生。分枝较少，主茎约第 6 节开始着生花序，花蓝紫色。荚浅绿白色，荚长 62 厘米，横径 0.8 厘米，单荚重 27 克，双荚率高。种子肾形，黑色。

**特　　性** | 生长势强。早熟，播种至初收春植约 55 天、秋植 45 天，可延续采收 25~30 天。抗病性、抗逆性强，适应性好。纤维少，品质好，较耐贮藏。每公顷产量 22~28 吨。

**栽培要点** | 参照丰产 2 号。

## 穗丰 8 号

Suifeng No.8 asparagus bean

**品种来源** | 广州市农业科学研究院育成，2003 年通过广东省农作物品种审定。

📍 **分布地区** | 全市各区。

**特　　征** | 蔓生。主茎第 6~8 节开始着生花序，花紫色。荚长 55~60 厘米，横径 0.8 厘米，荚浅绿色，荚面较平，荚形美观，单荚重 23 克。种子肾形，黑色。

**特　　性** | 生长势强。早中熟，播种至初收 64 天，可延续采收 25~30 天。较耐热，抗病性、抗逆性较强。荚肉紧实，纤维少，品质好。每公顷产量 24~30 吨。

**栽培要点** | 参照丰产 2 号。

# 穗丰 5 号

Suifeng No.5 asparagus bean

**品种来源**

广州市农业科学研究院 2011 年育成。

**分布地区** | 全市各区。

特　　征 | 蔓生。主茎第 5~6 节开始着生花序，花紫色。荚浅绿白色，荚面较平，荚形美观，有光泽，荚长 60 厘米，横径 0.9 厘米，单荚重 27 克。种子肾形，黑色。

特　　性 | 生长势强。中迟熟，播种至初收春植 65~70 天、夏秋植约 45 天，可延续采收 25~30 天。抗病性、抗逆性较强。荚肉紧实，纤维少，味甜，品质好，耐贮运。每公顷产量 23~30 吨。

栽培要点 | 参照丰产 2 号。

# 穗丰 9 号

Suifeng No.9 asparagus bean

**品种来源**

广州市农业科学研究院 2009 年育成。

**分布地区** | 全市各区。

特　　征 | 蔓生。主茎第 7~8 节开始着生花序，花浅紫白色。荚色油绿，荚面较平，荚形美观，有光泽，荚长 65 厘米，横径 0.8 厘米，单荚重 26.3 克。种子肾形，红白两色。

特　　性 | 生长势强。中迟熟，播种至初收春植 65~70 天、夏秋植约 45 天，可延续采收 25~30 天。抗病性、抗逆性较强。纤维少，味甜，品质好，耐贮运。每公顷产量 24~30 吨。

栽培要点 | 参照丰产 2 号。

## 珠豇 1 号

Zhujiang No.1 asparagus bean

**品种来源** 广东金作农业科技有限公司育成，2004 年通过广东省农作物品种审定。

**分布地区** | 全市各区。

特　　征｜蔓生。主茎第 5~6 节开始着生花序，花紫色。荚色油绿，有光泽，荚面平整，外形美观，荚长 55~60 厘米，横径 0.8 厘米，单荚重 23~26 克。种子肾形，黑色。

特　　性｜生长势强。中迟熟，播种至初收春植 69 天、夏秋植 50 天，可延续采收 25~30 天。耐热性中等，耐寒性较强，耐涝性强，高抗枯萎病，耐锈病。荚肉紧实，品质好，纤维少，结荚多而集中。每公顷产量 24~30 吨。

栽培要点｜播种期春植 2—4 月、秋植 7—8 月。其余参照丰产 2 号。

## 珠豇 3 号

Zhujiang No.3 asparagus bean

**品种来源** 广东金作农业科技有限公司育成，2004 年通过广东省农作物品种审定。

**分布地区** | 全市各区。

特　　征｜蔓生。主茎第 6~7 节开始着生花序，花浅紫色。荚面平整，尾部圆滑，绿白色有光泽，荚长 70 厘米，横径 1.0 厘米，单荚重 32.5 克。种子肾形，

红白两色。

特　　性｜生长势强。早熟，播种至初收春植 55 天、夏秋植 43 天，可延续采收 25~30 天。抗病性、抗逆性较强。荚肉紧实，品质好，纤维少。每公顷产量 23~30 吨。

栽培要点｜参照丰产 2 号。

## 碧玉油白

Biyuyoubai asparagus bean

**品种来源** 广州市金苗种子有限公司育成。

**分布地区** | 全市各区。

特　　征｜蔓生。主茎第 5~6 节开始着生花序，花黄白色。荚绿白色，有光泽，荚长 62 厘米，横径 0.9 厘米，单荚重 27.5 克。种子肾形，白色。

特　　性｜生长势强。早熟，播种至初收春植 60 天、夏秋植 45 天，可延续采收 25~30 天。抗病性、抗逆性较强。肉厚，纤维少，品质好，结荚多而集中，耐老化，耐贮运。每公顷产量 23~30 吨。

栽培要点｜参照丰产 2 号。

## 中度珠仔豆

Zhongduzhuzai asparagus bean

> 品种来源

广州长合种子有限公司育成。

📍 分布地区 | 全市各区。

特　　征 | 蔓生。主茎约第7节开始着生花序，花浅紫白色。结荚多而集中，荚形整齐，荚表面微凸。荚绿白色，有光泽，荚长48厘米，横径1.1厘米，单荚重30.5克。种子肾形，黑白两色。

特　　性 | 生长势强。早中熟，播种至初收春植60天左右、夏秋植约45天，可延续采收25~30天。耐热性、耐涝性、耐旱性较强。品质好，豆味浓，肉厚，纤维少，耐老化，耐贮运。每公顷产量20~25吨。

栽培要点 | 参照丰产2号。

## 全能王油白

Quannengwangyoubai asparagus bean

> 品种来源

广州广南农业科技有限公司育成。

📍 分布地区 | 全市各区。

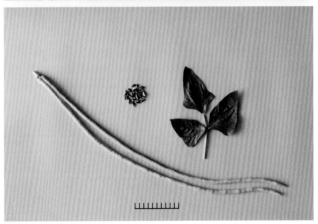

特　　征 | 蔓生。侧蔓2~3条，主茎第6~7节开始着生花序，花浅紫白色。荚绿白色，有光泽，荚长63.5厘米，横径0.9厘米，单荚重28.5克。种子肾形，红白两色。

特　　性 | 生长势强。早熟，播种至初收春植60天左右、秋植约45天，可延续采收25~30天。抗病性、抗逆性较强。肉厚，纤维少，不易老化，风味好，品质优。每公顷产量23~30吨。

栽培要点 | 参照丰产2号。

## 长丰黑仁油青

Changfeng black seed asparagus bean

**品种来源** | 广州市南星种业有限公司育成。

**分布地区** | 全市各区。

**特　　征** | 蔓生。主茎约第7节开始着生花序，花紫色。荚浅油绿色，荚长60厘米，横径1.0厘米，单荚重25克，双荚率高。种子肾形，黑色。

**特　　性** | 生长势强。早熟，播种至初收春植55天、秋植45天，可延续采收25~30天。较抗枯萎病、锈病。肉厚，不易露仁，纤维少，品质较好。每公顷产量24~30吨。

**栽培要点** | 播种期春植2—4月、秋植7—8月。其余参照丰产2号。

## 穗青豆角

Suiqing asparagus bean

**品种来源** | 广州市农业科学研究院1996年育成。

**分布地区** | 全市各区。

**特　　征** | 蔓生。侧蔓2~3条，主茎第8~9节开始着生花序，花紫色。荚深绿色，有光泽，先端白色，荚长55~58厘米，横径0.8厘米，单荚重18.5克。种子肾形，红色。

**特　　性** | 生长势强。中迟熟，播种至初收春植65~70天、夏秋植45天，可延续采收25~30天，抗病性、抗逆性较强。荚肉紧实，纤维少，味甜，品质好，耐贮运。每公顷产量20~25吨。

**栽培要点** | 播种期4—8月，直播或育苗移栽，株距15厘米，行距40厘米。施足基肥，适当淋水，前期预防徒长，后期控制早衰。及时搭架引蔓。

## 八月角

Bayue asparagus bean

**品种来源** | 地方品种。

**分布地区** | 增城等区有少量栽培。

**特　　征** | 蔓生。侧蔓2~3条，主茎第12~13节开始着生花序，花浅紫色。荚绿白色，先端白色，荚长25~30厘米，横径0.8~0.9厘米，单荚重14.0克。种子肾形，红色。

**特　　性** | 生长势强。迟熟，播种至初收55天，可延续采收25~30天，抗病性、抗逆性较强。荚肉紧实，品质优良，耐贮运。每公顷产量10~15吨。

**栽培要点** | 播种期8月，直播或育苗移栽，株距15厘米，行距40厘米。施足基肥，适当淋水，前期预防徒长，后期控制早衰。及时搭架引蔓。

# 菜豆

豆科菜豆属中的栽培种，一年生缠绕性草本植物，学名 *Phaseolus vulgaris* L.，别名四季豆、芸豆、芸扁豆、玉豆、龙牙豆、京豆等，染色体数 $2n=2x=22$。

菜豆起源于中南美洲，于17世纪时经欧洲传入亚洲。我国的栽培历史悠久，南北均广为栽培，广州地区已有100多年栽培历史，现年栽培面积约2 000公顷。

**植物学性状**：根系较发达，须根多，再生能力弱。茎细弱，被短茸毛。第一片真叶为对生单叶，呈心脏形；第二片真叶以后均为三出复叶，互生。总状花序，每花梗着生花2~8朵，最多可达十余朵。花为蝶形花，开花前龙骨瓣包裹着雌雄蕊，呈螺旋状弯曲。雌蕊尖端弯曲，呈环状，花柱较长，柱头上密生茸毛，呈毛刷状。花色有白色、淡紫色、紫色等。果实为荚果，为圆棍形或扁条形。嫩荚颜色有绿色、淡绿色、紫红色、紫红色花斑及黄色。自花授粉。种子形状多为肾形，少数为扁平或细长，也有椭圆形、圆球形等。种皮有白色、乳白色、灰色、黄色、棕色、蓝色、黑色或具有不同颜色的条纹或斑纹。种脐为白色。千粒重100~700克。

**类型**：按其生长习性可分为蔓生种、半蔓生种和矮生种3类。按豆荚纤维化的程度可分为软荚种和硬荚种，软荚种荚果肉质，粗纤维少，荚充分长成后仍柔嫩，为主要食用部分；硬荚种豆荚壁薄，纤维多，易老化，种子发育快，为主要食用部分。我国生产的都属于软荚种。按豆荚颜色可分为绿荚、黄荚、紫荚和紫色斑纹种，广州地区目前栽培的多为蔓生种中的绿荚种，如12号菜豆、35号玉豆等。

**栽培环境与方法**：喜温但不耐高温和低温霜冻，生长适温18~25℃。忌酸性土，pH以6.0~7.0为宜。宜选择肥沃、排水良好、2~3年未种过豆科作物的壤土或沙壤土种植。播种期春植为1—2月，可直播或用营养杯育苗；秋植为8月中旬至10月，宜直播。每667米² 用种量3~4千克。

**收获**：豆荚充分长粗，种子刚刚膨大，荚壁未硬化时应及时采收。

**病虫害**：病害主要有白粉病、炭疽病、花叶病、细菌性疫病、根腐病、锈病等，虫害主要有豆荚螟、小地老虎、螨类、美洲斑潜蝇、豆蚜等。

**营养及功效**：富含蛋白质、碳水化合物、维生素和矿物质等。味甘，性平，具滋补、解热、利尿、消肿等功效。

# Kidney bean

*Phaseolus vulgaris* L., an annual twining herb. Family: Leguminosae. Synonym: French bean, Common bean, Bush bean, Haricot bean, Snap bean. The chromosome number: $2n=2x=22$.

Kidney bean is originated in central and South America, introduced from Europe to Asia in the 17th century. Kidney bean has long be cultivated in China, and for more than 100 years in Guangzhou. At present, the cultivation area is about 2000 $hm^2$ per annum.

Botanical characters: Kidney bean has stronger root system. Stems thin, with trichomes. The first pair of true leaves is simple, heart shaped, and the followings are alternating tri-foliolate. Inflorescence, axillary racemes. 2~8 flowers. Flower papilionaceous, white, light purple and purple in color. Pod, round or flat. Young pods green, light green, white, amaranth, or variegated amaranth, yellow, etc. Self-pollination. Seeds reniform, few flat, slender, oval and round ball in shape. Black, white, ivory-white, grey, yellow, brown, blue and variegated, etc. in color. The hilium white. The weight per 1000 seeds is 100~700 g.

Types: According to the growth character, kidney bean can be divided into three types: vine, semi-vine and dwarfed. According to the fibrosis of pod, it can be divided into soft and hard pod types. The soft pod is fleshy, with few crude fiber, is the main edible part. The hard pod has thin pericarp, more fiber, the seed develop fast. The soft pod is the main type of kidney beans in China. According to the pod color, kidney beans can be divided into green, yellow, purple, and purple variegated types. At present, the main type in Guangzhou is the green vine type, including No.12 and No.35 kidney bean.

Cultivation environment and methods: The warm weather is favourable to the growth of kidney bean, which has no tolerance to high temperature, low temperature and frost. The suitable temperature for growth is 18~25℃. The optimum soil pH is 6.0~7.0. The fertile, good drainage sandy loam or loam soil are optimum soil. Sowing dates: spring, January to February, direct seedling or seedling transplanting; autumn, mid-August to October, usually direct seedling. 45~60 kg seeds per ha.

Harvest: The harvest occurs when pods are fully grown, the pericarp is not hardened.

Nutrition and efficacy: Kidney bean is rich in proteins, carbohydrates, vitamins, minerals, etc.

## 12 号菜豆

No.12 kidney bean

**品种来源**

广州市农业科学研究院育成，1992 年通过广东省农作物品种审定。

📍 分布地区 | 全市各区。

特　　征 | 蔓生，分枝力强。主蔓第 8~10 节开始着生花序，花白色，每花序开 6~9 朵花，结荚 2~5 条，荚长 18 厘米，宽 1.2 厘米，厚 1.0 厘米，浅绿色，单荚重约 13 克。种子肾形，白色。

特　　性 | 中熟，播种至初收春植约 75 天、秋植 50 天左右，可延续采收 25~30 天。耐寒，抗锈病力较强，翻花力强。品质优。每公顷产量 18~25 吨。

栽培要点 | 春植播种期 1—2 月，可直播或用营养杯育苗，秋植播种期为 8 月中旬至 10 月，直播。起深沟高畦，株距 9~12 厘米，行距 40 厘米。施足基肥，合理追肥，适当淋水，及时插架引蔓。

## 穗丰 4 号

Suifeng No.4 kidney bean

**品种来源**

广州市农业科学研究院育成，2005 年通过广东省农作物品种审定。

📍 分布地区 | 全市各区。

特　　征 | 蔓生。主蔓第 8~10 节着生花序，花白色。结荚 4~6 条。荚淡绿色，长扁形，长 20.6 厘米，宽 1.3 厘米，厚 0.6 厘米。种子肾形，白色。

特　　性 | 中熟，播种至初收春植约 75 天、秋植 50 天左右，可延续采收 25~30 天。抗锈病力较强，翻花力强。品质好。每公顷产量 18~25 吨。

栽培要点 | 参照 12 号菜豆。

## 35 号玉豆

No. 35 kidney bean

**品种来源**

番禺区大石蔬菜科学研究所 1988 年育成。

分布地区｜全市各区。

特　征｜蔓生，分枝力中等。主蔓第 4~6 节开始着生花序，花白色，每花序开 10~15 朵花，结荚 3~5 条。荚长 17.5 厘米，宽 1.1 厘米，厚 0.8 厘米，绿色，单荚重约 12 克。种子肾形，黄褐色。

特　性｜早中熟，播种至初收春植约 70 天、秋植 45 天左右，可延续采收 25~30 天。抗逆性较强。耐贮运，品质好。每公顷产量 15~23 吨。

栽培要点｜参照 12 号菜豆。

## 丰田 39 号

Fengtian No. 39 kidney bean

**品种来源**

广州市华叶种苗科技有限公司育成。

分布地区｜全市各区。

特　征｜蔓生，分枝力强。主蔓第 8~10 节开始着生花序，每花序开 10~15 朵花，花紫红色，结荚 2~5 条。荚长 22 厘米，宽 1.2 厘米，厚 0.9 厘米，荚浅绿油色，单荚重 14.5 克。种子肾形，黑色。

特　性｜中熟，播种至初收春植约 75 天、秋植 50 天左右，可延续采收 25~30 天。耐热，耐寒，抗锈病，翻花力强。品质好。每公顷产量 18~25 吨。

栽培要点｜参照 12 号菜豆。

## 新丰玉豆 2202

Xinfeng 2202 kidney bean

**品种来源**

广州市华艺种苗有限公司育成。

分布地区｜全市各区。

特　征｜蔓生，分枝力强。主蔓第 7~9 节开始着生花序，每花序开 10~15 朵花，花紫红色，结荚 2~5 条。荚长 23 厘米，宽 1.2 厘米，厚 0.9 厘米，荚浅绿油色，单荚重 15.5 克。种子肾形，黑色。

特　性｜中熟，播种至初收春植约 75 天、秋植 50 天左右，可延续采收 25~30 天。耐热，耐寒，抗锈病，翻花力强。品质好。每公顷产量 18~25 吨。

栽培要点｜参照 12 号菜豆。

# 豌豆

豆科豌豆属一年生或二年生攀援性草本植物，学名 *Pisum sativum* L.，别名荷兰豆、胡豆、淮豆、青豆、雪豆、寒豆、麦豆等，染色体数 $2n=2x=14$。

豌豆起源于亚洲中部、地中海沿岸、埃塞俄比亚、小亚细亚和外高加索地区。我国菜用豌豆的栽培历史较短，广州栽培历史近 200 年，年栽培面积约 2 500 公顷。

**植物学性状**：主根发达，侧根较多，根瘤发达，固氮能力较强。茎中空而质脆，圆筒形，绿色，茎、托叶及小叶表面都有白色蜡粉，呈革质。植株丛生或蔓生。叶互生，基部 1~3 节为单生叶，4~6 节以上为羽状复叶，小叶 2~3 对，顶端有 1~3 小叶变为卷须，攀援他物。花为短总状花序，花柄自叶腋伸出，先端着生 1~3 朵花，为蝶形花，花冠白色、粉红色或紫红色。荚果幼时扁平而长，成熟时近圆筒形，嫩荚浓绿色或黄绿色，种子成熟时黄褐色。自花授粉。种子圆形，表皮光滑或皱缩，颜色有黄、绿、白、青灰、褐、褐斑等，千粒重 150~800 克。

**类型**：有菜用豌豆和软荚豌豆 2 个变种。菜用豌豆由粮用品种演化而来，以鲜豆粒及嫩梢（又名荷兰豆尖、龙须菜、荷兰豆苗）供食用。食用嫩梢品种有美国手牌豆苗。软荚豌豆是在菜用品种的基础上选育而成的，其幼荚口感清脆，鲜嫩香甜，如改良 11 号、奇珍 76 甜豌豆。

**栽培环境与方法**：喜冷凉干燥，较耐寒而不耐热，广州地区一般 9 月中旬至 11 月下旬播种。忌连作，不耐涝。选择 pH 6.0~7.2 的较肥沃的沙壤土作栽培田块，施足基肥，合理追肥。直播，单行植或双行植。

**收获**：嫩荚在花后 12 天左右，豆荚停止伸长而种子刚开始发育时采收；甜豌豆要在豆荚饱满、不露仁时采收。采收嫩豆粒，在开花后 15 天左右，豆粒长到充分饱满，荚仍为深绿色或开始变浅绿色时采收。采收嫩梢（即豆苗）可在播种后 20~25 天开始第一次采收，以后每隔 5~7 天采收一次。

**病虫害**：病害主要有白粉病、褐斑病、豌豆尖镰刀菌萎凋病、根腐病等，虫害主要有豌豆潜叶蝇、豆秆蝇、豆蚜等。

**营养及功效**：富含蛋白质、糖分、维生素、矿物质和多种氨基酸。味甘，性平，具治寒热、止泻痢、益中气、消痈肿、下乳汁等功效。

# Vegetable pea

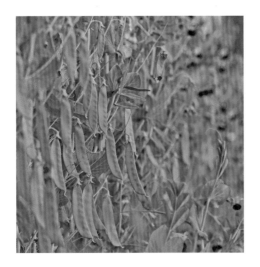

*Pisum sativum* L., an annual or biennial twining herb. Family: Leguminosae. Synonym: Pea, Wando. The chromosome number: $2n=2x=14$.

Vegetable pea was cultivated more than 200 years in Guangzhou. At present, the cultivation area is about 2500 $hm^2$ per annum.

Botanical characters: Vegetable pea has a strong tap-root, more root nodule, which has nitrogen fixation ability. The stem hollow and fragility, cylindrical, green. White wax powder on the surface of stem, leaf and lobular. Stem: vine or dwarfed. Leaves: simple on first to third nodes in base, and paripinnate on upper nodes, with 2~3 pairs of leaflets and branching tendril. Flower, shorted raceme with 1~3 papilionaceous flowers, white, pink and red-violet corolla. The young pods flat, long, dark green or yellow green, mature pod nearly cylindrical, brown in color. Self-pollination. The seed is round, smooth or wrinkled, yellow, green white, grey, brown and brown variegated. The weight per 1000 seeds is 150~800 g.

Types: There are two types of vegetable pea: edible pea and edible young pod type. The edible pea type, fresh peas and tender shoot are eaten as vegetable. The edible young pod type has crisp, fresh and sweet young pod, including Improved No.11, Qizhen No.76.

Cultivation environment and methods: The cold and dry weather is suitable for the growth of vegetable pea, which has cold tolerance while no heat tolerance. Sowing dates: mid-September to November. The optimum soil pH is 6.0~7.2. The fertile sandy loam is suitable for vegetable pea cultivation. Direct seeding, single or double row planting.

Harvest: The young pods are harvested about 12 days after anthesis. The tender peas are harvested about 15 days after anthesis. The tender shoot can be harvested 20~25 days after sowing, then harvested every 5~7 days.

Nutrition and efficacy: Vegetable pea is rich in proteins, carbohydrates, minerals and amino acids.

## 改良 11 号

Improved No.11 pea

> 品种来源

蔡兴利菜种行有限公司 20 世纪 90 年代育成。

📍 分布地区｜全市各区。

特　　征｜软荚型。蔓生，蔓长 150~180 厘米，侧蔓较多，节间长 5~9 厘米。叶长 18 厘米，深绿色。主蔓第 13~15 节开始着生花序，花紫红色，双生或单生。荚长 9.5 厘米，宽 1.6 厘米，单荚重 2.5 克，荚形较平直，绿色。

特　　性｜中熟，播种至初收 70~80 天，可延续采收 70~80 天。较耐白粉病。豆荚脆嫩，纤维少，品质优。每公顷产量 8~10 吨。

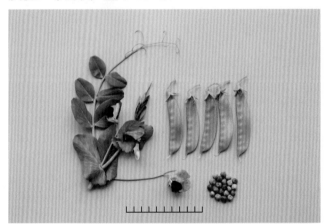

栽培要点｜适播期 9—11 月，直播。起深沟高畦，株距 4~6 厘米，行距 40 厘米。施足基肥，合理追肥，适当淋水，及时插架绑蔓。苗期注意防治豆秆蝇，生长期间注意防治白粉病。

## 大荚荷兰豆

Large pod pea

> 品种来源

汕头市金韩种业有限公司育成。

📍 分布地区｜全市各区。

特　　征｜软荚型。蔓生，蔓长 200~250 厘米，侧蔓较多，叶深绿色。主蔓第 11~12 节开始着生花序，花紫红色，双生或单生。荚长 12.1 厘米，宽 2.5 厘米，单荚重 9.1 克，荚形较平直，绿色。

特　　性｜中熟，播种至初收 70~80 天，可延续采收 50 天。不耐白粉病。豆荚脆嫩，纤维少，品质优。每公顷产量 10~12 吨。

栽培要点｜参照改良 11 号。

## 红花中花

Red flower mid-pod pea

> 品种来源

引自饶平县。

📍 分布地区 | 全市各区。

特　　征 | 软荚型。蔓生，蔓长 200~250 厘米，侧蔓 3~4 条，主蔓具红色纵线条。叶深绿色。主蔓第 13~15 节开始着生花序，花紫红色，双生或单生。荚长 10.2 厘米，宽 1.8 厘米，单荚重 8.2 克，荚形较平直，绿色。

特　　性 | 迟熟，播种至初收 85 天左右，可延续采收 60 天。不耐白粉病。豆荚脆嫩，品质中等。每公顷产量 9~11 吨。

栽培要点 | 参照改良 11 号。

## 奇珍 76 号甜豌豆

Qizhen No.76 sweet pea

> 品种来源

日昇种子有限公司育成。

📍 分布地区 | 全市各区。

特　　征 | 软荚型。蔓生，蔓长 180~200 厘米，侧蔓 2~3 条，节间长 5~7 厘米。叶长 18 厘米，深绿色。主蔓第 12~14 节开始着生花序，花白色，双生或单生。荚长 8 厘米，宽 1.2 厘米，厚 1.0 厘米，单荚重 7 克，荚形肥厚，绿色。

特　　性 | 中熟，播种至初收 65~75 天，可延续采收 70~80 天。不耐白粉病。豆荚脆嫩，纤维少，风味甜，品质优，生食、凉拌及炒食俱佳。每公顷产量 9~11 吨。

栽培要点 | 参照改良 11 号。

## 农普甜豌豆

Nongpu sweet pea

**品种来源** | 广州市农业科学研究院于 1996 年育成。

**分布地区** | 全市各区。

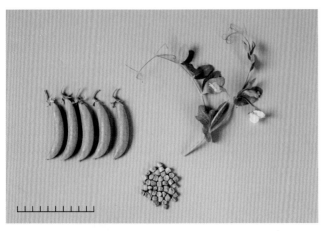

特　　征 | 软荚型。蔓生，蔓长 180~200 厘米，侧蔓 2~3 条，节间长 5~9 厘米。叶深绿色。主蔓第 13~15 节开始着生花序，花白色，双生或单生。荚长 7 厘米，宽 1.1 厘米，厚 0.9 厘米，单荚重 6.5 克，荚形肥厚，绿色。

特　　性 | 中熟，播种至初收 65~75 天，可延续采收 70~80 天。不耐白粉病。豆荚脆嫩，纤维少，风味甜，品质优，生食、凉拌及炒食俱佳。每公顷产量 9~11 吨。

栽培要点 | 参照改良 11 号。

## 美国手牌豆苗

Meiguo shoupai pea shoot

**品种来源** | 20 世纪 80 年代从美国引进。

**分布地区** | 全市各区。

特　　征 | 叶用型。植株半蔓生，分枝多。株高 20~25 厘米，节间长 3~5 厘米，叶长 16~20 厘米。具 10~12 节时开始采摘未张开嫩梢，叶片宽厚，深绿色。

特　　性 | 播种后约 22 天即可采收嫩梢，可延续采收 100~120 天。嫩梢叶片纤维少，品质优。每公顷产量 4~5 吨。

栽培要点 | 播种期 9 月至翌年 1 月，适播期 10—12 月，收获期 10 月至翌年 3 月。条播，行距 25 厘米，每公顷播种量 200 千克。施足基肥，勤追肥，苗期注意防治豆秆蝇，生长期间注意防治白粉病。

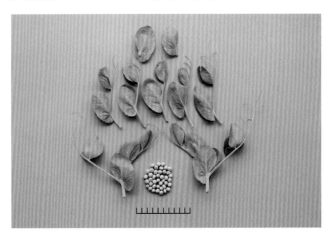

## 红花麦豆

Red flower pea

**品种来源** | 地方品种。

**分布地区** | 全市各区。

特　　征 | 硬荚型。蔓生，侧蔓多。叶绿色，托叶与茎相连部分紫红色，主蔓第 12~13 节开始着生花序，花紫红色，单花或双花。荚长 6.3 厘米，宽 0.8 厘米，鲜种子千粒重 200 克。

特　　性 | 迟熟，播种至初收约 90 天，可延续采收 15 天，荚硬，种皮较厚，纤维多，含淀粉多，豆粒供蔬食。耐寒，抗白粉病中等。每公顷鲜荚产量 5~6 吨。

栽培要点 | 适播期 11—12 月，直播。起深沟高畦，株距 5~6 厘米，行距 40 厘米。其余参照改良 11 号。

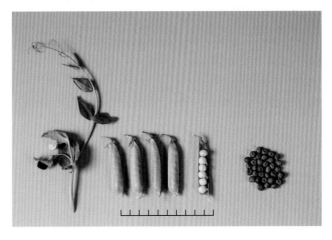

# 四棱豆

豆科四棱豆属一年生或多年生草本植物，学名 *Psophocarpus tetragonolobus* (L.) DC.，别名四角豆、杨桃豆、翼豆等。

原产非洲及东南亚，目前我国云南、广西、广东、海南、台湾等地有种植。

**植物学性状**：根系发达。植株长势强，其蔓可攀缘高达4米以上，多分枝，茎无毛，绿色或紫色。三出复叶，光润卵圆形，叶腋抽生花梗，总状花序，花白色或淡蓝色。嫩荚黄绿色，有4个棱角，每一棱角上有锯齿状翼。种子近球形，白色、黄色、棕色、黑色或杂以多种颜色。

**栽培环境与方法**：喜温暖湿润气候，生长发育适温为20~25℃，不耐霜冻，不耐久旱。属于短日照作物，对光照长短反应敏感，对土壤要求不严，较耐贫瘠，以肥沃的微酸性土壤为佳。多用种子繁殖，也可用块根、枝条扦插繁殖。种子坚硬，浸种催芽后育苗移栽或直播。

**收获**：采收黄绿色嫩荚供食用。

**病虫害**：病虫害少，主要防治地老虎、蚜虫和豆荚螟。

**营养及功效**：富含蛋白质、维生素、钙、铁等。嫩荚可炒食、盐渍或制酱菜等。嫩茎叶可炒食或煮汤。地下块根可炒食、制干或制淀粉，干种子可榨油或加工成豆制品。全株均可入药，常食对泌尿系统、心脑血管等疾病有一定的辅助疗效。

# Winged bean

*Psophocarpus tetragonolobus* (L.) DC., a perennial twining herb. Family: Leguminosae. Synonym: Goa bean, Four-angled bean, Four-cornered bean, Manila bean, Mauritius bean.

Winged bean is originated in Africa and Southeast Asia, and is cultivated in Yunnan, Guangxi, Guangdong, Hainan, and Taiwan.

Botanical characters: Winged bean has stronger root system. Climbing vine with branches, more than 4 m in height, green or purple in color. Leaf trifoliolate with ovate leaflet. Inflorescence, axillary racemes with white or pale blue flowers. The pod is yellow-green and has four wings with frilly edges running lengthwise. The skin is waxy and the flesh partially translucent in the young pods. Seeds nearly round in shape. White, yellow, brown, black and variegated, etc. in color.

Cultivation environment and methods: The winged bean thrives in hot weather and favours humidity. The suitable temperature for growth is 20~25℃. Winged bean is sensitive to day length, is a short-day plant. The fertile, sub-acid soil is favour for winged bean growth. Usually propagated by seeding, also by cuttage of tuber and shoot. Due to the hard seed coat, soaking and pregermination are required before seeding.

Harvest: The yellow-green young pods are harvested as vegetable.

Nutrition and efficacy: Winged bean is rich in proteins, vitamins, calcium, iron, etc. Leaves can be eaten like spinach, tubers can be eaten raw or cooked, seeds can be used in similar ways as the soybean.

# 菜用大豆 Vegetable soybean

豆科大豆属的栽培品种，一年生草本植物，学名 *Glycine max* (L.) Merr.，别名黄豆、毛豆、枝豆，染色体数 $2n=2x=40$。

菜用大豆原产于中国，约有 4 000 多年的历史，广州栽培历史悠久，但多为零星种植，以农家种为主。

**植物学性状**：根系发达，主根明显，再生能力弱。子叶出土，第 1 对真叶为单叶，以后为三出复叶。花梗从叶腋抽生，花小，白色或紫色，总状花序，每花序上有 8~10 朵花，结 2~5 荚果，嫩荚绿色，被灰白色或棕色茸毛。嫩荚种子绿色。

**栽培环境与方法**：喜温暖，生育期适温 20~25℃。选择 2~3 年未种过豆科作物、土层深厚、肥沃、地势高、排水良好的微酸性（pH 6.5~7.0）土壤种植。3—9 月播种，直播。

**收获**：豆荚鼓粒达 80%，荚色仍翠绿时采收。

**病虫害**：病害主要有病毒病、褐斑病、灰斑病等，虫害主要有豆荚螟、豆秆黑潜蝇、叶蝉、豆蚜等。

**营养及功效**：富含蛋白质、碳水化合物、纤维素、各种维生素和矿物质等。大豆味甘，性温，具宽中下气、消水健脾等功效。

*Glycine max* (L.) Merr., an annual herb. Family: Leguminosae. The chromosome number: $2n=2x=40$.

The vegetable soybean is originated in China, has been cultivated for about 4000 years, and grown sparsely for a long time in Guangzhou.

Botanical characters：Vegetable soybean, with a strong tap-root system. The first pair of true leaf is simple, the other leaves are trifoliolate. Axillary raceme with small white or violet flowers. There are 8~10 flowers in each inflorescence, 2~5 green pods with gray or brown trichomes. Young seeds green.

Cultivation environment and methods：The optimum temperature for the growth of vegetable soybean is 20~25℃. Warm weather, deep, fertile, good drainage and acidic (pH 6.5~7.0) soil are favourable to the growth of vegetable soybean. Direct sowing from March to September.

Harvest: Young pods are harvested when pods are still green and the seed are nearly in full size.

Nutrition and efficacy: Vegetable soybean is rich in proteins, carbohydrates, fibers, vitamins and minerals.

## 黄豆（毛豆）

Soybean

**品种来源**

地方品种。

**分布地区**｜花都、从化、增城等区零星种植。

**特　　征**｜株高 97 厘米，主茎节数 14 节，分枝数 4 个。叶卵圆形，绿色，长 10~12 厘米，宽 7 厘米。花白色，腋生，每花序结荚 1~3 个。荚长 6.5 厘米，宽 1.5 厘米，嫩荚绿色，成熟时黄白色。种子卵圆形，青绿色，千粒鲜重 900 克。

**特　　性**｜中熟，生长期 90 天左右。生长势强，抗性强，耐热，耐旱，病害少。以嫩豆粒供菜用，成熟种子供加工。每公顷种子产量 1.5~2.5 吨。

**栽培要点**｜播种期 3—8 月，直播，株行距 10~40 厘米，每穴播种 2~3 粒种子。施足基肥，5~6 叶时进行除草和培土，及时防治蚜虫、螨类及叶蝉。

# 七、茄果类
## SOLANACEOUS FRUITS

- 茄子
- 辣椒
- 番茄

# 茄子

茄科茄属以浆果为产品的一年生草本植物，在热带为多年生植物，学名 *Solanum melongena* L.，别名矮瓜、茄瓜、落苏，染色体数 $2n=2x=24$。

茄子起源于东南亚、南亚热带地区，公元 4—5 世纪传入我国，广州栽培历史悠久，是春、夏、秋季主要蔬菜之一，目前年种植面积约 1 200 公顷。

**植物学性状**：根系发达，纵型直根。茎木质化，直立，粗壮，分枝呈灌木状，假轴分枝。叶片长椭圆形至倒卵形，叶缘有不规则浅裂，叶深绿色或紫绿色，叶背被茸毛。两性花，花冠呈紫色、淡紫色或白色。自花授粉。浆果，长条形至扁卵形，有紫、绿、白等色，果肉白色至淡绿色。种子黄褐色，扁平，呈肾形，表面光滑，坚硬，千粒重 4~5 克。

**类型**：根据果实颜色分为 4 类。

1. **紫茄类型**　主栽品种有紫荣 8 号、农丰、园丰、润丰 3 号、紫丰 2 号、绿霸新秀、长丰 2 号、长丰 3 号、公牛等。

2. **青茄类型**　主栽品种有翡翠绿、观音手指等。

3. **白茄类型**　主栽品种有象牙白、白玉等。

4. **花茄类型**　主栽品种有玫瑰紫花茄等。

**栽培环境与方法**：喜温，耐热，喜光，不耐涝。结果期间的适宜温度为 25~30℃，17℃以下生长缓慢，会引起落花，5℃以下会受冻害。育苗移栽，在广州一年四季都能生长，冬季要注意防寒。生长期间需水量大，对矿物质肥的要求以氮、钾为主。

**病虫害**：病害主要有青枯病、绵疫病和褐纹病等，虫害主要有螨、蓟马、烟粉虱等。

**营养及功效**：富含胡萝卜素、硫胺素、核黄素、各种矿物质元素和维生素 P。味甘，性凉，具清热解毒、散血、消肿、宽肠等功效。

# Eggplant

*Solanum melongena* L., annual fruit vegetable with a strong taproot system. Family: Solanaceae. Synonym: Garden egg, Brinjal, Aubergine, Guinea squash. The chromosome number: $2n=2x=24$.

Eggplant is originated in tropics of southeast and south Asia, introduced into China in 4~5th century. Eggplant has long been cultivated in Guangzhou. It is one of the main vegetables in spring, summer and autumn. At present, the cultivation area is about 1200 $hm^2$ per annum in Guangzhou.

Botanical characters: Stem woody, erect, false-podial branching. Leaf, oblong to obovate, leaf margin irregularly lobed, dark green or purple green, trichomes on the dorsal side. Flowers, hermaphroditic, corolla purple, lavender or white. Self-pollination. Berry ovate-round to long-ovate, violet, green or white in color. Pulp white to light green in color. Seed, brown, flat, reniform, smooth, hard. The weight per 1000 seeds is 4~5 g.

Types: According to fruit color, eggplant can be divided into 4 types.

1. Violet fruit type: including Zirong No.8, Nongfeng, Yuanfeng, Runfeng No.3, Zifeng No.2, Lübaxinxiu, Changfeng No.2, Changfeng No.3, Gongniu, etc.

2. Green fruit type: including Feicuilü, Guanyinshouzhi, etc.

3. White fruit type: including Xiangyabai, Baiyu, etc.

4. Variegated violet fruit type: including Meiguizihua, etc.

Cultivation environment and methods: Eggplant growth require warm and sunny weather, with heat tolerance, while no water-logging tolerance. Suitable temperature for fruit development is 25~30 ℃. Flower abortion, fruit development slowly below 17℃. It may be freezing injury below 5℃. Seedling transplanting. Eggplant can be cultivated in Guangzhou year-round. The growth of eggplant requried suffficient irrigation and nitrogen and potassium.

Nutrition and efficacy: Eggplant is rich in carotene, thiamine, riboflavin, minerals and vitamin P.

## 长丰 2 号

Changfeng No. 2 eggplant

**品种来源**

广州农达种子科技有限公司育成的杂种一代。

**分布地区**｜南沙、番禺、从化、增城、花都、白云等区。

**特　征**｜叶片深绿色带紫色，叶缘波状浅缺刻。茎节密，花浅紫色。果长棒形，头尾均匀，果身直，尾部圆，果紫红色，果面平滑，着色均匀，光泽度好，果上萼片紫绿色，果长 30.0~32.0 厘米，横径 4.5~5.0 厘米，单果重 250~300 克，果肉白色，肉质紧实。

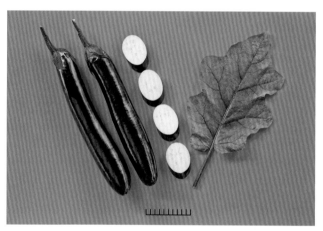

**特　性**｜生长势强。适应性强。中熟，播种至初收春植约 107 天，秋植约 84 天，延续采收 60~120 天。坐果率高，商品外形美观，高温期仍能保持良好的紫红色。耐贮运，品质优。较抗病，田间表现耐热性、耐涝性和耐旱性强。每公顷产量 70~80 吨。

**栽培要点**｜宜选土层深厚、肥沃的地块种植，注意轮作。适播期春植 12 月至翌年 2 月，秋植 6 月至 8 月上旬。每 667 米² 种植 800~1 000 株。施足基肥，追肥遵循"轻施苗肥、稳施花肥、重施果肥"的原则，每隔 10 天追施一次，注重氮、钾配合施用。不能缺水，但忌积水，干旱时应及时灌溉。加强整枝摘叶，防止群体过度荫蔽。门茄坐果时，保留 2 个主枝，摘除分叉以下所有侧枝，老叶要分批摘除。注意防治青枯病、绵疫病、褐纹病以及蓟马和螨类。

## 长丰 3 号

Changfeng No. 3 eggplant

**品种来源**

广州农达种子科技有限公司育成的杂种一代。

**分布地区**｜南沙、番禺、从化、增城、花都、白云等区。

**特　征**｜叶片深绿色带紫色，叶缘波状浅缺刻。茎节密，花浅紫色。果长棒形，头尾均匀，果身直，尾部圆，果深紫红色，果面光滑亮丽，着色均匀，果上萼片紫绿色，果长 32.0~34.0 厘米，横径 4.5~5.0 厘米，单果重 250~300 克，果肉白色，肉质紧实、细嫩。

**特　性**｜生长势强，植株紧凑，适宜密植，适应性广，生长后期表现更好。中晚熟，播种至初收春植约 108 天、秋植约 86 天，延续采收 60~120 天。坐果率高，整齐度好，商品外形美观，高温不易变色。耐贮运，品质优。对青枯病及褐纹病有较强的抗性，田间表现耐寒性较强。每公顷产量 70~80 吨。

**栽培要点**｜参照长丰 2 号。

## 紫荣 8 号

Zirong No.8 eggplant

**品种来源**
广州市农业科学研究院 2011 年育成的杂种一代。

**分布地区**｜南沙、番禺、从化、增城、花都、白云等区。

**特　征**｜叶片长卵形，深绿色，叶缘波状浅缺刻。花浅紫色。果长棒形，果深紫红色，果面平滑，光泽佳，头尾匀称，尾部圆，果长 33.0~35.5 厘米，横径 5.0~5.2 厘米，单果重 330 克左右，果肉白色，肉质紧实。

**特　性**｜生长势强，适应性较强。中熟，播种至初收春植约 105 天，秋植约 85 天，延续采收 60~120 天。坐果均匀。商品性状优良，采收全期果形一致性好，商品率高，高温时果色不易发白。耐老，耐贮运，品质优。抗青枯病能力较强，耐热性、耐寒性较好。每公顷产量 70~80 吨。

**栽培要点**｜参照长丰 2 号。

## 农丰

Nongfeng eggplant

**品种来源**
广东省农业科学院蔬菜研究所 2010 年育成的杂种一代，2012 年通过广东省农作物品种审定。

**分布地区**｜南沙、番禺、从化、增城、花都、白云等区。

**特　征**｜叶片长卵形，深绿色，叶缘波状浅缺刻。花浅紫色。果长棒形，果深紫红色，果面平滑，着色均匀、有光泽，头尾均匀，果长 30.3~31.9 厘米，横径约 5.1 厘米，单果重 284~302 克，果肉白色，肉质紧密。

**特　性**｜生长势强。中熟，播种至初收春植约 113 天、秋植约 88 天，延续采收 60~120 天。坐果均匀，商品率高，品质优。耐老，耐贮运。中抗青枯病，田间表现耐热性、耐寒性、耐涝性和耐旱性强。每公顷产量 70~80 吨。

**栽培要点**｜适播期春植 12 月至翌年 2 月，秋植 6 月至 8 月上旬。其余参照长丰 2 号。

## 园丰

Yuanfeng eggplant

**品种来源**

广东华农大种业有限公司育成的杂种一代。

📍 **分布地区** | 南沙、番禺、从化、增城、花都、白云等区。

**特　　征** | 叶片深绿色带紫色，叶缘波状浅缺刻。花浅紫色。果长棒形，外形较直，果紫红色，果面平滑，着色均匀，光泽度好，头尾均匀，果长31.0厘米左右，横径5.5厘米左右，单果重约350克，果肉白色，肉质紧实。

**特　　性** | 生长势强。早熟，播种至初收春植约105天、秋植约85天，延续采收60~120天。商品外形美观，品质优，耐贮运。较抗青枯病。每公顷产量70~80吨。

**栽培要点** | 参照长丰2号。

## 公牛

Gongniu eggplant

**品种来源**

广州亚蔬园艺种苗有限公司育成的杂种一代。

📍 **分布地区** | 南沙、番禺、从化、增城、花都、白云等区。

**特　　征** | 叶片深绿色带紫色，叶缘波状浅缺刻。茎节密，花浅紫色。果长棒形，头尾均匀，果身微弯，尾部圆，果皮紫红色，果面平滑，着色均匀，光泽度好，果上萼片紫绿色，果长31.5~33.0厘米，横径4.9~5.1厘米，单果重250~310克，果肉白色，肉质紧实。

**特　　性** | 生长势强。中熟，播种至初收春植约107天、秋植约84天，延续采收60~120天。坐果率高，商品外形美观。耐贮运，品质优。田间表现耐寒、耐热，较抗病。每公顷产量70~80吨。

**栽培要点** | 参照长丰2号。

## 润丰 3 号

Runfeng No. 3 eggplant

**品种来源** 广州华绿种子有限公司育成的杂种一代。

**分布地区** | 南沙、番禺、从化、增城、花都、白云等区。

**特 征** | 叶片深绿色带紫色，叶缘波状浅缺刻。茎节密，花浅紫色。果长棒形，外形较直，果深紫红色，果面平滑，着色均匀，光泽度好，头尾均匀，果长 32.0~35.0 厘米，横径 4.5 厘米左右，单果重 250~300 克，果肉白色，肉质紧实。

**特 性** | 生长势强。中熟，播种至初收春植约 110 天、秋植约 87 天，延续采收 60~120 天。商品外形美观，高温期仍能保持良好的紫红色，肉质细滑，品质优，耐贮运。较抗青枯病。每公顷产量 70~80 吨。

**栽培要点** | 参照长丰 2 号。

## 紫丰 2 号

Zifeng No. 2 eggplant

**品种来源** 广州兴田种子公司育成的杂种一代。

**分布地区** | 南沙、番禺、从化、增城、花都、白云等区。

**特 征** | 叶片深绿色带紫色，叶缘波状浅缺刻。茎节密，花浅紫色。果长棒形，头尾均匀，果身微弯，尾部圆，果皮紫红色，果面平滑，着色均匀，光泽度好，果上萼片紫绿色，果长 30.9~32.3 厘米，横径 5.1~5.3 厘米，单果重 295~309 克，果肉白色，肉质紧实。

**特 性** | 生长势强。中熟，播种至初收春植约 107 天、秋植约 84 天，延续采收 60~120 天。坐果率高，商品外形美观，高温期仍能保持良好的紫红色。耐贮运，品质优。对青枯病抗性中等，田间表现耐热性、耐涝性和耐旱性强。每公顷产量 70~80 吨。

**栽培要点** | 参照长丰 2 号。

## 绿霸新秀

Lübaxinxiu eggplant

> 品种来源

广州市绿霸种苗有限公司育成的杂种一代。

📍 分布地区 | 南沙、番禺、从化、增城、花都、白云等区。

特　　征 | 叶片深绿色带紫色，叶缘波状浅缺刻。茎节密，花浅紫色。果长棒形，头尾均匀，果身微弯，尾部圆，果皮紫红色，果面平滑，着色均匀，光泽度好，果上萼片呈紫绿色，果长 30.4~31.8 厘米，横径 4.9~5.1 厘米，单果重 275~289 克，果肉白色，肉质紧实。

特　　性 | 生长势强。中熟，播种至初收春植约 107 天、秋植约 84 天，延续采收 60~120 天。坐果率高，商品外形美观，高温期仍能保持良好的紫红色。耐贮运，品质优。对青枯病抗性中等，田间表现耐热性、耐涝性和耐旱性强。每公顷产量 70~80 吨。

栽培要点 | 参照长丰 2 号。

## 玫瑰紫花茄

Meiguizi eggplant

> 品种来源

广州市农业科学研究院 2012 年育成的杂种一代。

📍 分布地区 | 南沙、番禺、从化、增城、花都、白云等区。

特　　征 | 叶青绿色，叶缘波状浅缺刻。花浅紫色。果上萼片绿色，果长棒形，果色青中带紫，果面平滑，头尾匀称，果长 29.0~32.0 厘米，横径 4.2~4.4 厘米，单果重约 250 克，果肉绿白色，紧实。

特　　性 | 生长势强。早中熟，播种至初收春植约 98 天、秋植约 83 天，延续采收 60~120 天。坐果均匀，采收全期果形一致，商品率高，肉质细嫩，风味口感独特，品质优。中抗青枯病，耐寒性、耐阴性较好。每公顷产量 70~80 吨。

栽培要点 | 参照长丰 2 号。

## 观音手指（盐步青茄）

Guanyinshouzhi eggplant

**品种来源**

地方品种。

📍 **分布地区** | 南沙、番禺、从化、增城、花都、白云等区。

**特　　征** | 茎叶青绿色，叶缘波状深缺刻。花浅紫色。果上萼片绿色，果长条形，稍弯曲，果浅绿色，果面稍粗糙，无光泽，果尾尖细，果长30.0~32.0厘米，横径约2厘米，单果重约100克，果肉绿白色。

**特　　性** | 生长势强。早熟，播种至初收春植约96天、秋植约82天，延续采收60~120天。肉质柔滑，风味口感独特，品质优。耐湿热，抗病性一般。每公顷产量50~60吨。

**栽培要点** | 参照长丰2号。注意防治青枯病及绵疫病。

## 翡翠绿

Feicuilü eggplant

**品种来源**

广州市农业科学研究院2012年育成的杂种一代。

📍 **分布地区** | 南沙、番禺、从化、增城、花都、白云等区。

**特　　征** | 茎叶青绿色，叶缘波状浅缺刻。花浅紫色。果上萼片绿色，果长棒形，果青绿色，果面平滑，光泽度佳，头尾较匀称，果长29.5~32.1厘米，横径4.2~4.4厘米，单果重239~250克，果肉绿白色，肉质紧实。

**特　　性** | 生长势强。早中熟，播种至初收春植约99天、秋植约84天，延续采收60~120天。坐果均匀，采收全期果形一致，商品率高，肉质细嫩，风味口感独特，品质优。耐寒性、耐阴性较好。每公顷产量70~80吨。

**栽培要点** | 参照长丰2号。注意防治青枯病。

## 象牙白

Xiangyabai eggplant

**品种来源**

广州市农业科学研究院 2012 年育成的杂种一代。

📍 **分布地区** | 南沙、番禺、从化、增城、花都、白云等区。

特　　征 | 茎叶青绿色，叶缘波状浅缺刻。花浅紫色。果上萼片绿色，果长棒形，果色乳白，果面平滑，光泽佳，头尾较匀称，果长 28.3~31.1 厘米，横径 4.4~4.6 厘米，单果重 250~260 克，果肉白色，肉质紧实。

特　　性 | 生长势强。早中熟，播种至初收春植约 97 天、秋植约 84 天，延续采收 60~120 天。坐果均匀，采收全期果形一致，商品率高，品质优。耐寒性、耐阴性较好，中抗青枯病。每公顷产量 70~80 吨。

栽培要点 | 参照长丰 2 号。

## 白玉

Baiyu eggplant

**品种来源**

广东省农业科学院蔬菜研究所育成，2007 年通过广东省农作物品种审定。

📍 **分布地区** | 南沙、番禺、从化、增城、花都、白云等区。

特　　征 | 茎叶青绿色。花浅紫色。果长棒形，果乳白色，果面平滑，果尾偏尖，果长约 26.0 厘米，横径约 4.6 厘米，单果重 192 克，果肉白色，肉质紧实。

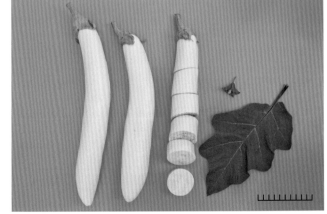

特　　性 | 生长势强。早中熟，播种至初收春植约 105 天，秋植约 86 天，延续采收 60~120 天。坐果均匀，商品率高，品质优。中抗青枯病，田间表现耐热性和耐寒性强，耐涝性较强，耐旱性中等。每公顷产量 70~80 吨。

栽培要点 | 参照长丰 2 号。

# 辣椒

茄科辣椒属一年生或多年生草本植物，学名 Capsicum annuum L.，别名番椒、海椒、辣子、甜椒等，染色体数 $2n=2x=24$。

辣椒起源于中南美洲热带地区的墨西哥、秘鲁等地，16世纪引入中国。最早记载辣椒的史料，见于明代高濂 1591 年著的《草花谱》，在书中他把辣椒称为"番椒"。全国各地均有栽培，广州年栽培面积约 1 500 公顷。

**植物学性状**：根系浅。茎直立，基部木质化，上部半木质化。茎的分枝一般为双叉分枝，也有三叉分枝。叶片单生、互生，卵圆形、披针形或椭圆形。花为完全花，常异花授粉，单生或簇生，花冠一般呈乳白色，个别品种为绿白色或紫白色。浆果灯笼形、扁圆形、牛角形、羊角形、长条形、短指形、樱桃形等，果皮肉质，一般有 2~4 个心室，嫩果多为绿色或黄色、白色、紫色等，成熟时多为红色或黄色。种子肾形，淡黄色，千粒重 5~8 克。

**类型**：按果实特征，可分为牛角椒、线椒、朝天椒、甜椒等类型。

1. **牛角椒类型**　广州栽培的主要类型，占栽培面积的 80% 以上。果实较大，辣味中等至微辣。果实牛角形或羊角形，果肉厚。品种主要有辣优 4 号、辣优 15 号、汇丰 2 号、湘研 158、东方神剑、茂椒 4 号等。

2. **线椒类型**　果实长条形，味辣而香，皮薄肉薄。品种主要有香妃、金田 3 号、永利等。

3. **朝天椒类型**　果实末端朝上，单生或簇生。广州种植以单生为主，味极辣，香，肉薄。品种主要有广良 5 号、艳红等。

4. **甜椒类型**　主要在温室大棚种植。植株高达 2 米以上，果实灯笼形，有红、黄、绿、奶白、紫等颜色，味甜，肉厚，可生食或熟食。品种主要有卡尔顿、黄欧宝等。

**栽培环境与方法**：喜温暖，忌高温，怕霜冻、寒冷，生长适温为 25~30℃。对光照的要求不太严格，较耐弱光。较耐旱，怕涝，不耐渍；较耐肥，不耐瘠瘦。对土壤要求不太严格。育苗移栽，春植播种期 1 月上旬至 2 月上旬，定植期 3 月上旬至 4 月上旬，采收期 5—7 月；秋植播种期 8 月上旬，定植期 9 月，采收期 10—12 月。每 667 米$^2$用种量 40 克左右。

**收获**：一般在花凋谢 20~25 天后采收青果，红果宜在果实八九成熟后采收。

**病虫害**：病害主要有猝倒病、立枯病、疫病、青枯病、病毒病、炭疽病、疮痂病、软腐病、枯萎病、灰霉病、白粉病等，虫害主要有小地老虎、烟青虫、斜纹夜蛾、蚜虫、茶黄螨、蓟马、红蜘蛛等。

**营养及功效**：富含维生素 C、辣椒素、胡萝卜素、辣椒红素、柠檬酸、苹果酸等。性热，味辛，具温中下气、开胃消食、散寒除湿等功效。

# Pepper

*Capsicum annuum* L., an annual or perennial vegetable with vigorous root system. Family: Solanaceae. Synonym: Chili, Hot pepper, Sweet pepper. The chromosome number: $2n=2x=24$.

Pepper is originated in Mexico, Peru and other tropical regions of Central and South America. It was introduced into China in 16th century. At present, the cultivation area is about 1500 hm$^2$ per annum in Guangzhou.

Botanical characters: Pepper, stems erect, woody on base, semi-woody on upper. Stems usually dichotomous branching, some trichotomous. Leaves is solitary or alternate. Leaf oval, lanceolate or ellipse. Often cross-pollinated crops. Complete flower, solitary or clustered. Corolla is ivory, few white-green or pink in color. Berries lantern, flat round, long horn, short horn, horn, elongated, short fingers, cherry, etc. in shape. Pericarp is fleshy. Fruit has 2~4 locules. Young fruit is green or yellow, white, purple and other colors, red or yellow when ripe. Seeds reniform, light-yellow. The weight per 1000 seeds is 5~8 g.

Types: According to fruit characteristics, it can be divided into Ox horn pepper, Thin cayenne pepper, Pod pepper, sweet pepper, etc.

1. Ox horn pepper: The main type in Guangzhou, accounting for more than 80% cultivation area. Larger fruit, moderate to mild pungent. Fruit is horn or long horn in shape. The cultivars include: Layou No.4, Layou No.15, Huifeng No.2, Xiangyan 158, Dongfangshenjian, Maojiao No.4, etc.

2. Thin cayenne pepper: Long fruit with thin pericarp pungent and fragrant. The cultivars include: Xiangfei, Jintian No.3, Yongli, etc.

3. Pod pepper: The fruit end upright, solitary or clustered, more solitary in Guangzhou. Strong pungent, fragrant. The cultivars include: Guangliang No.5, Yanhong, etc.

4. Sweet pepper: It is mainly planted in greenhouse in Guangzhou. The plant is large, more than 2 m in height. Fruit is lantern-shaped, thick pericarp, red, yellow, green, milky white, and purple etc. in color. It can be eaten raw or cooked. The cultivars include Kaerdun, Huangoubao, etc.

Cultivation environment and methods: Pepper grow well in warm weather, less tolerant to high temperature, frost and cold. The optimum temperature for growth is 25 ~ 30℃. Pepper is tolerant to weak light, drought, while not to waterlogged. Seedling transplanting, about 600 g seeds per ha. In spring, sowing dates: early January to early February, transplanting dates: early March to early April, harvesting dates: May to July. In autumn, sowing dates: early August, transplanting dates: September, harvesting dates: October to December.

Harvest: Green fruit can be harvested 20~25 days after anthesis. Red fruit should be harvested when fruit has 90% maturity.

Nutrition and efficacy: Pepper is rich in vitamin C, capsaicin, carotene, capsanthin, citric acid, malic acid, etc.

# 牛角椒 Ox horn pepper

## 辣优 4 号
Layou No.4 pepper

**品种来源**

广州市农业科学研究院育成的杂种一代，1999 年通过广东省农作物品种审定，2002 年通过全国农作物品种审定。

**分布地区** | 番禺、增城等区。

**特　　征** | 生长势强。株形较平展，株高 52 厘米，开展度 76~80 厘米。叶色浓绿。始花节位第 9~10 节。果实长牛角形，果长 15~20 厘米，横径 3.3 厘米，果皮较光滑，绿色，肉厚 0.3 厘米，2~3 个心室。单果重 35~50 克。

**特　　性** | 早熟，播种至初收春植 86 天、秋植 65 天，延续采收 45~82 天。耐高温高湿和低温弱光，抗病性强，田间表现抗疫病、青枯病和病毒病。味辣，品质优，商品性好。每公顷产量 44 吨左右。

**栽培要点** | 早春栽培：播种期 11 月至翌年 1 月，苗期 50~60 天，株行距 30 厘米 ×40 厘米，每 667 米² 种植 3 500~4 000 株；施足基肥，追肥以复合肥为主，全期每 667 米² 追肥 30~50 千克，兑水浇施为好；及时打掉门果以下侧芽；排除田间积水，及时防治病虫害。秋冬露地及温室栽培：7—10 月播种，8—11 月定植，定植后 2 周内要适当遮阴保苗，加强肥水管理，促进发棵封垄，这是高产的关键。定植方法、田间管理等参考早春栽培。

## 辣优 15 号
Layou No.15 pepper

**品种来源**

广州市农业科学研究院育成的杂种一代，2009 年通过广东省农作物品种审定。

**分布地区** | 番禺、增城、从化、花都等区。

**特　　征** | 生长势强，株高 58.3~66.5 厘米。第 1 朵花着生节位第 9.0~9.9 节。青果绿色，熟果大红色。果实长羊角形，果面光滑，有光泽，无棱沟，果顶部细尖，果长 17.8 厘米，横径 2.8 厘米，肉厚 0.3 厘米。单果重 36.9~38.6 克。

**特　　性** | 早中熟，播种至初收春植 103 天、秋植 76 天，延续采收 48~84 天。中抗青枯病，抗疫病。田间表现耐热性、耐寒性、耐涝性和耐旱性强。味辣，感观及品质优。每公顷产量 45 吨左右。

**栽培要点** | 适合春、秋、冬季露地、地膜覆盖栽培及再生辣椒越冬栽培。适播期 7 月至翌年 2 月。其余参照辣优 4 号。

## 汇丰 2 号

Huifeng No.2 pepper

**品种来源**

广东省农业科学院蔬菜研究所育成的杂种一代，2009年通过广东省农作物品种审定。

**分布地区** | 番禺、增城、从化、花都等区。

**特　征** | 生长势强，株高54.9~61.5厘米。第1朵花着生平均节位第10.7~11.7节。青果绿色，熟果大红色。果实羊角形，果面光滑，有光泽，无棱沟，果顶部细尖。果长18.0~18.2厘米，横径2.5~2.6厘米，肉厚0.3厘米左右。单果重36.4~39.4克。

**特　性** | 中熟，播种至初收春植104天、秋植76天，延续采收40~84天。感青枯病，中抗疫病。田间表现耐热性、耐寒性、耐涝性和耐旱性强。味辣，感观及品质优。每公顷产量43吨左右。

**栽培要点** | 适宜春、秋季种植。春植11月至翌年1月播种，秋植7—10月播种，育苗移栽，双行植，株行距30厘米×40厘米，每667米²种植4 000株左右。注意防治青枯病、疫病、病毒病、螨类、蚜虫和烟青虫等病虫害。

## 湘研 158

Xiangyan 158 pepper

**品种来源**

湖南湘研种业有限公司育成的杂种一代，2006年通过湖南省农作物品种审定。

**分布地区** | 番禺、增城、从化、花都等区。

**特　征** | 生长势强，株高55.3~62.5厘米。第1朵花着生节位第10~12节。青果绿色，熟果大红色。果实长羊角形，果面光滑，有光泽，果顶部渐细尖。果长19~22厘米，果宽3.2~3.5厘米，肉厚0.3厘米。大果型。单果重55~62克。

**特　性** | 中熟，播种至初收春植86天、秋植65天，延续采收春植45天、秋植68天。田间表现耐热，抗病毒病、疮痂病。味辣，品质优。每公顷产量43吨左右。

**栽培要点** | 适宜春、秋季种植。栽培管理参照汇丰2号。

## 东方神剑

Dongfangshenjian pepper

**品种来源**

广州市绿霸种苗有限公司育成的杂种一代，2008 年通过广东省农作物品种审定。

**分布地区** | 番禺、增城、从化、花都等区。

特　　征 | 生长势强，株高 40.1~52.5 厘米。始花节位第 9.4~11.3 节。果实羊角形，青果绿色，熟果大红色。果面平滑，无棱沟，有光泽。果长 14.5~17.4 厘米，横径 2.4~2.9 厘米，肉厚 0.3 厘米左右。单果重 31.2~44.8 克。

特　　性 | 中熟，播种至初收春植 98 天、秋植 78 天，延续采收 41~72 天。中抗青枯病，感疫病。田间表现抗病毒病和炭疽病，耐热性和耐旱性强，耐寒性和耐涝性中等。微辣，感官及品质优。每公顷产量 42 吨左右。

栽培要点 | 适宜春、秋季露地和保护地栽培。苗期春植 55~60 天，秋植 25~30 天；施足有机质基肥，增施磷钾肥；采用双行种植，规格为 50 厘米×33 厘米，每 667 米$^2$种植 2 600~3 000 株；采果前 3~5 天重施一次水肥，每 667 米$^2$施复合肥 20 千克；及时摘除侧枝，盛产期搭架护果；注意防治疫病。

## 茂椒 4 号

Maojiao No.4 pepper

**品种来源**

茂名市茂蔬种业科技有限公司育成的杂种一代，2010 年通过广东省农作物品种审定。

**分布地区** | 番禺、增城、从化、花都等区。

特　　征 | 生长势中等。第 1 朵花着生节位第 7.8~9.3 节。果实羊角形，青果黄绿色，熟果大红色。果面光滑，有棱沟。果长 15.6~18.5 厘米，横径 2.3~2.8 厘米，果肉厚 0.3 厘米。单果重 27.9~43.8 克。

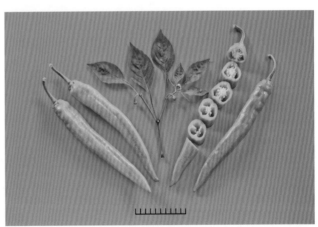

特　　性 | 中早熟，播种至初收春植 99 天、秋植 76 天，延续采收 40~73 天。感青枯病和疫病。田间表现炭疽病和病毒病发病较轻，耐热性、耐寒性、耐涝性和耐旱性强。味较辣，品质良。每公顷产量 45 吨左右。

栽培要点 | 适宜春、秋季种植。秋植 8—9 月播种，苗龄 30~35 天；春植 11—12 月播种，苗龄 50~55 天，采用营养杯基质育苗。起高畦种植，畦宽 1.5 米包沟，双行植，每 667 米$^2$种植 2 500~3 000 株。注意防治青枯病和疫病。

## 线椒 Thin cayenne pepper

### 香妃
Xiangfei thin cayenne pepper

**品种来源**

广州市农业科学研究院 2007 年育成的杂种一代。

**分布地区** | 番禺、增城、从化、花都等区。

**特　征** | 生长势强，株高 60 厘米。青果浅绿色，熟果鲜红色，果实长条形，果面光滑，有光泽。长 20.2~25.6 厘米，横径 1.5~1.8 厘米，肉厚 0.2 厘米。单果重 23.3~28.4 克。

**特　性** | 早中熟，播种至初收春植 108 天、秋植 79 天，延续采收 42~86 天。田间表现抗青枯病和疫病，耐雨水，耐热。鲜食脆嫩，味特香辣，品质优，耐贮运，连续挂果能力强，且上下果一致。每公顷产量 35 吨左右。

**栽培要点** | 适合春、秋季露地、地膜覆盖栽培。适播期 8 月上旬至翌年 2 月，培育适龄壮苗，株行距 30 厘米 ×40 厘米，每 667 米$^2$ 种植 3 000~3 500 株，秋季可适当密植。施足基肥，开花结果期勤施追肥。注意防治病虫害。

### 金田 3 号
Jintian No.3 thin cayenne pepper

**品种来源**

广东省农业科学院蔬菜研究所育成的杂种一代。

**分布地区** | 番禺、增城、从化、花都等区。

**特　征** | 生长势强，株高 61.5~69.5 厘米。第 1 朵花着生节位第 9.4~9.8 节。青果浅绿色，熟果大红色。果实长条形，果面光滑，果顶部渐细尖。果长 17.8~20.8 厘米，横径 1.2~1.5 厘米，肉厚 0.2 厘米。单果重 17.0~20.6 克。

**特　性** | 中熟，播种至初收春植 91 天、秋植 77 天，延续采收春植 42 天、秋植 82 天。田间表现耐热性、耐寒性、耐涝性和耐旱性强。味辣，品质优。每公顷产量 37 吨左右。

**栽培要点** | 适宜春、秋季种植。栽培管理参照香妃。

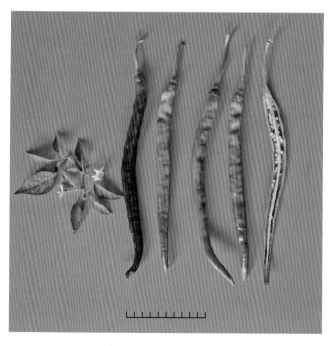

## 永利

Yongli thin cayenne pepper

> **品种来源**

深圳市永利种业有限公司育成的杂种一代。

📍 分布地区｜番禺、增城、从化、花都等区。

特　　征｜生长势强，株高 49.5~55.5 厘米。第一朵花着生节位第 9.7~10.6 节。青果绿色，熟果大红色。果实长条形，果面光滑，有光泽，有棱沟，果顶部渐细尖。果长 22.1~26.3 厘米，横径 1.7~2.1 厘米，肉厚 0.2 厘米。单果重 25.0~29.1 克。

特　　性｜中熟，播种至初收春植 91 天、秋植 77 天，延续采收春植 42 天、秋植 84 天。田间表现耐热性、耐寒性、耐涝性和耐旱性强。味辣，品质优。每公顷产量 38 吨左右。

栽培要点｜适宜春、秋季种植。栽培管理参照香妃。

## 朝天椒 Pod pepper
## 广良 5 号

Guangliang No.5 pod pepper

> **品种来源**

广东省良种引进服务公司从泰国引进的杂种一代。

📍 分布地区｜番禺、增城、从化、花都等区。

特　　征｜生长势强，株高 110 厘米。第 1 朵花着生节位第 12.1~12.8 节。青果绿色，熟果鲜红色。朝天椒类型，果面光滑，有光泽，果实着生方向向上。果长 6 厘米，横径 1.1 厘米，肉厚 0.1 厘米。单果重 5 克。

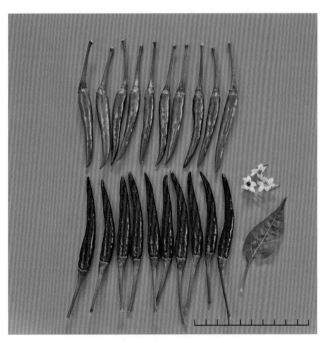

特　　性｜中熟，播种至初收春植 115 天、秋植 93 天，延续采收 96 天左右。田间表现耐热性、耐寒性、耐涝性和耐旱性强。味辣，品质优。连续采收果实仍保持较好的商品性。每公顷产量 30 吨左右。

栽培要点｜适宜春、秋季种植。适播期为 1—3 月和 7—8 月，重视苗期管理，选择非重茬地种植，适当疏植，每 667 米$^2$ 种植 1 100~1 500 株。加强肥水管理和病虫害防治。

## 艳红

Yanhong pod pepper

**品种来源**

海南林忠民菜种行有限公司从泰国引进的杂种一代。

📍 分布地区｜番禺、增城、从化、花都等区。

特　　征｜生长势强，株高120厘米。第1朵花着生节位第13.1~13.6节。青果浅绿色，熟果鲜红色。朝天椒类型，果面光滑，有光泽，果实着生方向向上。果长6.5厘米，横径1.0厘米，肉厚0.1厘米。单果重5克左右。

特　　性｜迟熟，播种至初收春植123天、秋植95天，延续采收90天。田间表现耐热性、耐寒性、耐涝性和耐旱性强。味辣，品质优，连续采收果实仍保持较好的商品性。每公顷产量29吨左右。

栽培要点｜参照广良5号。

## 甜椒 Sweet pepper

## 卡尔顿

Kaerdun sweet pepper

**品种来源**

法国科劳斯蔬菜种子公司。

📍 分布地区｜南沙、番禺、白云等区。

特　　征｜株高150~200厘米。叶长28厘米，叶宽10厘米，深绿色。花白色。果肉厚，4心室率高，成熟后果色鲜红，果形方正，10厘米×10厘米。单果重200克左右。

特　　性｜播种100天左右可采收青椒，120天左右采收成熟彩椒。采收期长，可达4~5个月。品质优良，商品性佳。每公顷产量50~60吨。

栽培要点｜宜采用温室大棚基质栽培。播种期8—9月，育苗移栽，每667米$^2$种植2 000株。栽培要做好基质和设施的消毒，实行水肥一体化管理，前期淡肥水，结果期加大肥水供应，增施钙肥；三秆整枝，及时引蔓。注意防治病虫害。

## 黄欧宝

Huangoubao sweet pepper

**品种来源**

从荷兰引进。

📍 分布地区｜南沙、番禺、白云等区。

特　　征｜株高150~200厘米。叶长18厘米，叶宽9厘米，深绿色。花白色。果肉中厚，肉厚0.50厘米，4心室率高，成熟后果色转为黄色，果形9.5厘米×9厘米。单果重150克左右。

特　　性｜中早熟，播种90天左右可采收青椒，120天左右采收成熟彩椒。连续坐果能力强，采收期长，可达4~5个月。品质优良，商品性佳。每公顷产量45~55吨。

栽培要点｜参照卡尔顿。

## 辣椒叶 Leaf-edible pepper

### 农普辣椒叶

Nongpu leaf-edible pepper

**品种来源**

广州市农业科学研究院1984年育成。

📍 分布地区｜全市各区。

特　　征｜食用部位为嫩茎叶和成长叶，是专用辣椒叶品种。生长势强，株形平展，株高70厘米，开展度80厘米。叶片青绿色，卵状披针形，叶长8~9厘米，叶宽4~5厘米。

特　　性｜中熟，播种至初收辣椒叶约45天。辣椒叶味甘鲜、嫩滑，口感佳，品质优。较耐肥，较耐热，耐病毒病。每公顷产量15~30吨。

栽培要点｜以采收叶为主，及早摘除嫩果。广州地区可周年种植，直播每667米²用种量150~250克，育苗移栽则为100克左右。其余参照辣椒品种。

# 番茄

茄科番茄属一年生草本植物，学名 *Lycopersicon esculentum* Mill.，别名西红柿、洋柿子、蕃柿、六月柿，染色体数 $2n=2x=24$。

番茄起源于南美洲，16 世纪末或 17 世纪初明代万历年间传入中国，广州番茄栽培始于 20 世纪初，至今已逾百年，年栽培面积约 6 500 公顷。

**植物学性状**：根系较发达。茎半蔓生或半直立。叶互生，羽状复叶或羽状复叶深裂。茎、叶密被腺毛，散发特殊气味。总状花序或聚伞花序，完全花，黄色，自花授粉为主。果实为多汁浆果，成熟果多为红色或粉红色，也有黄、橙、绿、紫等颜色，还有带彩色条纹的。种子肾形，灰黄色，表面被茸毛，千粒重 3 克左右。

**类型**：通常分为普通、樱桃、大叶、梨形、直立番茄 5 个变种。按植株生长习性可分为有限生长（自封顶、高封顶）和无限生长（非自封顶）两类型；按果实大小可分为大果型（150 克以上）、中果型（100~149 克）、小果型（100 克以下）。

1. **大、中型果类型**　单果重 100 克以上，一般烹调后食用。主要品种有金丰 1 号、年丰、益丰、新星 101、托美多、金石、拉菲、先丰等。

2. **小型果类型**　又称为樱桃番茄，单果重 100 克以下，一般用作生食。主要品种有红艳、金艳、绿樱 1 号、黄樱 1 号、华喜珍珠等。

**栽培环境与方法**：喜温，喜光。生长适宜的昼温为 24~28℃，夜温为 15~18℃。采用育苗移栽，每 667 米$^2$用种量 8~10 克，浸种 2~4 小时，催芽露白即播种，培育壮苗，加强田间管理。

**收获**：白熟期采收适宜长期贮藏及远途运输；黄熟期采收适于近期上市及短期贮运；成熟期和完熟期采收适于立即上市及鲜食，不宜贮运。

**病虫害**：病害主要有青枯病、病毒病、叶斑病、早疫病、晚疫病、脐腐病等，虫害主要有蚜虫、棉铃虫、烟青虫、烟粉虱、美洲斑潜蝇等。

**营养及功效**：富含多种维生素、矿物质、碳水化合物、有机酸、番茄红素及少量的蛋白质。味甘、酸，性平，具健脾开胃、生津止渴、除烦润燥等功效。

# Tomato

*Lycopersicon esculentum* Mill., annual fruit vegetable with a strong root system. Family: Solanaceae. The chromosome number: $2n=2x=24$.

Tomato is originated in South America, was introduced into China in 16~17th century. Tomato cultivation began in Guangzhou in the early 20th century. At present, the cultivation area is about 6500 hm$^2$ per annum in Guangzhou.

Botanical characters: Stem semi-vine or semi-erect. Leaves alternate, pinnate or pinnate parted. Both stem and leaf have dense short glandular hair. Inflorescence: racemose or cymes, complete flower, yellow corolla, self-pollinated. Mature fruit most red or pink, some yellow, orange, green, purple and other colors. Seeds reniform, pale yellow, with silver grey trichomes on the surface. The weight per 1000 seeds is about 3 g.

Types: Usually tomatoes were divided into common, cherry, big leaf, pear, erect tomato five varieties. According to plant growth habit, it can be divided into determinate and indeterminate growth type. According to the size of fruit, it can be divided into three types, large fruit (above 150 g), medium fruit (100~149 g) and small fruit (100 g or less).

1. Large and medium-sized fruit type: Single fruit is above 100 g, used for cooking. Main cultivars include: Jinfeng No.1, Nianfeng, Yifeng, Xinxing 101, Tuomeiduo, Jinshi, Lafei, Xianfeng, etc.

2. Small fruit type: Cherry tomatoes. Single fruit is less than 100 g, usually used as raw food. Main cultivars include: Hongyan, Jinyan, Lüying No.1, Huangying No.1, Huaxizhenzhu, etc.

Cultivation environment and methods: Tomato growth require warm and sunny weather. The optimum temperature for growth is 24~28 ℃ (day), 15~18 ℃ (night). Seedling transplanting, about 120~150 g seeds per ha.

Harvest: Fruit at breaker stage is suitable for storage and transport; fruit at turning stage is suitable for market and short-term storage; fruit at pink and ripe stage suitable for local market, not for storage.

Nutrition and efficacy: Tomato is rich in vitamins, minerals, carbohydrates, organic acids, lycopene, proteins, etc.

# 大番茄 Large fruit tomato

## 金丰 1 号
Jinfeng No.1 tomato

**品种来源**

广州市农业科学研究院育成的杂种一代，1992 年通过广东省农作物品种审定，2003 年通过全国蔬菜品种鉴定委员会鉴定。

📍 **分布地区**｜南沙、白云、增城等区。

**特　　征**｜无限生长类型。叶长 53 厘米，宽 43 厘米。第 1 花序着生于第 8~9 节，以后每隔 2~4 节着生 1 花序。每花序结果 3~6 个。果实高圆形，高 6.3~6.6 厘米，横径 6.0~6.2 厘米，青果深绿肩，熟果鲜红，4 心室，肉厚 0.9~1.1 厘米。单果重 120 克左右。

**特　　性**｜中熟，播种至初收 100~105 天，延续采收 40~60 天。田间表现不耐热，耐青枯病及病毒病。耐裂果，耐贮运，品质优。每公顷产量 40~50 吨。

**栽培要点**｜以秋、冬植为主，播种期 8 月至翌年 2 月。育苗移栽，双行植，株行距 45 厘米 ×60 厘米，每 667 米² 种植 2 200~2 500 株，及时整枝、搭架、绑蔓，适当疏花疏果。施足基肥，前期注意控制肥水，促进植株健壮，开花结果后多施追肥，注意氮、磷、钾配合，田间切忌过干过湿，及时排除积水。冬季温度过低，可用 2,4-D 10~20 毫克/升涂花防止落花。抓好轮作。前、中期注意防治蚜虫及烟粉虱，抑制病毒病传播蔓延；中、后期注意防治叶斑病、早疫病等，发现青枯病株应及时拔除，并在病穴上撒石灰。12 月至翌年 4 月采收。

## 年丰
Nianfeng tomato

**品种来源**

广州市农业科学研究院育成的杂种一代，2001 年通过广东省农作物品种审定。

📍 **分布地区**｜花都、白云、增城等区。

**特　　征**｜自封顶类型。株高 130~150 厘米，开展度 55~65 厘米。叶长 36 厘米，宽 41 厘米。第 1 花序着生于第 8~9 节，以后每隔 1 节着生 1 花序。每花序结果 4~6 个。果实扁圆形，高 5.0~5.5 厘米，横径 6.0~6.5 厘米，青果无绿肩，熟果鲜红，3 心室，肉厚 0.8 厘米。单果重 120 克左右。

**特　　性**｜早熟，播种至初收 90~95 天，延续采收 30~40 天。耐热，高抗青枯病，中抗病毒病。耐裂果，熟果软化快，耐贮性略差，品质中上。每公顷产量 45~50 吨。

**栽培要点**｜以春、夏植为主，播种期 6 月至翌年 2 月。育苗移栽，株行距 45 厘米 ×60 厘米，每 667 米² 种植 2 200~2 500 株，及时整枝、搭架、绑蔓，适当疏花疏果。施足基肥，定植 7 天淋提苗肥，第 1 穗果膨大时施 1 次催果肥，初收后一般每 7~10 天追肥 1 次。注意防治脐腐病，可采用根外追肥的方法进行防治，用 1% 过磷酸钙、0.1% 氯化钙或 0.1% 硝酸钙，每 15 天喷 1 次，连续 2 次；病毒病应以预防为主，苗期及定植后生长的中早期注意防蚜虫及烟粉虱，加强栽培管理，防止农事操作接触传毒，及时处理病株。6—10 月采收。

## 益丰

Yifeng tomato

> 品种来源

广州市农业科学研究院育成的杂种一代，2004年通过广东省农作物品种审定，2006年通过全国蔬菜品种鉴定委员会鉴定。

📍 分布地区｜南沙、白云、花都、从化、增城等区。

特　　征｜无限生长类型。叶长34.5厘米，宽31厘米。第1花序着生于第8~9节，以后每隔1~2节着生1花序。每花序结果3~6个。果实圆形，高7.0~7.5厘米，横径6.8~7.2厘米，青果浅绿肩，褪绿肩快，熟果鲜红，4~5心室，肉厚0.9~1.1厘米。单果重150克左右。

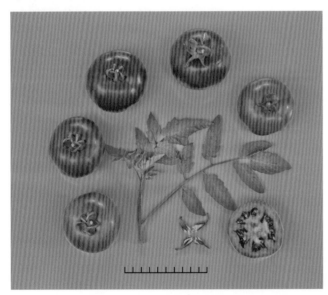

特　　性｜早熟，播种至初收春植105~110天、秋植90~100天，延续采收30~60天。耐热，抗青枯病、抗TMV病毒病。不耐裂果，果实硬实，耐贮运，品质中等。每公顷产量45~55吨。

栽培要点｜可春、秋植，播种期8月至翌年2月。育苗移栽，株行距45厘米×60厘米，每667米$^2$种植2 200~2 500株，及时整枝、搭架、绑蔓，适当疏花疏果。施足基肥，开花结果后多施追肥，适当增加钾肥用量，田间切忌过干过湿，及时排除积水。前、中期注意防治蚜虫及烟粉虱，抑制病毒病传播蔓延；中、后期注意防治疫病。11月至翌年6月采收。

## 金石

Jinshi tomato

> 品种来源

华南农业大学科技实业发展有限公司育成的杂种一代，2012年通过广东省农作物品种审定。

📍 分布地区｜南沙、番禺等区。

特　　征｜无限生长类型。叶长50厘米，宽38厘米。第1花序着生于第9~10节，以后每隔2~3节着生1花序。每花序结果4个。果实扁圆形，高5.5~6.0厘米，横径7.0~7.5厘米，青果无绿肩，熟果鲜红，3心室，肉厚0.9~1.0厘米。单果重150克左右。

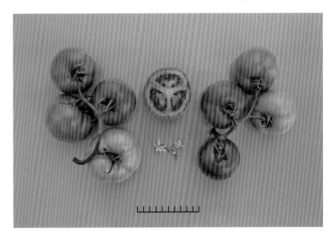

特　　性｜中熟，播种到初收106~116天，延续采收40~60天。耐热性、耐寒性好。田间表现不抗青枯病和病毒病。耐裂果，熟果硬实，耐贮运，品质优。每公顷产量50~60吨。

栽培要点｜以春、秋植为主，播种期春季1—2月，秋季8月。注意防治青枯病和病毒病。其余参照金丰1号。

## 新星101

Xinxing 101 tomato

**品种来源**

广东省农业科学院蔬菜研究所育成的杂种一代，2001年通过广东省农作物品种审定。

📍 **分布地区** | 花都、白云、增城等区。

特　　征 | 无限生长类型。叶长40厘米，宽36厘米。第1花序着生于第11节，以后每隔2~4节着生1花序。每花序结果3~7个。果实高圆形，高6.6~6.8厘米，横径6.4~6.5厘米，青果深绿肩，熟果鲜红，3~4心室，肉厚0.8~1.0厘米。单果重120克左右。

特　　性 | 中熟，播种到初收春植约106天、秋植约95天，延续采收40~60天。耐热性一般，耐寒性好。田间表现抗青枯病及病毒病。耐裂果，耐贮运，品质优。每公顷产量45~55吨。

栽培要点 | 以春、秋植为主，播种期8月至翌年2月。栽培管理参照益丰。

## 托美多

Tuomeiduo tomato

**品种来源**

广州兴田种子有限公司引进的杂种一代。

📍 **分布地区** | 南沙、番禺等区。

特　　征 | 无限生长类型。叶长45厘米，宽31厘米。第1花序着生于第9节，以后每隔2~3节着生1花序。每花序结果3~5个。果实扁圆形，高6.0~6.5厘米，横径8.0~8.2厘米，幼果偏绿白，青果无绿肩，熟果鲜红，4心室，肉厚0.9~1.0厘米。单果重200克左右。

特　　性 | 中熟，播种至初收约100天，延续采收40~60天。耐热性一般，耐寒性好。田间表现不抗青枯病，耐病毒病。耐裂果，熟果硬实，很耐贮运，品质优。每公顷产量50~60吨。

栽培要点 | 以秋、冬植为主，播种期8—10月。其余参照金丰1号。

## 拉菲

Lafei tomato

> 品种来源

广州亚蔬园艺种苗有限公司引进的杂种一代。

📍 分布地区 | 南沙、番禺等区。

特　　征 | 无限生长类型。叶长42厘米，宽38厘米。第1花序着生于第8~9节，以后每隔2~3节着生1花序。每花序结果4~6个。果实扁圆形，高5.8~7.0厘米，横径6.9~7.2厘米，青果深绿肩，熟果鲜红，2心室，肉厚0.8~1.0厘米。单果重150克左右。

特　　性 | 中熟，播种到初收106~116天，延续采收40~60天。耐热性、耐寒性好。田间表现不抗青枯病和病毒病。耐裂果，熟果硬实，耐贮运，品质优。每公顷产量50~60吨。

栽培要点 | 以春、秋植为主，播种期春季1—2月，秋季8月。注意防治青枯病和病毒病。其余参照金丰1号。

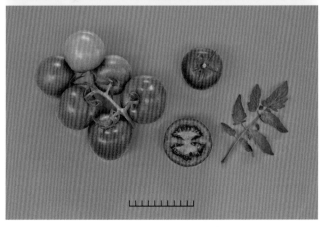

## 先丰

Xianfeng tomato

> 品种来源

广州南蔬农业科技有限公司育成的杂种一代，2013年通过广东省农作物品种审定。

📍 分布地区 | 花都、白云、增城等区。

特　　征 | 无限生长类型。叶长56厘米，宽44厘米。第1花序着生于第9~11节，以后每隔2~3节着生1花序。每花序结果3~6个。果实扁圆形，高6.3~6.6厘米，横径6.8~7.2厘米，青果无绿肩，熟果鲜红，4心室，肉厚1.1~1.2厘米。单果重130克左右。

特　　性 | 中熟，播种至初收102~106天，延续采收33~53天。耐热性、耐寒性强，感青枯病和病毒病。耐裂果，耐贮性好，品质中等。每公顷产量45~60吨。

栽培要点 | 以秋、冬植为主，播种期8—10月。其余参照金丰1号。

# 樱桃番茄 Cherry tomato

## 红艳 Hongyan cherry tomato

**品种来源**

广州市农业科学研究院育成的杂种一代。

**分布地区** | 花都、白云、南沙、增城等区。

特　　征 | 无限生长类型。叶长52厘米，宽45厘米。第1花序着生于第10~11节，以后每隔2~3节着生1花序。每花序结果12~25个。果实椭圆形，高4.2~4.3厘米，横径3.5~3.9厘米，脐尖，青果有绿肩，熟果鲜红，2心室，肉厚0.5厘米，可溶性糖7%~8%。单果重20克左右。

特　　性 | 早熟，播种至初收94~103天，延续采收35~55天。丰产，田间表现耐热性、耐寒性、耐旱性强，抗青枯病、病毒病。较耐裂果，果硬，耐贮性好，品质优。每公顷产量45~50吨。

栽培要点 | 以春、秋植为主，播种期8月至翌年2月。育苗移栽，株行距45厘米×60厘米，每667米²种植2 000株，及时整枝、搭架、绑蔓，适当疏花疏果。施足基肥，定植后7天淋提苗肥，第1穗果膨大时施1次催果肥，初收后一般每7~10天追肥1次。防治青枯病：应选择地势高、排水良好的田块，水旱轮作；拔除田间病株，在病穴及周围撒生石灰，用200毫克/升链霉素，每7~10天1次，连续3~4次喷根颈部；用1∶1∶150倍波尔多液、30%氧氯化铜800倍液灌根，每7~10天1次。病毒病应以预防为主，苗期及定植后生长的中早期注意防蚜虫及烟粉虱，加强栽培管理。

## 金艳 Jinyan cherry tomato

**品种来源**

广州市农业科学研究院育成的杂种一代。

**分布地区** | 花都、白云、南沙、增城等区。

特　　征 | 无限生长类型。叶长53厘米，宽39厘米。第1花序着生于第12节，以后每隔3节着生1花序。每花序结果10~30个。果实椭圆形，高4.4~4.5厘米，横径3.6~3.7厘米，青果有绿肩，熟果橙黄色，2~3心室，肉厚0.5厘米，可溶性糖7%~8%。单果重20克左右。

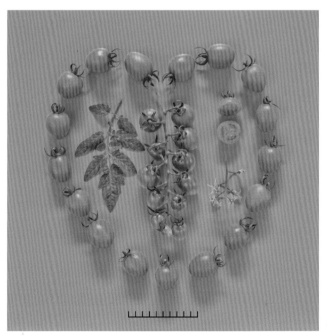

特　　性 | 早熟，播种至初收95~105天，延续采收35~55天。丰产性好，田间表现耐热性、耐寒性、耐旱性强，抗青枯病、病毒病。较耐裂果，果硬，耐贮性好，品质优。每公顷产量45~50吨。

栽培要点 | 以春、夏植为主，播种期8月至翌年2月。其余参照红艳。

## 黄樱 1 号

Huangying No.1 cherry tomato

> 品种来源

广州市农业科学研究院育成的杂种一代。

分布地区｜花都、白云、南沙、增城等区。

特　　征｜无限生长类型。叶长 40 厘米，宽 33 厘米。第 1 花序着生于第 10 节，以后每隔 3 节着生 1 花序。每花序结果 10~50 个。果实圆形，高 3.1~3.5 厘米，横径 3.5~3.9 厘米，青果有绿肩，熟果米黄色，3 心室，肉厚 0.3~0.4 厘米，可溶性糖 7%~8%。单果重 20 克左右。

特　　性｜早熟，播种至初收 94~103 天，延续采收 35~55 天。丰产性好，田间表现耐热性、耐寒性、耐旱性强，抗青枯病、病毒病。较易裂果，果软，皮薄，品质优。每公顷产量 45~50 吨。

栽培要点｜以春、秋植为主，播种期 8 月至翌年 2 月。应注意水分均衡供应，要及时采收，采收时应将萼片一起采摘。其余参照红艳。

## 绿樱 1 号

Lüying No.1 cherry tomato

> 品种来源

广州市农业科学研究院育成的杂种一代。

分布地区｜花都、白云、南沙、增城等区。

特　　征｜无限生长类型。叶长 42 厘米，宽 52 厘米。第 1 花序着生于第 10 节，以后每隔 3 节着生 1 花序。每花序结果 12~50 个。果实圆形，高 3.5~3.7 厘米，横径 3.7~3.9 厘米，青果有绿肩，熟果绿色，3 心室，肉厚 0.4~0.5 厘米，可溶性糖 7%~8%。单果重 20 克左右。

特　　性｜早熟，播种至初收 94~103 天，延续采收 35~51 天。丰产性好，田间表现耐热性、耐寒性、耐旱性强，抗病毒病。较易裂果，果软，皮薄，品质优，风味佳。每公顷产量 45~50 吨。

栽培要点｜以春、秋植为主，播种期 8 月至翌年 2 月。应注意水分均衡供应，要及时采收，采收时应将萼片一起采摘。其余参照红艳。

## 华喜珍珠

Huaxizhenzhu cherry tomato

**品种来源**

广东华农大种业有限公司育成的杂种一代。

📍 分布地区 | 花都、白云、南沙、增城等区。

特　　征 | 无限生长类型。叶长44厘米，宽41厘米。第1花序着生于第10节，以后每隔3节着生1花序。每花序结果10~11个。果实长圆形，高4.2~4.3厘米，横径3.0~3.2厘米，青果有绿肩，熟果粉红色，2心室，肉厚0.5~0.7厘米，可溶性糖7%~8%。单果重20克左右。

特　　性 | 早熟，播种至初收94~103天，延续采收35~55天。丰产性好，田间表现耐热性、耐寒性、耐旱性强，抗病毒病。耐裂果，果硬，耐贮性较好，品质优。每公顷产量45~50吨。

栽培要点 | 以春、秋植为主，播种期8月至翌年2月。其余参照红艳。

## 红箭

Hong jian cherry tomato

**品种来源**

广东省农业科学院蔬菜研究所育成的杂种一代。

📍 分布地区 | 花都、白云、南沙、增城等区。

特　　征 | 有限生长类型。株高152厘米，开展度88厘米，叶长40厘米，宽31厘米。第一花序着生于第10节，每间隔1~2节着生1花序。每花序结果10~26个。果实长圆形，高3.5~3.7厘米，横径2.2~2.5厘米，青果有绿肩，熟果鲜红，2~3心室，肉厚0.2~0.3厘米，可溶性糖6%~7%。单果重13克左右。

特　　性 | 早熟，播种至初收94~103天，延续采收35~51天。丰产性好，田间表现耐热性、耐寒性、耐旱性强，抗病毒病。较易裂果，果中硬，品质良好。每公顷产量40~45吨。

栽培要点 | 以春、秋植为主，播种期8月至翌年2月。其余参照红艳。

# 八、根茎类
## ROOT VEGETABLES

- 萝卜
- 胡萝卜

# 萝卜

十字花科萝卜属二年生草本植物，学名 *Raphanus sativus* L., 别名萝白、莱菔，染色体数 $2n=2x=18$。

萝卜原产欧洲、中国和日本。《尔雅》（公元前 2 世纪）记载称萝卜为"葖、芦萉（菔）"，宋代苏颂等著的《本草图经》（公元 1061 年）提到莱菔"南北皆通有之……北土种之尤多"。广州已有千年以上栽培历史，年栽培面积约 1 800 公顷。

植物学性状：直根系，直根肥大。茎短缩，叶片长倒卵形或羽状深裂，淡绿色至浓绿色，叶面光滑或被硬毛。复总状花序，完全花，白色或浅紫色。异花授粉。种子近圆球形，略扁，黄褐色，千粒重 16~17 克。

类型：按生产季节分为：

1. 夏秋萝卜　5—9 月播种，7—11 月收获。主要有短叶 13 号萝卜、耙齿萝卜等品种。

2. 冬春萝卜　10—12 月播种，12 月至翌年 4 月收获。主要有南畔洲萝卜、白玉春萝卜等品种。

栽培环境与方法：半耐寒性作物。营养生长期需适度高温（以利于形成叶丛），而后温度逐渐降低有利于肉质根膨大（不能低于 6℃）。在肉质根形成期需要大量水分促进膨大，否则易开裂。另外，水分失调易出现肉质根"糠心"现象，需钾肥、磷肥比氮肥要多，最好种植在富含腐殖质、土层深厚、排水良好的沙壤土。

收获：肉质根充分膨大、根基部圆润，叶色转淡后即可采收。

病虫害：病害主要有霜霉病、病毒病、软腐病、黑腐病等，虫害主要有萝卜蚜、桃蚜、菜螟、黄曲条跳甲、菜青虫、小菜蛾等。

营养及功效：富含糖类、维生素、纤维素、淀粉酶、芥子油等。味辛、甘，性温，具消食、理气、化痰、止咳、清肺利咽、散瘀消肿等功效。

# Radish

*Raphanus sativus* L., a biennial herb. Family: Cruciferae. Synonym: Chinese radish. The chromosome number: $2n=2x=18$.

Radish, originated in Europe, China and Japan, has been grown for more than 1000 years in Guangzhou. At present, the cultivation area is about 1800 $hm^2$ per annum in Guangzhou.

Botanical characters: Radish, with a succulent tap root (the edible part). Stem dwarfed. Leaf long obovate or pinnate parted, light green to dark green in color, smooth or with hard hairs. Inflorescence: a compound raceme with complete flowers, white or light purple in color. Cross-pollination. Seeds: approximate round, flat and yellow brown in color. The weight per 1000 seeds is 16~17 g.

Types: The radish in Guangzhou can be divided into two types according to growth seasons:

1. Summer and autumn radish: sowing dates: May to September, harvesting dates: July to November. The main cultivars include Short leaf No.13, Pachi radish, etc.

2. Winter and spring radish: sowing dates: October to December, harvesting dates: December to April the following year. The main cultivars include Baisha nanpanzhou, White spring radish, etc.

Cultivation environment and methods: Radish is a cool season crop. At the stage of vegetative growth, foliage formation require moderate high temperature. And the succulent tap root growth need low temperature (not below 6℃), sufficient and evenly irrigation. The growth of radish required more potassium, phosphate than nitrogen. It should be planted in sandy loam soil which is rich in humus, deep soil and good drainage.

Harvest: The radish can be harvested when the succulent root is in full size.

Nutrition and efficacy: The radish is rich in sugars, vitamins, celluloses, amylase, glucosinolates.

## 短叶 13 号

Short leaf No.13 radish

> 品种来源

汕头市白沙蔬菜原种研究所育成，1982 年通过广东省农作物品种审定。

📍 分布地区｜全市各区。

特　　征｜株高 40~50 厘米，开展度 40~45 厘米。叶片倒长卵形，长 20 厘米，宽 7 厘米，绿色，全缘，边缘向内稍弯曲；叶柄浅绿色。肉质根长圆柱形，长 25~30 厘米，横径 3.5~4.8 厘米，露出地面 2/3，皮和肉均白色，单根重 300~500 克。

特　　性｜早熟，播种至收获 45~60 天。抗病性、耐热性较强。皮薄，质脆少渣，味甜美，品质优。可生食、熟食，也可加工。每公顷产量 30~45 吨。

栽培要点｜播种期 5—9 月。穴播，每 667 米$^2$ 用种量 650~800 克。每公顷施用腐熟基肥 15~22 吨。株行距 15 厘米 ×20 厘米。苗期少施肥，肉质根膨大期施重肥。注意防治黄曲条跳甲、蚜虫、菜螟等。

## 耙齿萝卜

Pachi radish

> 品种来源

引自江门市新会区。

📍 分布地区｜全市各区。

特　　征｜株高 60~65 厘米，开展度 38~42 厘米。叶片倒披针形，长 32 厘米，宽 8.5 厘米，绿色，波状叶缘，边缘向内稍弯曲；叶柄浅绿色。肉质根长圆柱形，长 20~25 厘米，横径 3.5~4.5 厘米，露出地面 1/2，皮和肉均白色，单根重 180~250 克。

特　　性｜早熟，播种至收获 50~60 天。耐热性、耐湿性强。肉质稍结实，纤维较多，味甜，品质优。每公顷产量 20~25 吨。

栽培要点｜播种期 5—7 月，每 667 米$^2$ 用种量穴播 500 克、条播 800 克。其余参照短叶 13 号萝卜。

## 南畔洲

Nanpanzhou radish

> 品种来源

汕头市白沙蔬菜原种研究所育成。

📍 分布地区 | 全市各区。

**特　　征** | 植株直立，株高 48 厘米，开展度 75~80 厘米。叶片长椭圆形，长 54.7 厘米，宽 13.5 厘米，深绿色；叶缘羽状全裂，侧裂叶 8 对；叶柄短，浅绿色。肉质根圆筒形，长 25~30 厘米，横径 6.5~8.0 厘米，露出地面 1/3，皮肉白色，表皮嫩薄、光滑，单根重 800~1 500 克。

**特　　性** | 迟熟，播种后 90~120 天收获。冬性强，抽薹迟。纤维少，味甜，品质好。鲜菜用、制萝卜干或脆渍。每公顷产量 45~60 吨。

**栽培要点** | 播种期 9 月中旬至 12 月中旬。穴播，每 667 米² 用种量 400~500 克。株行距 26 厘米×33 厘米，间苗 2 次。结合除草松土，保持土壤湿润，注意防治黄曲条跳甲和蚜虫。

## 白玉春

Baiyuchun radish

> 品种来源

北京世农种苗有限公司 1994 年引进的韩国农友 BIO 株式会社育成品种。

📍 分布地区 | 全市各区。

**特　　征** | 株高 39~46 厘米，开展度 70~75 厘米。叶片长椭圆形，长 50 厘米，宽 10.8 厘米，叶片数 21，深绿色；叶缘羽状全裂；叶柄短，浅绿色。肉质根圆筒形，偶有分叉，长 25~30 厘米，横径 6~10 厘米，露出地面 1/3，皮肉白色，表皮嫩薄光滑，肩部钝圆，单根重 1 500~2 000 克。

**特　　性** | 长势中等。中熟，播种至收获约 80 天。对黑腐病和霜霉病有较强的抗性。耐抽薹，糠心晚，质脆，味甜，风味好，品质优，商品性好，耐贮运。每公顷产量 60~70 吨。

**栽培要点** | 播种期 4—8 月、10—11 月。穴播，每穴 1~2 粒种子，每 667 米² 用种量 500 克，株行距 20 厘米×35 厘米，间苗 2 次。结合除草松土，保持土壤湿润，及时追肥。适宜生长温度为 12~32℃，夏季短时能耐 36℃高温，但在高温条件下较易发生软腐病、病毒病。

# 胡萝卜 Carrot

伞形科胡萝卜属二年生草本植物，学名 *Daucus carota* L. var. *sativa* DC.，别名金笋、红萝卜，染色体数 $2n=2x=18$。

胡萝卜原产亚洲西部，广州栽培历史悠久，年栽培面积约 800 公顷，主要分布于增城、从化区。栽培品种主要从国外引进，目前生产上以五寸人参为主。

**植物学性状**：直根系，直根肥大。茎短缩。叶丛生于短缩茎上，为二至三回羽状复叶，小叶狭披针形，绿色，叶面被茸毛，叶柄细长。复伞形花序，完全花，细小，白色。双悬果，含 2 粒种子。种子椭圆形，黄褐色，纵棱上密被刺毛。千粒重 1.2~1.5 克。

**营养及功效**：胡萝卜富含胡萝卜素、糖类及矿物质，可炒食、煲汤、腌渍等。味甘、辛，性微温，具下气补中、利胸膈、和肠胃、安五脏等功效。

*Daucus carota* L. var. *sativa* DC., a biennial herb. Family: Umbelliferae. The chromosome number: $2n=2x=18$.

Carrot originated in west Asia, has long been cultivated in Guangzhou. At present, the cultivation area is about 800 $hm^2$ per annum in Guangzhou. The radish cultivars in Guangzhou are introduced abroad, the main one is 5 feet carrot.

Botanical characters: Carrot, with a succulent taproot (the edible part). Stem dwarfed. Leaf borne on the dwarfed stem, 2~3 pinnate compound, leaflets long lanceolate, green in color, with hairs, petioles long. Inflorescence: a compound umbel with complete flowers, small, white in color. The fruits cremocarp, with 2 seeds. Seeds: elliptical, yellow brown in color, ridges with hooked spines. The weight per 1000 seeds is 1.2~1.5 g.

Nutrition and efficacy: The carrot is rich in carotene, sugars and minerals.

## 新黑田五寸人参

Xinheitian wucun carrot

**品种来源** 广州市大田园种子有限公司1990年引进的日本品种。

**分布地区** | 增城、从化等区。

**特　征** | 植株较直立，半开张。株高 40~55 厘米，开展度 50~60 厘米。三回羽状复叶，叶片长 24 厘米，绿色；叶柄长 20 厘米，宽 0.5 厘米。肉质根长圆锥形，浅红色，末端较圆，肉质根全部入土，长 17~20 厘米，横径 4~5 厘米，红心红肉；表皮橘红色，光滑有光泽。木质部较大，侧根少。单根重 300~350 克。

**特　性** | 中晚熟，播种至采收 100~120 天。稍耐热，耐寒，耐旱，抽薹迟。肉质根水分少，质脆味甜，固形物含量高，品质优良，宜生食和加工。每公顷产量 37.5~52.5 吨。

**栽培要点** | 播种期 8—10 月。适于沙壤土种植。撒播或条播，每 667 $米^2$ 用种量 250~300 克。基肥要充足，播后覆盖稻草，发芽期注意保湿，盛苗期适当控制水分，间苗 2 次，株距 12~14 厘米，追肥着重破肚期。收获期 12 月上旬至翌年 1 月。为防止裂口、影响光泽，收获前 7~10 天不宜浇水。

# 九、薯芋类
# TUBER CROPS

- 芋
- 马铃薯
- 沙葛
- 葛
- 姜
- 山药
- 大薯

广州蔬菜品种志

# 芋 Taro

天南星科芋属多年生块茎植物，常作一年生栽培，学名 Colocasia esculenta (L.) Schott，别名芋头、芋艿，染色体数 $2n=2x=28$，$2n=3x=42$。

芋起源于中国、印度、马来半岛等地区。中国最早在《管子》（公元前5世纪至前3世纪）中即有芋的相关记载，在珠江、长江、淮河流域及台湾等地均有种植。广州栽培历史悠久，300多年前记载称芋为"广芋"，年种植面积约750公顷。

**植物学性状**：须根系，根较浅。叶片阔卵形，互生，深绿色；叶柄长，基部半月形，绿色、紫色或紫黑色。主球茎（母芋）节上的腋芽可形成小球茎（子芋），依次可形成孙芋、曾孙芋等。球茎有圆形、卵圆形、椭圆形和长椭圆形等，皮黄褐色，肉白色或带紫红色斑纹。佛焰花序，很少开花。浆果，种子近卵形，紫色。

**类型**：广州的芋分为球茎用芋和叶柄用芋两个变种。

1. 球茎用芋 依球茎的生长习性分为：

（1）多子芋。子芋多且较大。品种有红芽芋。

（2）魁芋。母芋大，子芋少且较小。品种有槟榔芋。

2. 叶柄用芋 近年已很少种植。

**栽培环境与方法**：喜温暖潮湿，较耐阴，生长适温21~27℃，球茎膨大要求较高温度和较大昼夜温差。忌连作。选母芋中部的子芋作种芋直接种植或育苗移栽。种植期12月至翌年3月，收获期9月至翌年2月。

**收获**：芋叶变黄衰败时采收球茎。贮藏适温6~10℃。

**病虫害**：病害主要有疫病、软腐病等，虫害主要有斜纹夜蛾、蚜虫、小地老虎、蝼蛄等。

**营养及功效**：球茎富含淀粉、蛋白质、钾、钙、磷、铁、维生素A、维生素C及膳食纤维等。芋性平、滑，味辛，具宽肠胃、破宿血、补气健脾等功效。

Colocasia esculenta (L.) Schott, a perenial succulent tuber plant, as annual cultivation. Family: Araceae. Synonym: Dasheen, Cocoyam, Tari, Giant taro. The chromosome number: $2n=2x=28$, $2n=3x=42$.

Taro was originated in India, China and the Malay Peninsula regions. Taro has long been cultivated in Guangzhou, it was mentioned as "Guang Taro" in literatures 300 years ago. At present, the cultivation area is about 750 $hm^2$ per annum in Guangzhou.

**Botanical characters**: Shallow fibrous root. Leaf broad ovate, alternate, dark green; long petiole, lunette in basal part, green, purple or purple black in color. The adventitious buds are formed to produce daughter corms, called cormels. Corm round, oval, elliptical and oblong in shape, yellow brown, white or white with purple stripe. Inflorescences: spadix, rarely blossom.

**Types**: In Guangzhou, taros are classified into two types: edible corm and edible petiole types.

1. Edible corm type: According to the growth habit of the corm:

(1) Multi-cormels taro: cormel more and larger, such as Red bud taro.

(2) Large corm taro: Large mother taro, small and less cormels, such as Binglang taro.

2. Edible petiole taro type: In recent years, few edible petiole taro has been cultivated.

**Cultivation environment and methods**: Taro prefer mild and wet, with shade toleration, appropriate growth temperature 21~27℃, bulb enlargement requires higher temperature and diurnal temperature. Avoid continuous cropping. Select son-taro at middle of mother taro for planting. Planting season is from December to March the following year. The harvest date is from September to February the following year.

**Harvest**: Harvest corms when taro leaves turn yellow. Storage temperature 6~10℃.

**Nutrition and efficacy**: The corm is rich in starches, proteins, potassium, calcium, phosphorus, iron, vitamin A, vitamin C, dietary fibers, etc.

## 红芽芋

Red bud taro

品种来源

地方品种。

📍 分布地区 | 全市各区。

特　　征 | 株高 90~100 厘米，开展度 40~50 厘米。叶片阔卵形，长 35~40 厘米，宽 30~35 厘米，叶基心形，深绿色；叶柄紫红色。母芋较大，近圆形，每株子芋 7~12 个，芋皮较厚，褐色，肉白色，芽鲜红色。单株产量 850~1 100 克。

特　　性 | 中熟，生长期 210~240 天。耐贮运。含淀粉较多，品质优，鲜菜用或干制。每公顷产量 23~28 吨。

栽培要点 | 种植期 2—3 月。宜选壤土种植。株行距 50 厘米 ×70 厘米。施足基肥，适时追肥，增施钾肥。注意培土、排灌等田间管理工作。9—10 月采收。

## 槟榔芋（香芋）

Binglang taro

品种来源

地方品种。

📍 分布地区 | 全市各区，其中以花都区炭步镇槟榔香芋最出名，该产品于 2014 年 5 月获得国家农业部农产品地理标志登记。

特　　征 | 株高 110~160 厘米。叶簇直立。叶片阔卵形，长 40~50 厘米，宽 35~40 厘米，先端较尖，叶基心形，深绿色，叶片中央与叶柄相连部位及叶脉紫红色，叶柄绿色。母芋长椭圆形，子芋长卵形，棕黄色或棕褐色，肉白色，有紫红色斑纹。每株子芋 5~7 个。单株产量 2.5~3.5 千克。

特　　性 | 晚熟，生长期 240~280 天。耐旱性较差，耐贮运。淀粉含量高，香味浓，品质优。可作鲜菜、干制及制罐头。每公顷产量 45~55 吨。

栽培要点 | 种植期 2—3 月。选择易排灌、土层厚、保水性能好的壤土种植。株行距 50 厘米 ×90 厘米。植穴深 15~20 厘米，芋种斜放，覆土 5~7 厘米厚。多施有机肥作基肥，幼苗出土后开始追肥，全期追肥 5~6 次。生长期培土 2 次，第 1 次在 4 叶期，第 2 次在 7 叶期左右，结合培肥、除草。注意保持土壤湿润。11 月至翌年 1 月收获。

# 马铃薯

茄科茄属中能形成地下块茎的一年生草本植物，学名 Solanum tuberosum L., 别名薯仔、土豆、洋芋等，染色体数 $2n=4x=48$。

马铃薯起源于秘鲁和玻利维亚等国，100多年前从东南亚传入广东省，广州年种植面积约 1 500 公顷，以块茎繁殖，目前广州的马铃薯品种主要是粤引85-38。

**植物学性状**：须根系。地上茎有直立茎、半直立茎和匍匐茎3种，颜色有绿色、紫褐色等，横切面为圆形或三角形。奇数羽状复叶。花为两性花，有白色、粉红色、蓝色或紫色等，在广州少开花。地下块茎扁圆形、椭圆形、圆形或长筒形等，表皮黄色、黄白色、粉红色、紫色等，肉黄色、黄白色或紫色。果实为浆果，圆形，含 100~250 粒种子，种皮为浅褐色或淡黄色，千粒重 0.5~0.6 克，休眠期 5~6 个月。

**栽培环境与方法**：喜冷凉。茎叶生长适温为 17~21 ℃，低于 7 ℃或高于 42 ℃时生长停止；块茎形成和生长发育适温为 17~19 ℃，低于 2 ℃或高于 29 ℃时生长停止。喜光照；最适于在沙壤土、壤土种植；生长期间保持土壤湿润，接近收获时土壤要相对干爽。对矿物质肥的要求以氮、钾为主。

**病虫害**：病害主要有青枯病、黑茎病、环腐病、晚疫病和病毒病，虫害主要有蚜虫、螨类、小地老虎。

**营养及功效**：富含淀粉、钾、纤维素、多种维生素和微量元素等。马铃薯性平、寒，味甘，具和胃调中、益气健脾、强身益肾、活血消肿等功效。

# Potato

*Solanum tuberosum* L., an annual tuber herb. Family: Solanaceae. The chromosome number: $2n=4x=48$.

Potato was originated in Peru and Bolivia, and was introduced to Guangdong province from southeast Asia 100 years ago. At present, the cultivation area is about 1500 $hm^2$ per annum, and the main cultivars is Yueyin 85-38 in Guangzhou.

Botanical characters: Fibrous root system. Stems round or triangular, erect, semi-erect and creeping, green, purple-brown in color. Leaf, odd pinnate. Flowers white, pink, blue or purple in color. Potato rarely blossom in Guangzhou. Tuber (the edible part) flat, oval, round or tubular in shape, yellow, yellowish white, pink, purple in color. Flesh yellow, white or purple. Fruit: round berry, with 100~250 seeds. Seed light brown or light yellow. The weight per 1000 seeds is 0.5~0.6 g.

Cultivation environment and methods: Potato perfer cool season. The optimum temperature for growth of stems and leaves is 17~21℃, the growth stop below 7℃ or above 42℃. The optimum temperature for tuber formation and growth is 17~19℃, tuber growth stop below 2℃ or above 29℃. Potato growth requires sunny weather, moist sandy or loam soil, and more nitrogen and potassium.

Nutrition and efficacy: Potato is rich in starches, potassium, crude fibers, vitamins, trace elements, etc.

## 粤引 85-38

Yueyin 85-38 potato

**品种来源**

广东省农业科学院作物研究所引进推广。

**分布地区** | 从化、增城、南沙、番禺、花都、黄埔、白云等区。

特　征 | 株型直立，株高28厘米左右。叶片数16片左右，小叶3~5对，有茸毛，深绿色。茎绿色。薯形整齐，薯块长椭圆形，长12~16厘米，宽7~8厘米，黄皮黄肉，表皮光滑，芽眼浅且少，单薯重150~500克。

特　性 | 早熟，播种至初收约90天。分枝少，喜冷凉，怕霜冻，高抗病毒，中感晚疫病。

栽培要点 | 选择土层深厚、肥沃的沙壤土种植，注意轮作，适施基肥。种薯先进行消毒处理，纵切成25~50克的种块，每块带1~2个芽眼，种块切口要用草木灰或石灰蘸涂以杀菌，直接催芽或播种。广州地区宜在10月中下旬至12月上旬播种。双行种植，株距20厘米左右，行距25~30厘米，每667米²种植4 500~5 000株，需用种薯125~150千克。出苗后早培土、早追肥，保持土壤湿润，注意防治病虫害。

# 沙葛 Yam bean

豆科豆薯属一年生缠绕性藤本植物，以块根供食用，学名 *Pachyrhizus erosus* (L.) Urban.，别名豆薯、凉薯、地瓜、番葛，染色体数 $2n=2x=22$。

沙葛起源于墨西哥和中美洲，广州已有 100 多年栽培历史，目前年种植面积约 300 公顷，主要栽培品种是牧马山。

**植物学性状**：直根系，多须根，主根上部膨大成肉质块根，块根扁圆形或纺锤形，具数条纵沟，表皮淡黄色，肉白色。茎蔓长 2 米左右，被黄褐色茸毛，每节发生侧蔓。三出复叶，互生，绿色，表面光滑，被疏毛，具托叶。总状花序，自第 5~6 叶腋开始抽生，以后每节叶腋连续发生，花蝶形，浅紫蓝色或白色。自花授粉。种子近方形，扁平，褐色，种皮坚硬，千粒重 200~250 克。

**栽培环境与方法**：喜高温，30℃左右有利于种子的发芽及茎叶花果和块根生长发育。耐旱，耐瘠，以土层较深、透气性良好的土壤为佳。种子繁殖，2—7 月播种，7—12 月收获。

**病虫害**：病害主要有软腐病，虫害主要有金龟子等地下害虫。

**营养及功效**：富含纤维素、糖类、蛋白质、维生素 C 及淀粉。味甘，性凉，具清凉去热、生津止渴、化痰消积等功效。种子及茎叶含鱼藤酮，对人畜有毒，可作为杀虫剂。

*Pachyrhizus erosus* (L.) Urban., an annual lianoid herb. Family: Leguminosae. Synonym: Sha-kot. The chromosome number: $2n=2x=22$.

Yam bean was originated from Mexico and Central America. It has been cultivated 100 years in Guangzhou. At present, the cultivation area is about 300 $hm^2$ per annum, and the main cultivar is Mumashan in Guangzhou.

Botanical characters: Taproot system, with many fibrous roots. Tuber fleshy, flat round or fusiform with several longitudinal groove, yellow and flesh white in color. The vine is about 2 m in length, with yellow brown hair. Trifoliate leaves alternate, green, smooth surface, with thinning hair. Inflorescence: axillary raceme, papilionaceous. Self pollination. Seeds nearly square, flat, brown, hard seed coat. The weight per 1000 seeds is 200 ~ 250 g.

Cultivation environment and methods: Yam bean perfer high temperature. About 30℃ is favour for seed germination, growth and development of stem, leaf and tuber. Yam bean is tolerant to drought and barren. Deep and good permeability soil is appropriate. Seed propagation. Sowing dates: February to July. Harvesting dates: July to December.

Nutrition and efficacy: Yam bean is rich in celluloses, carbohydrates, proteins and vitamin C.

## 牧马山 Mumashan yam bean

**品种来源** ｜ 引自四川。

**分布地区** ｜ 从化、增城等区。

**特　征** ｜ 缠绕蔓生。茎圆形，绿色，节间长 10~15 厘米，被黄褐色茸毛。三出复叶，小叶近菱形，长 10~15 厘米，宽 12~19 厘米，深绿色。块根为扁纺锤形，纵径 8~10 厘米，横径 12~15 厘米，具纵沟 3~4 条，皮色乳黄，有须根，肉白色。单个重约 500 克。

**特　性** ｜ 生长势强。早熟。耐旱，耐瘠，耐热，不耐涝。全生育期 120~150 天。质地脆嫩多汁，纤维少，品质优。每公顷产量 60~70 吨。

**栽培要点** ｜ 适播期 2 月上旬至 7 月下旬。宜选土层深厚、排水良好的沙壤土地块种植，注意轮作，施足基肥，高畦种植。直播，株行距 13 厘米×15 厘米，每 667 $米^2$ 种植 12 000~14 000 株，及时插竹引蔓，株高 150 厘米打顶，摘除侧芽，定期追肥，土壤保持湿润，排除田间积水。采收期 7 月上旬至 12 月下旬。

# 葛

豆科葛属多年生缠绕性藤本植物，以块根供食用，学名 *Pueraria thomsonii* Benth.，别名粉葛，染色体数 $2n=2x=24$。

葛起源于亚洲东南部，广州栽培历史悠久，200多年前记载为岭南物产，年种植面积约200公顷，2012年南沙的庙南粉葛成为国家地理标志产品。

植物学性状：根分吸收根和贮藏根，吸收根为须根，贮藏根即块根，纺锤形或长棒形，表皮皱褶，黄褐色，肉白色。茎圆形，绿色，多侧枝，被黄褐色茸毛。三出复叶，绿色，被黄褐色茸毛。总状花序，花蝶形，紫蓝色。荚果长椭圆形，被红褐色粗毛，含种子8~12粒。

栽培环境与方法：耐热，喜光，耐旱，20~25℃生长迅速，较强光照有利于生长发育。对土壤要求不严格。

广州原有细叶粉葛和大叶粉葛，现在主要是细叶粉葛，以插条或葛仔繁殖。

病虫害：病害主要有软腐病，虫害主要有金龟子等地下害虫。

营养及功效：富含纤维素、维生素C、钙、磷、铁等营养物质，为药食两用植物，素有"南葛北参"和"南方人参"之美誉。味甘，性平，具润肠、清热等功效。

# Kudzu

## 细叶粉葛

Thin leaf kudzu

### 品种来源

地方品种。

📍 **分布地区** | 南沙、番禺、从化、增城、花都、黄埔、白云等区。

特　　征 | 缠绕蔓生，茎圆形，被黄褐色茸毛。三出复叶，深绿色；块根为纺锤形，皮色乳黄色至深褐色，有皱褶，具纵沟，有须根，单个重1.0千克以上，长24厘米，粗10厘米以上，肉质白色至乳黄色。

特　　性 | 耐热，耐旱，不耐涝，病虫害少。种苗定植至采收早熟栽培220天，传统栽培300天。肉质细嫩，纤维少，葛味浓。每公顷产量15~20吨。

栽培要点 | 宜选土层深厚、肥沃的地块种植，注意轮作，施足基肥。冬无严寒、灌溉水充足的番禺、南沙地区采用庙南粉葛的早熟栽培方法，即8月中旬至12月下旬开始繁殖种苗"葛仔"，12月底至翌年1月初定植。每667米²种植800~1 000株。及时插竹引蔓，株高150厘米以下摘除侧芽，每株留蔓1~2条，株高150厘米以上任其生长，定期追肥，土壤保持潮湿，田间不能积水。7月中旬至8月上旬采收上市。传统栽培在12月至翌年2月以葛藤育苗，3月开始移植，冬季采收。

*Pueraria thomsonii* Benth., a perennial lianoid herb. Family: Leguminosae. Synonym: Fan-Kot. The chromosome number: $2n=2x=24$.

Kudzu was originated in southeast Asia, and has long been cultivated in Guangzhou. Kudzu was mentioned as Lingnan products 200 years ago. At present, the cultivation area is about 200 $hm^2$ per annum in Guangzhou.

**Botanical characters:** The tuberous roots fusiform or long rhomboid in shape, brown and flesh white in color. Round green stems with many laterals, covered with yellowish brown hairs. Green trifoliolate leaf covered with yellowish brown hairs. Inflorescence: raceme, papilionaceous. Purple and blue in color. The long oval pod with reddish brown hair contains 8~12 seeds.

**Cultivation environment and methods:** Kudzu is tolerant to heat and drought. Sunny weather is favour for growth and development of the plant. It grow rapidly at 20~25℃. There are Small leaf kudzu and Large leaf kudzu in Guangzhou, the former is popularly cultivated. Propagation: stem cutting.

**Nutrition and efficacy:** Kudzu is rich in celluloses, vitamin C, calcium, phosphorus, iron, etc.

# 姜 Ginger

姜科姜属多年生草本植物，作一年生栽培，学名 *Zingiber officinale* Rosc.，别名生姜，染色体数 $2n=2x=22$。

广州栽培历史悠久，200多年前已有记载。

**植物学性状**：根浅生。叶片披针形，绿色，具平行叶脉。叶鞘长，抱合成假茎，直立，绿色。地下茎腋芽可不断分生，形成多分枝的肉质根茎，表皮淡黄色，肉浅黄白色，为食用器官。

**栽培品种**：有疏轮大肉姜和密轮细肉姜两个品种。

**栽培环境与方法**：喜温暖，不耐寒，不耐霜冻。茎叶生长适温25~28℃，根茎生长适温18~25℃。对水分要求严格。要求土壤疏松、透气、有机质丰富。广州地区种植期为2—3月。

**收获**：6—9月采收嫩姜，10—12月采收老姜。

**病虫害**：病害主要有茎腐病、斑点病、纹枯病、炭疽病等，虫害主要有姜螟、异型眼蕈蚊、猿叶虫、烟蓟马等。

**营养及功效**：富含挥发油、姜辣素、多种维生素及矿物质。味辛，性微温，具除风邪寒热、伤寒头痛鼻塞、咳逆上气等功效。

*Zingiber officinale* Rosc., a perennial herb as annual cultivation. Family: Zingiberaceae. Synonym: Ginger root. The chromosome number: $2n=2x=22$.

Ginger has long been cultivated in Guangzhou and it was mentioned in literatures 200 years ago.

**Botanical characters**: Shallow roots. Leaf lanceolate, green, with parallel venation. Long sheath form the erect and green pesudo-stem. Axillary buds of rhizome can continuously differentiate to development branching. Succulent rhizomes (the edible part) with light yellow rinds, light yellow white flesh.

**Cultivars**: There are two cultivars: Fat flesh ginger and Thin flesh ginger.

**Cultivation environment and methods**: Ginger prefer warm weather, is not tolerant to cold and frost. The suitable temperature for shoot growth is 25~28℃, and rhizomes growth is 18~25℃. Ginger growth requires loose soil with rich organic matters. The planting date is from February to March in Guangzhou.

**Harvest**: Young gingers are harvested from June to September, and elder gingers are harvested from October to December.

**Nutrition and efficacy**: Ginger is rich in volatile oils, gingerol, vitamins and minerals.

## 疏轮大肉姜

Fat flesh ginger

🏷️ **品种来源**

地方品种。

📍 **分布地区** | 全市各区。

特　　征 | 植株直立。株高 50~70 厘米。叶片披针形，长 18~23 厘米，宽 2.0~3.5 厘米，深绿色。地下肉质根茎单行排列，分枝较疏，长 30~40 厘米，嫩时黄白色，老熟时浅黄色，肉黄白色，嫩芽粉红色。单株重 1~3 千克。

特　　性 | 生长期 150~210 天。喜阴凉，宜间作。耐旱，忌过湿，抗软腐病力差。根茎肥大。肉脆，味辣，品质优。每公顷产量约 15 吨。

栽培要点 | 种植期 2—3 月。选择排水良好的沙壤土种植，株距 20~25 厘米，行距 40 厘米，选择无病姜种，每公顷用种量 750~1 200 千克。采收嫩姜的宜密植，每公顷用种量 3 000~3 500 千克。种后畦面覆盖，幼苗出土后薄追肥，忌中午淋水，以免引起腐烂，5—7 月培土 1~2 次。6—9 月采收嫩姜，10 月后采收老姜。

## 密轮细肉姜

Thin flesh ginger

🏷️ **品种来源**

地方品种。

📍 **分布地区** | 全市各区。

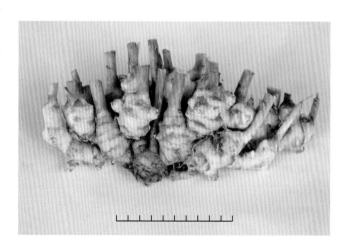

特　　征 | 植株直立。株高 60~80 厘米。叶片披针形，长 15~20 厘米，宽 2.0~2.5 厘米，绿色。地下肉质根茎分枝较密，成双行排列，表皮和肉均浅黄色，嫩芽紫红色。单株重 0.5~2.0 千克。

特　　性 | 生长期 150~210 天。喜阴凉，宜间作。耐旱，忌过湿，抗软腐病力较强。根茎分生多。肉质致密，辣味浓，品质优。每公顷产量 8~15 吨。

栽培要点 | 参照疏轮大肉姜。

# 山药 Chinese yam

薯蓣科薯蓣属一年生或多年生缠绕性藤本植物，学名 *Dioscorea batatas* Decne.，别名淮山，染色体数 $2n=4x=40$。

中国是山药重要的原产地和驯化中心，4世纪前期《山海经》中有山药的分布记载，广州有近百年的栽培历史，近年来，种植面积不断扩大。3—4月播种，11月至翌年3月收获，主要品种有桂淮2号、紫玉淮山。

**植物学性状**：根深，须根多。地上茎蔓生，长可达3米以上，有紫红色斑点，横切面圆形或多棱形。叶互生或对生，箭形，深绿色，叶腋生的零余子可作繁殖材料。薯块长棒形，表皮粗糙，褐色或紫褐色，肉白色或紫红色，为食用器官。

**栽培环境与方法**：喜温，生育适温为20~30℃，10℃块茎可以萌芽。地上茎叶不耐霜冻，温度降到10℃以下植株停止生长，5℃以下的低温会受冻害，短时间的0℃气温会冻死。要求强光照，短日照对地下块茎的形成和肥大有利，叶腋间的零余子也在短日照下出现。喜干燥，耐旱忌涝，耐阴。

**病虫害**：病害主要有黑斑病、炭疽病，虫害主要有蝼蛄、小地老虎、金龟子、斜纹夜蛾等。

**营养及功效**：富含淀粉、蛋白质、矿物质、多种维生素等营养物质，是一种药食两用植物。味甘，性温、平，具益肾气、健脾胃等功效。

*Dioscorea batatas* Decne., an annual or perennial lianoid plant. Family: Dioscoreaceae. Synonym: Yam, Chinese potato. The chromosome number: $2n=4x=40$.

Chinese yam was originated in China, and it was mentioned in literatures of early 4th century. Chinese yam has been cultivated in Guangzhou for more than 100 years. In recent years, the cultivation area expands continuously. Planting dates: March to April, and harvesting dates: November to March the following year. The main cultivars include Guihuai No.2 Chinese yam, Ziyu Chinese yam.

Botanical characters: Deep main roots and many fibrous root. Stems twing vine with purple spots, round or prismatic in shape, more than 3 m in length. Leaf alternation or opposite, sagittate, dark green. The bulbils borne at leaf axil, which could be used for propagation. Tuber, the edible part, clavate, rind coarse, brown or purple brown and flesh white or purple in color.

Cultivation environment and methods: Chinese yam prefers warm and sunny weather. The suitable temperature for growth is 20~30℃, and tuber sprout above 10℃. The plant growth stagnate below 10℃, and is freeze injury below 5℃. The short day is favour for tuber formation and development, axil bulbils formation. Chinese yam is tolerant to drought while not to waterlogged.

Nutrition and efficacy: Chinese yam is rich in starches, proteins, minerals, vitamins and other nutrients.

## 桂淮 2 号

Guihuai No.2 Chinese yam

**品种来源**

广西农业科学院经济作物研究所育成。

📍 分布地区 | 南沙、番禺、从化、增城、花都、黄埔、白云等区。

特　　征 | 茎右旋，圆棱形，主茎长 4~5 米。叶互生或对生，嫩叶紫红色，叶片呈卵状三角形至阔卵形，先端渐尖，基部心形，边缘全缘，叶色深绿，叶表光滑有光泽，蜡质层明显，叶脉网状，大脉数 7 条，呈紫红色，叶柄基部为紫红色，叶腋着生 1~3 个零余子。薯长圆柱形，长 50~100 厘米，单株重 0.6~1.5 千克，表皮棕褐色，根毛较少，薯块断面白色，肉质细腻。

特　　性 | 中晚熟，定植至初收 210~240 天，可延续采收 30~50 天。耐旱，不耐涝，每公顷产量约 30 吨。

栽培要点 | 选择排灌方便、土层深厚的沙壤土种植。选择无病虫的零余子或薯块作种，零余子可直接播种。种薯切块后要用草木灰或石灰蘸涂切口以杀菌，催芽播种。3—4 月播种，施足基肥。深沟高垄，采用专用定向种植槽栽培，株距 18~20 厘米，每 667 米² 种植 1 500~2 000 株。及时插竹引蔓，保留一条粗壮的芽条，其余的芽条剪除，主蔓长 150 厘米以下的侧蔓及时打掉，生长期保持土壤湿润。合理施肥，及时做好炭疽病、褐斑病、金龟子、斜纹夜蛾等病虫害的防治。

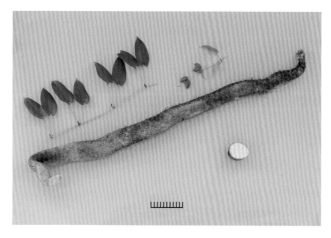

## 紫玉淮山

Ziyu Chinese yam

**品种来源**

增城农业科学研究所育成。

📍 分布地区 | 南沙、番禺、从化、增城、花都、黄埔、白云等区。

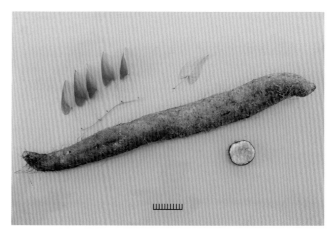

特　　征 | 茎蔓生，主茎长 4~5 米，侧蔓开展度 2~3 米，主茎四棱形。叶对生，叶片呈阔卵形，先端渐尖，基部阔心形，黄绿色，叶表光滑有光泽，叶脉网状，大叶脉数 7 条，叶柄基部和叶脉相连部分均为紫红色。少数植株叶腋间会长出少量的零余子。少数植株茎的末端会长出少量不结籽的扁三棱状花。薯块长圆柱形，粗壮顺滑，长 95 厘米左右，单株重 1.5~2.5 千克；薯皮上部棕褐色，芦头浅红色，肉紫红色，分布均匀。

特　　性 | 适应性广，耐热性、耐旱性强，耐涝性、耐寒性中等，对疫病和炭疽病抗性中等。生长势旺盛，顶部分枝多。中晚熟，定植至初收 210 天，可延续采收 60 天。每公顷产量 30~40 吨。

栽培要点 | 参照桂淮 2 号。

# 大薯 Yam

薯蓣科薯蓣属多年生缠绕草本植物，作一年生栽培，学名 *Dioscorea alata* L.，别名参薯、田薯、薯蓣，为淮山的变种，染色体数 $2n=3x=30$，$2n=8x=80$。

广州栽培历史悠久，300 年前已有记载，以远郊栽培为主，现年种植面积约 60 公顷。

**植物学性状**：须根系，叶片箭形，深绿色，对生。茎蔓生，断面多角形，具棱翼。叶腋生零余子，可作繁殖材料。薯块长棒形、纺锤形或块状，表皮紫褐色或浅褐色，肉白色或紫红色，为食用器官。

种植期 3—4 月，收获期 10 月至翌年 4 月。

类型：广州的大薯依肉色分为：

1. 白肉薯　肉白色，品种有早白薯、糯米薯。
2. 红肉薯　肉浅紫色，品种有紫薯。

*Dioscorea alata* L., a perennial lianoid herb as annual cultivation. Family: Dioscoreaceae. Synonym: Greater yam, Winged yam, Asiatic yam. The chromosome number: $2n=3x=30$, $2n=8x=80$.

Yam has long been cultivated in Guangzhou, it was mentioned in literatures 300 year ago. It is mainly grown in the far outskirts of city. The total area per annum is about 60 $hm^2$.

Botanical characters: Fibrous system. Opposite leaves sagittate and dark green. Stem trailing, with four wings, green in color. The leaf axils can develop lateral shoots or form aerial tuber which called bulbil. The tuber, the edible part, clavate, rind coarse, brown or purple brown and flesh white or purple in color.

Planting dates: March to April. Harvesting dates: October to April the following year.

Types: Yam in Guangzhou can be divided into two types based on the flesh color. White flesh yam, the cultivars include Zaobaishu yam and Nuomishu yam. Red flesh yam, light purple flesh, the cultivars include Purple yam.

## 早白薯

Zaobaishu yam

**品种来源** 地方品种。

**分布地区**｜南沙、番禺、从化、增城、花都、黄埔、白云等区。

**特　征**｜茎四棱，绿色，节间长 12~14 厘米。叶片箭形，长 12~14 厘米，宽 7~8 厘米，深绿色。薯块短柱形，长 20~25 厘米，粗 6 厘米左右，单个重 0.5 千克左右，表皮粗糙，灰褐色，须根少，肉白色。

**特　性**｜早熟，播种至采收 180~200 天。耐热，不耐涝。含淀粉多，嫩滑，品质优。每公顷产量 12~20 吨。

**栽培要点**｜种植期 3—4 月。宜选沙壤土种植，施足基肥，深沟高畦。单行植，株距 16~18 厘米。选择无病虫的零余子或薯块作种，零余子可直接播种。种薯切块后要用草木灰或石灰蘸涂切口以杀菌。催芽播种，选择出苗整齐的薯块种植，种植时芽向下或侧放，植后覆土，留苗 1~2 条。生长盛期培肥 3 次，注意保持土壤湿润。10—12 月采收。

## 糯米薯

Nuomishu yam

**品种来源** 地方品种。

**分布地区**｜南沙、番禺、从化、增城、花都、黄埔、白云等区。

**特　征**｜茎四棱，节间长约 10 厘米，青绿色。叶片箭形，长 14 厘米，宽 9 厘米，深绿色。薯块长棒形，顶端幼细而身粗圆，表皮黑褐色，须根少，长 50~55 厘米，粗约 8 厘米，单个重 1.5~2 千克，肉淡白色。

**特　性**｜中熟，播种至采收 180~220 天。耐热，不耐湿。含淀粉多，品质好。每公顷产量约 23 吨。

**栽培要点**｜参照早白薯。

## 紫薯

Purple yam

**品种来源** 地方品种。

**分布地区**｜南沙、番禺、从化、增城、花都、黄埔、白云等区。

**特　征**｜枝叶繁茂，茎四棱，棱翅高，紫红色，节间长约 14 厘米。叶片阔卵形，叶基心形，长 14 厘米，宽 8 厘米，绿色。薯块扁块状，长 35 厘米左右，粗 8~9 厘米，单个重 1.0~1.5 千克，有分叉，表皮粗糙，黑褐色，须根多，肉浅红色。

**特　性**｜晚熟，播种至采收 250 天。不耐湿。品质中等。每公顷产量约 23 吨。

**栽培要点**｜参照早白薯。12 月采收。

# 十、葱蒜类
# BULB CROPS

- 分葱
- 大葱
- 大蒜
- 薤
- 韭

# 分葱

百合科葱属多年生草本植物，作一年生栽培，学名 *Allium fistulosum* L. var. *caespitosum* Makino，别名香葱，染色体数 $2n=2x=16$。

广州栽培历史悠久，公元 4 世纪《广志》已有述及。

**植物学性状**：弦状根。管状叶，先端尖，青绿色。叶鞘抱合成假茎，假茎基部膨大，形成小鳞茎，丛生于短缩茎上，小鳞茎纺锤形或卵形，分蘖性强，鳞衣白色或紫红色，肉白色。不抽薹开花。

**栽培品种**：有疏轮香葱、密轮香葱和四季葱 3 个品种，是秋植冬春收的香菜之一。

**栽培环境与方法**：鳞茎繁殖。喜冷凉，较耐寒。耐肥，对土壤的适应性强。广州地区可周年栽培，但高温季节应注意遮阳、防旱。一般采用穴栽，栽后 1 个月培土一次以增加葱白长度。

**收获**：栽后约 2 个月可陆续采收。

**病虫害**：病害主要有霜霉病、锈病、软腐病等，虫害主要有葱蓟马、葱斑潜蝇、红蜘蛛等。

**营养及功效**：具香辛味，能增进食欲，有防治心血管疾病之功效。

# Bunching onion

# 疏轮香葱（玉葱）

Thin tiller bunching onion

邓彩联 摄

*Allium fistulosum* L. var. *caespitosum* Makino, a perennial herb as annual cultivation. Family: Liliaceae. Synonym: Welsh Onion, Green bunching onion, Spring onion. The chromosome number: $2n=2x=16$.

Bunching onion has long been cultivated in Guangzhou as mentioned in *Guangzhou Records* in the 4th century.

Botanical characters: String-shaped root system. Leaf pipe-shaped with acuminate tip, green-blue in color. Leaf sheaths closed to form pseudostems which thicken at the base to form bulblets, which cluster on the dwarfed stem. Bulblets are fusiform or ovate, strong tillering. The scale white or red purple in color. Flesh is white in color.

Cultivars: Main cultivars in Guangzhou include Thin tiller bunching onion, Dense tiller bunching onion, Four season onion, etc. Planting dates: autumn. Harvesting dates: winter to spring.

Cultivation environment and methods: Bunching onion is propagated by bulbs. In Guangzhou, bunching onion can be cultivated throughout the year, and shading is required in summer. Ridging once a month after planting in order to increase pseudo-stem length.

Harvest: Bunching onions can be harvested about 60 days after transplanting.

Nutrition and efficacy: Bunching onion is pungent, is benefit to appetite and preventing cardiovascular disease.

**品种来源**

地方品种。

分布地区｜全市各区。

特　　征｜株高40厘米，开展度30厘米。叶长30厘米，横径0.7厘米，青绿色，蜡粉少。葱白（假茎）长10厘米，横径1.8厘米，基部形成小鳞茎，纺锤形，鳞衣红褐色，肉白色。

特　　性｜生长势强，分蘖力较强，种植至初收60~70天。耐风雨，抗寒力强。香味浓，品质优。每公顷产量约30吨。

栽培要点｜种植期7—8月。可分株繁殖，株行距18厘米×24厘米。收获期9月至翌年5月，收获时每穴留2~3分蘖，使再分蘖生长，每隔30~35天收获1次。留种则在3月浅种于高畦，5月葱叶转黄软垂时便可采收葱头（鳞茎），晒干后挂在通风处贮存。

## 密轮香葱（火葱、红头葱）

Dense tiller bunching onion

### 四季葱

Four seasons bunching onion

**品种来源**

地方品种。

📍 分布地区｜全市各区。

**品种来源**

地方品种。

📍 分布地区｜白云、花都等区。

特　　征｜株高40厘米，开展度30厘米。叶长30厘米，横径0.8厘米，青绿色，蜡粉少。葱白（假茎）长10厘米，横径1.6厘米，基部形成小鳞茎，纺锤形，鳞衣淡紫红色，肉白色。

特　　性｜生长势强，分蘖力强。种植至初收60~70天。耐寒，耐湿。品质较优。每公顷年产量约70吨。

栽培要点｜种植期8月至翌年3月，以冬、春植为主。株行距15厘米×20厘米。其余参照疏轮香葱。

特　　征｜株高35厘米，开展度30厘米。叶较短而柔软，长25厘米，横径0.5厘米，青绿色，蜡粉少。葱白（假茎）长10厘米，横径1.5厘米，基部形成小鳞茎，纺锤形，鳞衣紫红色，肉白色。

特　　性｜生长势强，分蘖多而快。种植至初收55~65天。组织柔软，耐风雨力弱，耐寒。香味稍淡，品质中等。每公顷产量约25吨。

栽培要点｜参照疏轮香葱。

# 大葱  Welsh onion

百合科葱属二年生草本植物，作一年生栽培，学名 *Allium fistulosum* L. var. *giganteum* Makino，别名水葱，染色体数 $2n=2x=16$。

广州栽培历史悠久，公元 4 世纪已有述及。

**植物学性状**：弦状根。叶管状，先端尖，绿色，表面被蜡粉。叶鞘抱合成假茎，白色，着生于盆状短缩茎上，分蘖性强。冬季抽薹开花，伞形花序，花白色。种子近三角形，黑色。

**栽培品种**：有软尾水葱和硬尾水葱两个品种。

**栽培环境与方法**：喜冷凉，较耐热，不耐旱，不耐涝。要求土层深厚、排水良好、富含有机质的壤土。种子繁殖。2—4月播种，5—11月收获。常与其他蔬菜间种。

**病虫害**：病害主要有紫斑病、霜霉病、锈病、病毒病等，虫害主要有蓟马、葱斑潜蝇、葱种蝇等。

**营养及功效**：假茎（葱白）和嫩叶含碳水化合物、蛋白质、硫化丙烯等。味辛，性平，具发汗解表、健胃理气、消肿、利二便等功效。

*Allium fistulosum* L. var. *giganteum* Makino, a biennial herb. Family: Liliaceae. Synonym: Spring onion. The chromosome number: $2n=2x=16$.

Welsh onion has long been cultivated in Guangzhou. It was mentioned in ancient writings in the 4th century.

**Botanical characters**: String-shaped root system. Leaf pipe-shaped with acuminate tip, green in color, covered wax powder. Leaf sheaths forming a pseudostem, white in color, clustered at the basin-shaped dwarf stem, strong tillering. It bolts and flowers in winter. The flower is umbel and white in color. Seeds nearly triangular, black in color.

**Cultivars**: Main cultivars in Guangzhou include Soft leaf and Hard leaf welsh onion.

**Cultivation environment and methods**: Welsh onion prefers cool weather, is tolerant to heat while not to drought and waterlogged. It requires deep, good drainage, rich in organic matter loam. Seed propagation. Sowing dates: February to April, harvesting dates: May to November. Welsh onion usually interplant with other vegetables.

**Nutrition and efficacy**: Pseudo-stems and leaves are rich in carbohydrates, proteins, propylene sulfide, etc.

## 软尾水葱（尖尾水葱、细叶水葱）
Soft leaf welsh onion

**品种来源**

地方品种。

📍 **分布地区** | 全市各区。

**特　　征** | 株高48~50厘米。叶较细，长37~39厘米，横径0.6~0.7厘米，绿色，先端尖而软，有蜡粉。葱白（假茎）长9~11厘米，横径0.8~0.9厘米。12月至翌年1月抽出花茎，中空，花白色，花球直径5~6厘米。种子黑色。

**特　　性** | 生长较快，分蘖力强，播种至初收110~130天。耐热，不耐旱，不耐贮，易腐烂。品质柔软，味较淡。每公顷产量22~30吨。

**栽培要点** | 播种期2—4月。苗期50~80天，株行距15厘米×20厘米，每穴植2~3株，深种则葱白长，浅种则分蘖快。定植后50天可采收。采收期5—11月，留种期10—11月，翌年1—2月采种。

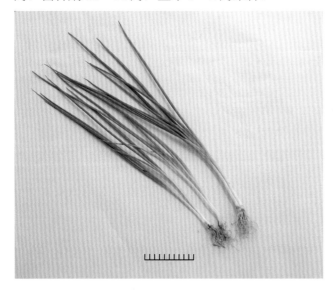

## 硬尾水葱（屈尾水葱、大叶水葱）
Hard leaf welsh onion

**品种来源**

地方品种。

📍 **分布地区** | 全市各区。

**特　　征** | 直立生长，株高41~43厘米。叶长33厘米，深绿色，被蜡粉，先端钝而硬。葱白（假茎）长9~10厘米，横径1.0~1.1厘米。12月至翌年1月抽出花茎，中空，花白色。种子黑色。

**特　　性** | 生长较慢，播种至初收130~150天。组织较坚硬，抗逆性强，分蘖力弱。品质粗硬，香味浓。每公顷产量20~23吨。

**栽培要点** | 播种期2—4月。苗期60~90天，株行距20厘米×20厘米。其余参照软尾水葱。采收期5—11月。

# 大蒜 Garlic

百合科葱属一、二年生草本植物，学名 *Allium sativum* L.，别名蒜，染色体数 $2n=2x=16$。

广州已有千年以上的栽培历史。

**植物学性状**：弦状根。叶带状，扁平对生，深绿色。叶鞘抱合成假茎，基部形成鳞茎（蒜头）。鳞茎由若干鳞芽（即蒜瓣）组成，鳞衣白色或紫红色。抽薹或不抽薹，花茎（蒜薹）细长。伞形花序，花小，紫色。种子退化。

**类型**：按叶片质地分为硬尾大蒜和软尾大蒜。

**栽培环境与方法**：喜冷凉，生长适温为18~20℃，超过26℃鳞茎停止生长。广州地区以采收蒜青为主，鳞茎繁殖，种植期9月至翌年1月，收获期12月至翌年4月。采收蒜头的种植期为9—11月，翌年3—4月收获。

**收获**：花茎顶部开始弯曲、总苞下部变白时采收蒜薹，鳞茎于充分肥大时收获。

**病虫害**：病害主要有露菌病、叶枯病、紫斑病、软腐病、花叶病毒病等，虫害主要有葱地种蝇、葱潜蝇、葱蓟马等。

**营养及功效**：鳞茎含维生素C、粗纤维、钙、磷、铁及大蒜素等。味辛，性温，具杀菌止痢、理胃温中等功效。

*Allium sativum* L., a biennial herb. Family: Liliaceae. The chromosome number: $2n=2x=16$.

Garlic has long been cultivated in Guangzhou more than thousand years.

**Botanical characters**: String-shaped root system. Leaves belt-shaped, flat, opposite, dark green in color. Leaf sheath inclose to form pseudo-stem, forms a bulb at the base. The bulb consists of a few bulb buds (namely cloves). The scale leaves of the garlic white or red-purple in color. It bolts or not. The flower stalk (garlic stalk) is slender. The flower is umbel and purple in color. Seeds aborted.

**Types**: According to leaf texture, it was divided into hard and soft leaf garlic.

**Cultivation environment and methods**: Garlic prefers cool weather, the optimum temperature for growth is 18~20℃. Bulb growth stagnate above 26℃. The green garlic plant is harvested mainly as products in Guangzhou. It is propagated by bulbs. Sowing dates: September to January the following year, harvesting dates: December to April the following year. For bulb production, sowing dates: September to November, harvesting dates: March to April the following year.

**Harvest**: The green garlic plant can be harvested about 90 days after planting. When the top of the stems begin to bend, the lower part of involucre begin to white, the garlic can be harvested. Bulbs is harvested in full size.

**Nutrition and efficacy**: Bulb is rich in vitamin C, crude fibers, calcium, phosphorus, iron, allicin, etc.

## 硬尾大蒜

Hard leaf garlic

**品种来源**

地方品种。

📍 分布地区 | 全市各区。

特　　征 | 植株较直立，株高 55~60 厘米。叶长 45 厘米，宽 1.8 厘米，叶厚挺拔，深绿色，有蜡粉。假茎长 17~19 厘米，横径 1.5~1.7 厘米。每个鳞茎有蒜瓣 10~11 个，鳞衣红色。

特　　性 | 晚熟，不抽薹，栽植至蒜青采收 90~120 天，采收蒜头需 130~150 天。抗性强。蒜叶较粗硬，质脆，味辣，品质优。每公顷产蒜青 22~30 吨，或干蒜头约 7 吨。

栽培要点 | 种植期 9 月中下旬。选择无病虫、充实的蒜瓣栽植，株行距 6 厘米×8 厘米。12 月可收蒜青，翌年 2 月采收蒜头。

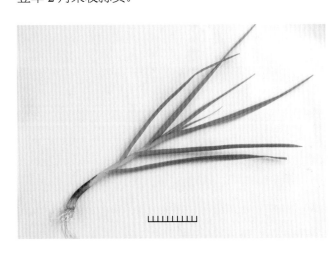

## 软尾大蒜

Soft leaf garlic

**品种来源**

地方品种。

📍 分布地区 | 白云、花都等区。

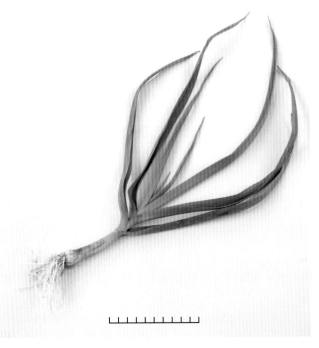

特　　征 | 株高 45~50 厘米。叶长 50 厘米，宽 2.8 厘米，浅绿色。假茎长 20~25 厘米，横径 1.3~1.5 厘米。蒜薹高 45~55 厘米，横径 0.5~0.6 厘米，绿白色，每个鳞茎有蒜瓣 7~12 个，鳞衣白色。

特　　性 | 早熟，生长快，易抽薹，植后 100~110 天收获蒜薹，120~130 天收获蒜头。品质优良。每公顷产蒜青 15~20 吨，或蒜薹 1.5 吨，或干蒜头 4~5 吨。

栽培要点 | 种植期 9 月中下旬。选择无病虫、充实的蒜瓣栽植，株行距 8 厘米×8 厘米。

# 薤 Scallion

百合科葱属多年生宿根草本植物，作二年生栽培，学名 *Allium chinense* G. Don.，别名荞，染色体数 $2n=3x=24$。

广州栽培历史悠久，地方志认为是广州物产之一。

**植物学性状**：弦状根。叶管状，细长，横切面呈三角形，浓绿色，薄被蜡粉。叶鞘抱合成假茎，基部形成小鳞茎，簇生于短缩茎上，卵形，白色。秋季抽出花茎。伞形花序，花小，紫红色，不结种子。

**栽培品种**：广州有丝荞和头荞两个品种，前者供蔬食，后者加工为酸荞头。

**栽培环境与方法**：喜冷凉，生长适温为16~25℃，较耐寒。薤对土壤适应性广，各种土质均可栽培，但以排水良好的沙壤土最宜。鳞茎繁殖。种植期8—11月，收获期12月至翌年6月。

**收获**：以叶和鳞茎供食用的，于1—4月陆续采收；收获鳞茎的，应在鳞茎充分膨大、叶片开始转黄后采收。

**病虫害**：病害主要有薤炭疽病、紫斑病等，虫害主要有葱蓟马、螨类等。

**营养及功效**：富含糖类、蛋白质和矿物质。味辛、苦，性温、滑，具行气导滞、温中散结等功效。

*Allium chinense* G. Don., a perennial herb, cultivated as a biennial. Family: Liliaceae. The chromosome number: $2n=3x=24$.

Scallion has long been cultivated in Guangzhou. It is indigenous to Guangzhou as mentioned by the topography.

**Botanical characters**: String-shaped root system, Pipe-shaped leaf slender, cross-section triangular, dark green in color, coated with wax. Leaf sheaths inclose to form pseudo-stem with bulblets borne at the base, ovate and white in color. It bolts a flower stalk in autumn. The flower is umbelliferous, small, and purple in color, no seed.

**Cultivars**: Two cultivars are cultivated in Guangzhou, Vegetable scallion and Head scallion. The former is used as fresh vegetable and the later is processed as sour head scallion.

**Cultivation environment and methods**: Scallion prefers cool weather, the optimum temperature for growth is 16~25℃. It requires good drainage sandy loam. It is propagated by bulbs. Sowing dates: August to November, harvesting dates: December to June the following year.

**Harvest**: Plant as vegetable is harvested in January to April. Bulbs are harvested in full size, when leaves began to turn yellow.

**Nutrition and efficacy**: Scallion is rich in carbohydrates, proteins and minerals.

## 丝荞

Vegetable scallion

### 品种来源

地方品种。

📍 分布地区｜增城、白云、花都等区。

特　　征｜丛生。叶长 35~40 厘米，横径 0.3 厘米，淡绿色，被蜡粉，柔软易倒伏，叶鞘后期紫红色。荞白（假茎）长 14~20 厘米，横径 0.8~1.2 厘米，基部形成小鳞茎，纺锤形，白色。10 月抽薹开花，花紫色，不结种子。

特　　性｜种植至初收 100~120 天。分蘖力很强。耐寒，忌高温及过湿。品质爽脆，味甜。每公顷产量 15~26 吨。

栽培要点｜种植期 8—9 月。选排水良好的沙壤土，穴植，12 月早收的株行距 15 厘米 ×28 厘米，每穴种植鳞茎 3~4 个，浅种，每公顷用种量 1.5~2.3 吨；迟收的株行距 15 厘米 ×33 厘米，每穴种植鳞茎 2 个，稍深植，每公顷用种量 1.13 吨。生长前期宜与其他蔬菜间种防热，后期注意培土，使荞白伸长。

## 头荞（木荞、燕荞）

Head scallion

### 品种来源

地方品种。

📍 分布地区｜增城、从化、白云、花都、黄埔等区。

特　　征｜丛生。叶长 33~45 厘米，横径 0.4 厘米，浓绿色，被蜡粉，叶稍硬而韧。荞白（假茎）长 10~15 厘米，横径 1.0~1.3 厘米，基部形成卵形鳞茎，高 3~4 厘米，横径 1.5~2.0 厘米，白色。花茎紫红色，不结种子。

特　　性｜种植至初收 150~180 天。生长势壮旺，分蘖力弱。忌高温及过湿。鳞茎肉厚而爽脆，味稍甜，供腌渍用。每公顷荞头产量 15~22 吨。

栽培要点｜种植期 10—11 月，收获期翌年 5—6 月。其余参照丝荞品种。

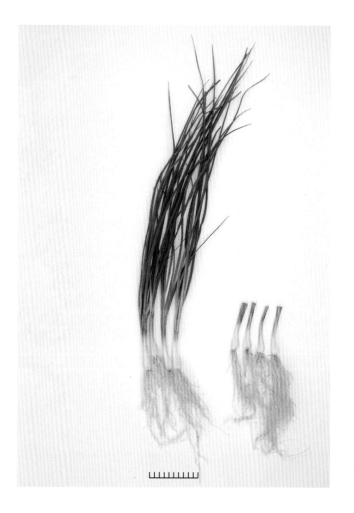

# 韭  Chinese chive

百合科葱属多年生宿根草本植物，学名 *Allium tuberosum* Rottl. ex Spr.，别名韭菜，染色体数 $2n=4x=32$。

广州栽培历史悠久，公元 13 世纪中叶已有述及。

**植物学性状**：弦状根。叶带状，扁平，深绿色。叶鞘抱合成白色假茎，着生于盆状短缩茎上。夏、秋季抽出花茎（即韭菜薹），细长，绿色。伞形花序，花白色。种子黑色。

**栽培品种**：栽培韭品种有细叶韭菜、大叶韭菜和阔叶韭菜 3 个品种。

**栽培环境与方法**：耐低温，不耐高温，生长适温 12~23℃。耐阴，在冷凉气候和中等光照强度条件下生长良好。播种期 11—12 月，翌年 3—4 月定植，第 1 年以间种为主，第 2 年为盛产年，收韭青、韭菜薹为主，第 3~4 年收 1 次韭青后可软化栽培 1 次韭黄。

**收获**：株高 30 厘米左右时收割叶片。采收韭菜薹的，则薹长约 35 厘米时采收。

**病虫害**：病害主要有疫病、锈病等，虫害主要有迟眼蕈蚊、韭萤叶甲等。

**营养及功效**：富含碳水化合物、蛋白质、多种维生素和挥发性物质硫化丙烯，具辛香味。味辛，性温，具除胃热、安五脏、活血壮阳等功效。

*Allium tuberosum* Rottl. ex Spr., a perennial herb. Family: Liliaceae. Synonym: Chinese leeks. The chromosome number: $2n=4x=32$.

Chinese chive has long been cultivated in Guangzhou. It was mentioned in ancient writings in the middle of 13rd century.

Botanical characters: String-shape root system. Leaf belt-shaped, flat, dark green in color. Leaf sheaths inclosed to form a white pseudo-stem, clustered at the basin-shaped dwarf stem. It bolts a slender, green flower stalk during summer and autumn. The flower is umbelliferous and white in color. Seed is black.

Cultivars: Main cultivars in Guangzhou include Large leaf Chinese chive, Small leaf Chinese chive and Broad leaf Chinese chive.

Cultivation environment and methods: Chinese chive is tolerant to low temperature, while not to high temperature. The optimum temperature for growth is 12~23℃. It grows well in cool and moderate sunlight conditions. Sowing dates: November to December, planting dates: March to April the following year. The green leaves and flower stalks are harvested as products. Harvesting of blanched Chinese chive in the third or fourth year.

Harvest: Leaves are harvested when the plant is 30 cm high. Flower stalk is harvested when flower stalk is 35 cm length.

Nutrition and efficacy: Chinese chive is rich in carbohydrates, proteins, vitamins, propylene sulfide, etc.

## 细叶韭菜

Small leaf Chinese chive

**品种来源** 地方品种。

**分布地区** | 全市各区。

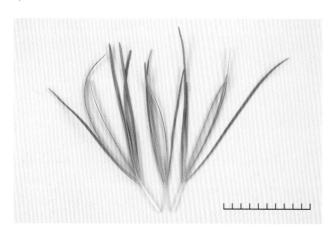

**特　征** | 株高 37 厘米。叶长 27 厘米，宽 0.5 厘米，较薄，上部弯垂，绿色。假茎长 10 厘米，横径 0.5 厘米，白色。花茎长 35 厘米，横径 0.4 厘米，绿色。

**特　性** | 播种至初收 270 天。生长较快，分蘖力强，抽薹早，6—10 月采收花茎。较耐热、耐寒、耐风雨能力强。品质柔软，纤维较多，味浓。每公顷年收韭青 40~45 吨、韭黄 8 吨、花茎 5 吨。

**栽培要点** | 播种期 11—12 月。苗期 120 天，定植期 3—4 月，株行距 25 厘米 ×40 厘米，每穴定植 3~4 株，植时对齐假茎顶端。追肥以薄施为宜。一般春季 30 天、夏秋季 60 天、冬季 45 天可采收 1 次韭青。如果软化韭黄，则韭青延长 15 天收割。每次收获后，即行中耕、施肥、培土。

## 大叶韭菜

Large leaf Chinese chive

**品种来源** 地方品种。

**分布地区** | 全市各区。

**特　征** | 株高 40 厘米。叶较宽厚而硬直，长 30 厘米，宽 1 厘米，深绿色。假茎长 10 厘米，横径 0.8 厘米，青白色。花茎较粗，长 35 厘米，横径 0.6 厘米，绿色。

**特　性** | 播种至初收 330 天。分蘖力中等，抽薹较迟，8—9 月采收花茎。较耐热、耐寒、耐风雨能力强。产量较高。品质脆嫩，纤维较少，味较淡。每公顷年收韭青 65~70 吨、韭黄 10 吨、花茎 3 吨。

**栽培要点** | 参照细叶韭菜。

## 年花韭菜

Nianhua Chinese chive

**品种来源** 20 世纪 90 年代从台湾引进。

**分布地区** | 白云、南沙、番禺等区。

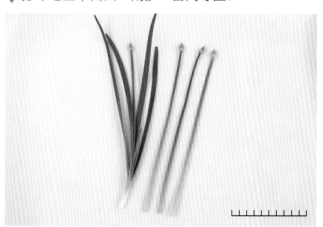

**特　征** | 株高 45 厘米。叶片宽厚而柔软，长 35 厘米，宽 1.5 厘米，深绿色。假茎长 10 厘米，横径 1.0~1.2 厘米，青白色。花茎粗，基部近方形，长 35 厘米，横径 0.8 厘米，绿色。

**特　性** | 生长快，分蘖力较强，抽薹早。以采收花茎为主，3—12 月都可采收花茎。较耐热、耐寒、耐风雨能力强。产量高。纤维少，柔软，味较淡。

**栽培要点** | 参照细叶韭菜。

# 十一、香辛类
## AROMATIC VEGETABLES

- 罗勒
- 紫苏
- 球茎茴香
- 艾叶
- 香茅
- 芫荽
- 莳萝
- 薄荷
- 山柰
- 欧芹
- 香花菜
- 迷迭香
- 香蜂草

# 罗勒 Basil

唇形科罗勒属一年生草本植物，学名 *Ocimum basilicum* L.，别名九层塔，染色体数 $2n=2x=48$。

罗勒引自欧洲，广州各地零星栽植。

**植物学性状**：茎直立，茎高50~60厘米，钝四棱形，绿色至紫色，多分枝。叶对生，具短柄，卵形或长椭圆形，先端锐尖，基部渐狭，略全缘。总状花序，花冠唇形，花瓣白色至淡紫色。坚果，椭圆形，黑褐色。

**栽培环境与方法**：耐热性强，高温下生长迅速，但不耐寒。喜光，不耐涝。多采用直播，也可育苗移栽。茎高20厘米后则可采摘幼嫩茎叶供食。病虫害少，易于栽培。

**营养及功效**：含有丁香油酚、丁香油酚甲醚和桂皮香甲酯等挥发性芳香油。嫩叶可食，亦可泡茶饮，具祛风、健胃及发汗等功效。

*Ocimum basilicum* L., an annual herb. Family: Labiatae. Synonym: Sweet basil, Common basil. The chromosome number: $2n=2x=48$.

It is introduced from Europe and cultivated sparsely in Guangzhou.

Botanical characters: Stem erect, 50~60 cm in height blunt quadrangular in shape, green to purple in color and many branching. Leaf opposite, with short petiole, oval or oblong in shape, entire margin. Inflorescens: raceme, labiate corolla, white to mauve in color. Nut, oval, black brown in color.

Cultivation environment and methods: Basil is tolerant to high temperature but not tolerant to cold and water logging. Basil perfer sunny weather. Mainly direct sowing, or seedling transplanting. Tender shoot can be harvested as vegetable when the plants are 20 cm in height.

Nutrition and efficacy: Stems, leaves and flowers of basil contain eugenol, eugenolmethyl ether and methyl cinnamate.

# 香茅 Citronella

禾本科香茅属多年生草本植物，学名 *Cymbopogon citratus* (DC. ex Nees) Stapf，别名柠檬草，染色体数 $2n=40, 60$。

广州各地零星种植。

**植物学性状**：茎直立，秆高达2米，分蘖性强，节下被白色蜡粉。叶鞘无毛，不向外反卷，内面浅绿色；叶舌质厚，长约1毫米；叶片长30~90厘米，宽0.5~1.5厘米，先端狭尖，平滑或边缘粗糙。

**栽培环境与方法**：周年可分株繁殖。生长势强，栽培容易。

**营养及功效**：全株具有柠檬之芳香气味，茎叶可提取柠檬香精油，药用有通络祛风之效。

*Cymbopogon citratus* (DC. ex Nees) Stapf, a perennial herb. Family: Poaceae. Synonym: Lemon grass. The chromosome number: $2n=40, 60$.

Citronella is sparsely cultivated in Guangzhou.

Botanical characters: Stem erect, 2 m in height, strong tillering, white powder on the base. Leaf sheaths glabrous, inner side light green. Ligule thick, 1 mm in length. Leaf 30~90 cm in length, 0.5~1.5 cm in width, apex narrowly acuminate, smooth or rough edges.

Cultivation environment and methods: The plant can be divided for propagation year round. Citronella grow vigorly and is easy to cultivate.

Nutrition and efficacy: The whole plant has lemon scented. The stem and leaf can be used for extraction of essential oil of lemon.

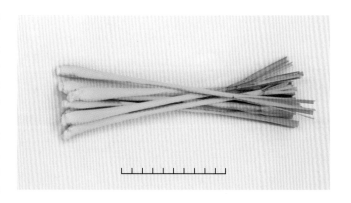

## 薄荷 Mint

唇形科薄荷属多年生宿根草本植物，学名 *Mentha haplocalyx* Brip. var. *piperascens* (Malinn) C. Y. Wu et H. W. Li.，别名青薄荷、香荷，染色体数 $2n=2x=12$。

广州栽培历史悠久，各地零星栽植。

**植物学性状**：根系浅。茎方形，多分枝。叶对生，卵形，先端锐尖，基部楔形至近圆形，叶缘锯齿状，绿色。茎叶被茸毛，以嫩茎叶作香菜。

**栽培环境与方法**：种子和茎可繁殖。栽植后可采收多年，收获期3—9月。

**营养及功效**：茎叶含有薄荷油，有香气，具解热祛风化痰等功效。

*Mentha haplocalyx* Brip. var. *piperascens* (Malinn) C. Y. Wu et H. W. Li., a perennial herb. Family: Labiatae. Synonym: Aromatic mint. The chromosome number: $2n=2x=12$.

Mint has long been cultivated sparsely in Guangzhou.

Botanical characters: Shallow root system. Stem square and with many branching. Leaf opposite, oval, green, serrate margin. Stems and leaves are covered with hairs. Tender stems and leaves are used as aromatic vegetable.

Cultivation environment and methods: Seed and stem cutting are used for propagation. The plant can be harvested for many years. The harvest date is from March to September.

Nutrition and efficacy: Stems and leaves contain mint oil.

## 香花菜 Spearmint

唇形科薄荷属多年生宿根草本植物，学名 *Mentha spicata* L.，别名留兰香、绿薄荷，染色体数 $2n=36,48,54,84$。

广州栽培历史悠久，各地零星种植。

**植物学性状**：根系浅。茎方形，分枝多，绿色。叶对生，卵圆形或披针形，深绿色，叶缘锯齿状，两面无毛或近无毛。轮伞花序，密集顶生成穗状，花紫色或白色。

**栽培环境与方法**：地上茎和种子均可繁殖。栽植后可收获多年，收获期3—9月。

**营养及功效**：茎叶有香气，常作调味香料食用，也可提炼留兰香油。

*Mentha spicata* L., a perennial herb. Family: Labiatae. Synonym: Green mint. The chromosome number: $2n=36, 48, 54, 84$.

Spearmint has long been cultivated sparsely in Guangzhou.

Botanical characters: Shallow root system. Green stem square with many branches. Leaf opposite, oval or lanceolate, dark green, with serrate margin. Inflorescence: whorled umbel, dense terminal to form spicate inflorescens. Flower purple or white in color.

Cultivation environment and methods: Stem cutting and seed are used for propagation. The plant can be harvested for many years. Harvesting date is from March to September.

Nutrition and efficacy: The stems and leaves with aromatic odor can be used for spearmint oil extraction.

# 紫苏 Perilla

唇形科紫苏属一年生草本植物，学名 *Perilla frutescens* (L.) Britt.，别名香苏，染色体数 $2n=2x=38$，40。

广州多为零星种植，有个别企业采用大棚生产供出口。

**植物学性状**：根系较发达。茎直立，四棱，分枝多，具茸毛，紫色或绿色。叶对生，阔卵形或卵圆形，叶缘钝锯齿，叶面紫色或绿色，被茸毛。总状花序，顶生或腋生，花冠紫红色。坚果卵形，灰褐色，含1粒种子。

**类型**：根据叶的颜色和形状分为紫苏和青紫苏两类，品种有大鸡冠紫苏和青紫苏。

**栽培环境与方法**：喜温暖湿润，开花期适温26~28℃。不耐干旱，较耐阴。对土壤适应性较广。一般春季播种育苗，栽培过程中注意及时摘心打杈。

**收获**：6—8月为采收高峰期，每次采摘2对叶片。

**病虫害**：病害主要有褐斑病、锈病等，虫害主要有蚜虫、斜纹夜蛾等。

**营养及功效**：富含维生素、多种矿物质及紫苏醛、紫苏醇等挥发油。味辛，性温，具解肌发表、散风寒、行气宽中、消痰利肺、和血温中、止痛、解鱼蟹毒等功效。

*Perilla frutescens* (L.) Britt., an annual herb. Family: Labiatae. The chromosome number: $2n=2x=38$, 40.

Perilla is usually sparsely cultivated in Guangzhou.

**Botanical characters**: Vigorous root system. Stem erect, four angular, covered with hairs, purple or green in color. Leaf opposite, broad ovate or oval in shape, with obtusely serrate margin, purple or green in color with hairs. Inflorescence: raceme, terminal or axillary, purple red flower. Fruit nut, oval, pale brown in color, with 1 seed.

**Types**: According to the color and shape of the leaf, perilla can be divided into perilla and green perilla. The cultivars include Cockscomb perilla and Green perilla.

**Cultivation environment and methods**: Perilla perfer warm and mosit weather, is not tolerant to drought but tolerant to shade. Sowing or transplanting in spring. High yield require pinching in time.

**Harvest**: Harvesting perilla occur from June to August.

**Nutrition and efficacy**: Perilla is rich in vitamins, minerals, perillaldehyde, perilla alcohol, etc.

## 大鸡冠紫苏

Cockscomb perilla

> 品种来源

地方品种。

📍 分布地区 | 各区零星种植。

特　　征 | 株高 80~90 厘米，开展度 70 厘米。茎具浅沟纹，淡紫色。叶阔卵形，长 14 厘米，宽 13 厘米，先端短尖，叶缘深锯齿状，叶面皱，叶面紫绿色，叶背紫红色，有茸毛；叶柄长 4.5 厘米，紫红色。花序顶生或腋生，花紫白色。种子褐色。

特　　性 | 播种至初收 65~75 天，延续采收 120 天。生长势强。耐热怕冷，耐涝，病虫害少。香味稍淡。

栽培要点 | 1—2 月播种。每公顷用种量 3.0~4.5 千克。苗期 40~45 天，3~4 片叶移植。施足基肥，松土追肥。一般在 7 节上留侧枝 4 条，每隔 15 天摘叶 1 次。花期 9—10 月，11 月采种。

## 青紫苏

Green perilla

> 品种来源

地方品种。

📍 分布地区 | 全市各区。

特　　征 | 株高 38 厘米，开展度 23~31 厘米。茎具浅沟纹，青紫色。叶阔卵形，顶端锐尖，长约 10 厘米，宽 8.5 厘米，青绿色，叶缘深锯齿状，叶背被白色茸毛；叶柄长 4.5~6.8 厘米。花序顶生或腋生。

特　　性 | 生长势强。耐热怕冷，耐涝，病虫害轻。香味浓郁。

栽培要点 | 花期 8—9 月。参照大鸡冠紫苏。

# 芫荽 Coriander

伞形科芫荽属一年生草本植物，学名 *Coriandrum sativum* L.，别名香菜，染色体数 $2n=2x=22$。

植物学性状：根系浅。茎短缩。叶互生，半直立，一至二回羽状复叶，小叶卵圆形，浅裂或深裂，具长柄，浅绿色。伞形花序，花小，白色。双悬果，圆球形，黄褐色，果面有棱，含种子2粒。

栽培品种：广州栽培历史悠久，有沙滘芫荽、大叶芫荽和留香芫荽3个品种。

栽培环境与方法：喜冷凉，生长适温为17~20℃。属长日照作物。要求肥沃而保水力强的壤土或沙壤土。多采用直播，一般在秋冬播种。

收获：株高20~25厘米时即可分次采收。

病虫害：病害主要有叶斑病、白粉病等，虫害主要有蚜虫、红蜘蛛等。

营养及功效：富含维生素C、烟酸及钾、钙、铁等。味辛，性温，具利大小肠、消食下气、发汗透疹、止血等功效。

*Coriandrum sativum* L., an annual herb. Family: Umbelliferae. The chromosome number: $2n=2x=22$.

Botanical characters: Shallow root. Stem dwarf. Leaf alternate, semi-erect, 1~2 pinnate, leaflets ovate, lobed or parted and light green in color, with long petioles. Umbelliferous, small white flower. Cremocarp round ribbed, yellow brown, containing 2 seeds.

Cultivars: Coriander has long been cultivated in Guangzhou. The cultivars include Shajiao, Large leaf and Liuxiang coriander.

Cultivation environment and methods: Coriander prefers to cool season, suitable temperature for growth is 17~20 ℃. Long day plant. Fertile loam or sandy loam with moisture holding capacity is required. Direct seeding in autumn and winter time.

Harvest: Coriander can be harvested when the plants are 20~25 cm in height.

Nutrition and efficacy: Coriander is rich in vitamin C, niacin, calcium, potassium, iron, etc.

## 沙滘芫荽 Shajiao coriander

**品种来源** 地方品种。

**分布地区** | 白云区。

**特　　征** | 株高20~23厘米，开展度15~20厘米。叶长20~23厘米，小叶3~4对，近圆形，长约1.5厘米，宽约2厘米，绿色，叶缘波状浅裂；叶柄绿白色。复伞形花序，花白色或微紫色。果实近球形，有棱。单株重8克。

**特　　性** | 早熟，以秋冬种植为主，播种至初收秋植50~60天、冬植60~80天。生长势较强，抽薹较早。耐寒，不耐热，病虫害少。香味浓郁，品质优。每公顷产量15~18吨。

**栽培要点** | 播种期8月至翌年1月。选沙壤土种植，条播或撒播。播前将果实压破，使种子分离，以利于萌发。每公顷用种量45~60千克。高温季节可进行低温催芽和遮阳网覆盖栽培，周年可播种。施足基肥，播后保持土壤湿润，勤施薄肥。

## 大叶芫荽 Large leaf coriander

**品种来源** 引自泰国。

**分布地区** | 全市各区。

**特　　征** | 株高22~30厘米，开展度15~20厘米。叶面绿色，叶缘深绿色，叶柄白绿，有3~4对小叶，叶卵圆形，叶柄长15~20厘米，柄基近白色。复伞形花序，花小，白色。果实近球形，有棱。单株重12~15克。

**特　　性** | 以秋冬种植为主，播种至初收秋植45~55天、冬植60~70天。生长势较强，抽薹迟。耐热，耐寒，病虫害少。质地柔嫩，香味浓。每公顷产量18~22吨。

**栽培要点** | 周年均可播种，以10月至翌年3月为最适播期。夏季高温季节，宜采用低温催芽和遮阳网覆盖栽培。其余参照沙滘芫荽。

## 留香芫荽 Liuxiang coriander

**品种来源** 广东省良种引进服务公司引进推广。

**分布地区** | 全市各区。

**特　　征** | 株高28~30厘米，开展度15~20厘米。羽状复叶，有3~4对小叶，叶卵圆形，叶柄长15~20厘米，柄基近白色。复伞形花序，花小，白色。果实近球形，有棱。单株重12~15克。

**特　　性** | 以秋冬种植为主，播种至初收秋植45~50天、冬植60~70天。生长势较强，抽薹迟。耐热，耐寒，病虫害少。质地柔嫩，香味浓。每公顷产量18~20吨。

**栽培要点** | 周年可以播种。在4—8月高温多雨季节要采取遮阳覆盖防雨防热。其余参照沙滘芫荽。

# 山奈 Rhizoma kaempferiae

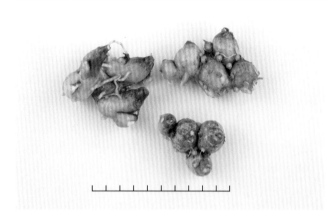

姜科山奈属多年生草本植物，作一年生栽培，学名 *Kaempferia galanga* L.，别名沙姜。

广州栽培历史悠久，多为零星种植。

**植物学性状**：根茎块状，单个或数个相连，具芳香味。无地上茎。叶2~4片贴近地面生长，近卵圆形或阔卵圆形，全缘，幼叶被软柔毛。穗状花序，花白色，基部有紫斑，有香味。蒴果，长椭圆形。

**栽培环境与方法**：喜温暖。湿润。生长适温为18~28℃。不耐涝，不耐旱，不耐寒。一般春种，秋冬季收获。

**营养及功效**：块茎中富含钾、挥发油等，可鲜食或制成粉、片，用作香料或调味品。

*Kaempferia galanga* L., a perennial herb. Family: Zingiberaceae.

Rhizoma kaempferiae has long been cultivated sporadically in Guangzhou.

**Botanical characters:** The rhizome single or several linked, aromatic. No overground stem. 2~4 leaves grow close to the ground. Leaf is nearly ovate or broadly ovate, entire margin. The young leaves are covered with hairs. Inflorescences spicate, flowers are white with purple spots on the base, fragrant. Capsule is oblong.

**Cultivation environment and methods:** Rhizoma Kaempferiae perfers warm and moist weather. Suitable temperature for growth is 18~28℃. It is not tolerant to waterlogging, drought and cold. Planting in spring, harvesting in autumn and winter.

**Nutrition and efficacy:** The tuber is rich in potassium, volatile oil, etc. It can be eaten freshly or made for powder or spices.

# 迷迭香 Rosemary

唇形科迷迭香属多年生灌木，学名 *Rosmarinus officinalis* L.，别名艾菊。

广州有零星种植。采用枝条扦插或种子繁殖。

**植物学性状**：茎直立，横断面方形。叶对生，线形，叶面主叶脉凹陷，叶背银白色，密布细茸毛。全株皆有芳香味。唇形花，蓝紫色、白色或粉红色。坚果，卵圆形，褐色。

**营养及功效**：全株芳香，茎叶可提取精油或作调味品。味辛，性温，具健胃、镇静安神等功效。

*Rosmarinus officinalis* L., a perennial shrub. Family: Labiatae.

Rosemary is cultivated sparsely in Guangzhou. Cuttings or seed are used for propagation.

**Botanical characters:** Stem erect, square. Leaf opposite, linear, leaf main vein sag, silvery white back of leaf, with fine hairs. The whole plant is aromatic. Labiate flower blue purple, white or pink. Nuts is ovoid in shape and brown in color.

**Nutrition and efficacy:** The whole plant of rosemary is aromatic. The stem and Leaves can be fried or made for condiment.

# 球茎茴香 Fennel

伞形科茴香属的一个变种，为一、二年生草本植物，学名 *Foeniculum vulgare* Mill. var. *dulce* Batt. et Trab.，别名大茴香。

广州有零星种植。

**植物学性状**：株高 70~100 厘米。叶为具三回羽状深裂的细裂叶，裂叶细如丝状，叶面光滑，无毛。球茎扁球形，高约 10 厘米，宽 6~7 厘米，厚 3~4 厘米。单株重 800~1 000 克，球茎重 300~500 克。球茎品质柔嫩，纤维少，香味特殊。

**栽培环境与方法**：喜冷凉，但适应性强，能耐寒，稍耐热。以球茎和嫩叶供食用。

*Foeniculum vulgare* Mill. var. *dulce* Batt. et Trab., an annual or biennial herb. Family: Umbelliferae. Synonym: Sweet fennel, Finocchio.

Fennel is sporadically cultivated in Guangzhou.

Botanical characters: Plant 70~100 cm in height . Leaf 3 pinnate compound, glossy, glabrous. Bulb oblate, 10 cm in height, 6~7 cm in width and 3~4 cm in thickness. A single plant weight is 800~1000 g and bulb weight is 300~500 g. Bulb tender, little fibre and aromatic.

Cultivation environment and methods: Fennel prefers to cool season, and is tolerant to cold and heat. Bulb and tender leaves are edible.

# 莳萝 Dill

伞形科莳萝属一、二年生草本植物，学名 *Anethum graveolens* L.，别名小茴香、刁草，染色体数 $2n=2x=10$。

广州地区有少量种植。

**植物学性状**：以嫩叶供食用。根系浅，茎短缩。株高 20~50 厘米，叶轮生，三回羽状分裂，裂片狭长成线状，绿色，具较长的叶柄。伞形花序，花淡黄色，花期较短。

**栽培环境与方法**：喜温暖湿润的气候条件，生长适温为 20~25℃。

**营养及功效**：全株芳香，叶可提取精油或作调味品。味辛，性温，具下气利膈、顺气止痛、生津解渴等功效。

*Anethum graveolens* L., an annual or biennial herb. Family: Umbelliferae. Synonym: Fennel, Dill weeds. The chromosome number: $2n=2x=10$.

Dill has been marginally cultivated in Guangzhou.

Botanical characters: The leaf is edible as vegetable. Root system shallow. Stem dwarf. Plant 20~50 cm in height. Leaf whorl, 3 pinnate compound, green, long petiole. Inflorescences umbel, pale yellow.

Cultivation environment and methods: Dill perfers warm and moist weather. The suitable temperature for growth is 20~25℃.

Nutrition and efficacy: The whole plant of dill is aromatic. The leaves can be fried or made for condiment.

## 欧芹 Parsley

伞形科欧芹属一、二年生草本植物，学名 *Petroselinum crispum* (Mill). Nym. ex. A. W. Hill，又名洋芫荽，染色体数 $2n=2x=22$。

**植物学性状**：叶为浓绿色，具长柄，叶缘锯齿状，有卷缩型及平坦型。花小，淡绿。

**栽培环境与方法**：喜冷凉湿润，不耐热，生长适温为15~20℃。广州极少量种植。

*Petroselinum crispum* (Mill). Nym. ex. A. W. Hill, an annual or biennial herb. Family: Umbelliferae. The chromosome number: $2n=2x=22$.

Botanical characters: Leaf dark green, long petioles, serrate margin, curly or flat surface. Flower small, light green.

Cultivation environment and methods: Parsley prefers cool and moist weather and is not tolerant to heat. The suitable temperature for growth is 15~20℃. Parsley is rarely cultivated in Guangzhou.

## 香蜂草 Lemon balm

唇形科多年生草本植物，学名 *Melissa officinalis* L.，别名蜜蜂花。

**植物学性状**：植株直立。茎横断面四棱形，多分枝。叶对生，近心脏形，绿色，两面疏生茸毛及透明腺点，带有柠檬香味。轮伞形花序，顶生或腋生，花白色或淡紫色。坚果，长圆形，黑褐色有光泽。

**栽培环境与方法**：广州有零星种植。采用种子、分枝或枝条扦插繁殖。

*Melissa officinalis* L., a perennial herb. Family: Lamiaceae. Synonym: Balm mint, Bee balm, Blue balm.

Botanical characters: Stem erect, four angular, with many branch. Leaf opposite, nearly heart in shape, green in color. Thin hair and transparent glandular spots on two sides of leaf, with lemon scent. Inflorescences verticillaster, terminal or axillary, flower white or pale purple in color. Nut oblong, dark brown in color.

Cultivation environment and methods: Lemon balm is cultivated sporadically in Guangzhou, and propagated by seeds, branching and cuttings.

## 艾叶 Mugwort

菊科蒿属多年生草本植物，学名 *Artemisia argyi* H.，别名艾草、艾蒿等。

**植物学性状**：植株有浓烈香气。主根明显，略粗长，直径达1.5厘米，侧根多。茎单生，高80~150厘米，有明显纵棱，褐色或灰黄褐色，基部稍木质化，上部草质。叶互生，羽状裂叶，叶柄基部呈翼状，叶面绿色，叶背密生白色茸毛。

**栽培环境与方法**：野生或各地有零星栽植。新鲜嫩茎叶可做菜用。

*Artemisia argyi* H., a perennial herb. Family: Asteraceae. Synonym: Chinese mugwort.

Botanical characters: Strong aroma. Root obvious, the diameter is 1.5 cm, with many branch. Stem with obvious ridge, 80~150 cm in height, brown or yellowish-brown in color. Leaf alternate pinnate, green, alary base of petiole. The blade back is covered with white hairs.

Cultivation environment and methods: Mugwort is wild or sporadicly cultivated in Guangzhou. Tender shoot is used as vegetable.

# 十二、水生蔬菜类
## AQUATIC VEGETABLES

- 莲藕
- 茭白
- 慈姑
- 菱

- 荸荠
- 豆瓣菜
- 水芹

广州蔬菜品种志

# 莲藕

睡莲科莲属多年生水生草本植物，学名 *Nelumbo nucifera* Gaertn.，别名藕，染色体数 $2n=2x=16$。

广州地区栽培已有 300 多年历史，新垦莲藕以品质好而闻名，已成为国家地理标志产品，除供应珠江三角洲以外，每年出口量达 1 万吨以上。目前广州莲藕栽培面积 800 公顷左右。

植物学性状：节上发不定根。茎分匍匐茎和根状茎。种藕顶芽抽出形成匍匐茎，条形，称莲鞭。莲鞭生长到一定程度，先端数节积累养分形成肥大根状茎，称母藕。母藕节上又发生侧莲鞭，积累养分形成子藕，子藕节上又可形成孙藕。叶单生，圆盘形，全缘，绿色，叶脉呈放射状。花单生，两性，白色或淡红色，子房上位，花谢后成为莲蓬，心皮多数，每心皮有种子 1 粒，椭圆形，浅黄色。

类型：按食用产品器官分为藕莲和子莲 2 种类型。藕莲以收藕为主，主要品种有海南洲、京塘丝藕、鄂莲 5 号、鄂莲 6 号等。子莲以采收莲子为主，有白莲和红莲等品种，现已少有种植。

栽培环境与方法：宜栽培于浅水条件下，采用种藕种植，每 667 米$^2$ 用种量 200 千克，2—5 月种植，6—12 月收获，翌年可自行留种。施足基肥，适当追肥。保持浅水，不能过深。

收获：6 月可以采收嫩藕，10 月后采收成熟藕。

病虫害：病害主要有腐败病、褐斑病，虫害主要有蚜虫、斜纹夜蛾。

营养及功效：富含淀粉、蛋白质、维生素 C、糖和多酚化合物等。味甘，性平，具交心肾、厚肠胃、固精气、强筋骨、补虚损、利耳目、除寒湿、止脾泄久痢等功效。

# Lotus root

*Nelumbo nucifera* Gaertn., a perennial aquatic herb. Family: Nymphaeaceae. The chromosome number: $2n=2x=16$.

The lotus root has been cultivated for more than 300 years in Guangzhou. Now the lotus root of Xinken in Nansha District becomes most renowned due to the fine quality, and the amount of exports is over ten thousand tons annually. At present, the cultivation area is about 800 $hm^2$ per annum in Guangzhou.

Botanical characters: Adventitious roots are borne on the nodes. Stem is divided into stolon and rhizome. Stolon developing from the terminal bud of the mother lotus root is linear in shape, called lotus tail. When the growth of the lotus tails reaches a certain stage, several terminal nodes which accumulate nutrients become thick to form a rhizome, the mother lotus root. Lateral lotus tails are borne on the nodes of the mother lotus root, due to nutrient accumulation, which form secondary lotus roots. Leaf, solitary, round basin-shaped, entire, with shooting veins, green. Flower solitary, bisexual, white or light red in color with a superior ovary. After fertilization of flowers, lotus seeds are formed. Numerous carpels, with one seed in each carpel, oval, light white yellow in color.

Types: Lotus roots are divided into two types according to the edible part: rhizome type and seed type. The rhizome type is cultivated for the lotus root. The cultivars include Hainanzhou, Jingtang Elian No.5, Elian No. 6. Seed type is cultivated chiefly for seed. The cultivars include white lotus root and red lotus root, etc. Now this type is rarely cultivated.

Cultivation environment and methods: Lotus root perfer in shallow water. Propagated by lotus roots, 3000 kg per ha are required. Planting dates: February to July and harvesting dates: June to December. Sufficient basal fertilizer and topdressing are required.

Harvest: The young lotus root can be harvested in June, and the elder lotus root can be harvested after October.

Nutrition and efficacy: Lotus root is rich in starchs, proteins, vitamin C, sugars, polyphenol compounds, etc.

## 海南洲（猫头）

Hainanzhou lotus root

> 品种来源

地方品种。

📍 **分布地区** ｜ 南沙、番禺等区。

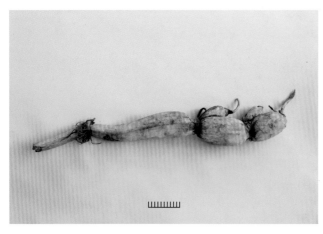

特　　征｜株高 150~180 厘米。叶直径 70 厘米，深绿色，叶脉明显，叶窝较深，被蜡质；叶柄黄绿色，有刺。花瓣白色。母藕长 80 厘米，具 4~5 节，横径 7 厘米，藕褐色，短圆筒形，孔道较大，单藕重 1.5~2 千克。

特　　性｜中晚熟，生长期 150 天左右。主藕轻微弯曲，藕皮较厚，有锈斑。嫩藕爽脆。成熟藕粉香，品质优，耐贮运，保存期较长。该品种较耐盐碱，较耐深水，非常适宜咸淡水交界地区种植。每公顷产量 15~18 吨。

栽培要点｜播种期 2—4 月。用种藕种植，株行距 150 厘米×200 厘米。施足基肥，定植 25 天后开始追肥，全期共追肥 2~3 次。注意防治蚜虫、斜纹夜蛾。收获期 9—12 月，采收完毕留小量藕种作明年用种，但留种不要超过 3 年。

## 京塘丝藕

Jingtang lotus root

> 品种来源

地方品种。

📍 **分布地区** ｜ 花都区北兴京塘村。

特　　征｜株高 100~150 厘米。叶直径 65 厘米，绿色，被蜡质，有光泽，叶肉薄而脆，叶脉明显，叶窝较浅；叶柄青黄色，有刺。花细，花瓣白色，边缘带浅红色。母藕细长，长 1.5~2 米，具 4~5 节，节间瘦长，横径 4~5 厘米，孔道细，单藕重 1.5 千克。

特　　性｜晚熟，生长期 200~240 天。生长势旺盛，藕深生。耐贮运。淀粉多，品质优。每公顷产量 7.5 吨。

栽培要点｜每年利用藕田留下的藕在翌年 3—4 月萌发新株。11 月排干田水，冬至前后采收。其余参照海南洲。

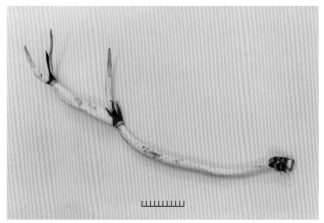

## 鄂莲 5 号

Elian No. 5 lotus root

> 品种来源

武汉市蔬菜科学研究所育成，又名 3735。

📍 **分布地区**｜南沙、番禺等区。

**特　　征**｜株高 160~180 厘米。叶直径 70~80 厘米，绿色，被蜡质。花瓣白色。母藕长 100~120 厘米，具 5~6 节，横径 7 厘米，横断面椭圆形，表皮白色，藕肉厚实，孔道较小，单藕重 2~3 千克。主藕入泥深 30 厘米。

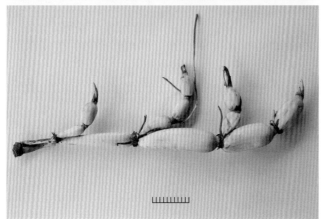

**特　　性**｜早中熟，生长期 125 天左右。抗逆性较强，较耐腐败病。肉质肥厚，品质优，耐贮藏。产量高，每公顷产量 30~37.5 吨。

**栽培要点**｜注意定植后早追肥，封行前每 667 米$^2$ 施复合肥 50 千克，藕开始膨大时，视其长势可每 667 米$^2$ 施复合肥 20~40 千克。该品种较早熟，7 月可采青荷藕，8 月下旬可收成熟藕。其余参照海南洲。

## 鄂莲 6 号

Elian No. 6 lotus root

> 品种来源

武汉市蔬菜科学研究所育成。

📍 **分布地区**｜南沙、番禺等区。

**特　　征**｜株高 160~180 厘米。叶近圆形，叶片直径 72 厘米左右。花瓣白色，开花较多。藕节间为短筒形，表皮黄白色，肩部圆钝，节间均匀，一般主藕 5~7 节，长 90~110 厘米，横径 8 厘米，单藕重 3.5~4 千克。主藕入泥深 25~30 厘米。

**特　　性**｜生长势强。早中熟，生长期 125 天左右。抗逆性较强。肉质肥厚，品质优。主藕入泥较浅，容易采收。产量高，每公顷产量 30~37.5 吨。

**栽培要点**｜注意定植后早追肥，封行前每 667 米$^2$ 施复合肥 50 千克，藕开始膨大时，视其长势可每 667 米$^2$ 施复合肥 20~40 千克。该品种较早熟，7 月可采青荷藕，8 月下旬可收成熟藕。其余参照海南洲。

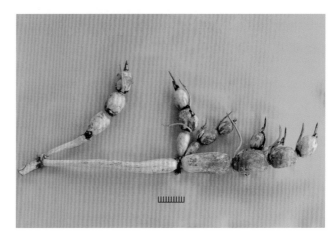

# 茭白 Water bamboo

禾本科菰属多年水生草本植物，学名 *Zizania caduciflora* (Turcz.) Hand.-Mazz.，别名茭笋，染色体数 $2n=2x=34$。

广州栽培历史悠久，地方志把茭白列为地产蔬品之一，现南沙、番禺、白云等区有小面积种植。

**植物学性状**：须根发达。茎分地上短缩茎和地下根状茎。地上短缩茎分蘖形成株丛，进入生殖生长后，拔节伸长，同时有黑穗菌（*Ustilago esculenta*）寄生，刺激先端数节膨大，形成肉质茎。地下根状茎在土中匍匐生长，其侧芽和顶芽向上生长成为分株。叶片条形，叶鞘长，相互抱合形成假茎。圆锥花序，紫色花，在广州一般不开花结籽。食用部分为肉质茎，呈纺锤形，黄白色，肉白色，味道鲜美。

**栽培环境与方法**：宜栽培于浅水条件下，采用分株繁殖，夏季种植，秋季收获。

**营养及功效**：富含蛋白质、维生素C、矿物质及多种氨基酸。味甘，性寒，具解热毒、除烦渴、利二便、除烦热、消渴等功效。

*Zizania caduciflora* (Turcz.) Hand.-Mazz., a perennial aquatic herb. Family: Gramineae. The chromosome number: $2n=2x=30$.

The water bamboo was mentioned in topography as one of local vegetable more than 100 years ago. It is cultivated sparsely in Nansha, Panyu and Baiyun districts.

Botanical characters: The water bamboo with vigorously fibrous roots. Stem is divided into above-ground dwarf stem and under-ground rhizome. Tillers are borne on the dwarf stem, forming clustered plants. The elongation of the dwarf stem and tillers begin at the reproductive growth stage. Due to the stimulating effects of parasitic fungi (*Ustilago esculenta*), the terminal parts become swollen to form a fleshy stem (the edible part), which is fusiform with yellow-white skin and white flesh. The rhizome grows prostratly in soil, while its lateral and terminal buds growing upright, becomes a division. Leaf linea, leaf sheath long, forming a pseudostem. Inflorescence, panicle, purple in color. It generally does not flowering or seeding in Guangzhou.

Cultivation environment and methods: Water bamboo perfers shallow water. Propagated by plant division in summer, and harvesting in autumn and winter.

Nutrition and efficacy: Water bamboo is rich in proteins, vitamin C, minerals and amino acids.

## 软尾茭笋 Soft leaf water bamboo

**品种来源** 地方品种。

**分布地区** | 南沙、番禺、白云等区。

**特　征** | 株高150~180厘米。叶长130厘米，宽2.5厘米，柔软弯垂。肉质茎长纺锤形，长10~12厘米，横径3~4厘米，具3~4节，表皮黄白色，肉白色，单个重90~110克。

**特　性** | 定植至初收90天。每公顷产量20~25吨。

**栽培要点** | 4月育苗，7月定植，株行距100厘米×165厘米。注意施足基肥，适当追肥，前期灌浅水，后期适当加深水位，勤培土使茭笋嫩且洁白。注意防治蚜虫，摘除老叶。收获期9—10月，采收完毕把母株集中留种田，次年4月分株繁殖。

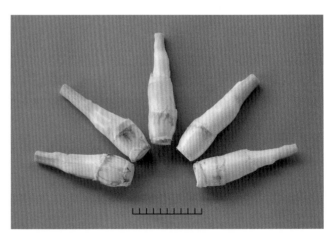

# 慈姑 Chinese arrowhead

泽泻科慈姑属多年水生草本植物，学名 *Sagittaria sagittifolia* L., 别名乌芋，染色体数 $2n=2x=22$。

广州 300 多年前已有栽培，现南沙、番禺、白云、增城等区有小面积种植，年栽培面积约 250 公顷。

*Sagittaria sagittifolia* L., a perennial aquatic herb. Family: Alismataceae. The chromosome number: $2n=2x=22$.

Chinese arrowhead has been cultivated in Guangzhou for more than 300 years. It is cultivated sparsely in Nansha, Panyu, Baiyun and Zengcheng. At present, the cultivation area is about 250 $hm^2$ per annum in Guangzhou.

Botanical characters: Fibrous, flesh roots, which are without any root hair. The stem is divided into 3 types: the dwarf stem, the stolon and the corm. The stolon is formed from the elongation of axillary bud on the dwarf stem. The terminal bud of stolon is a leaf bud and it elongats to form a new plant. The terminal part of the stolon is thickened due to nutrient accumulation to form a corm (the edible part), which is ovate or spherical in shape and white or white-yellow in color. Leaves are hastate, alternate, green, petiole long, with shallow grooves. Inflorescence, a raceme, monoecious, and white. No seeding in Guangzhou generally.

植物学性状：须根系，肉质，无根毛。茎分短缩茎、匍匐茎和球茎。短缩茎上腋芽伸长成为匍匐茎，匍匐茎顶芽出土生叶成为分株。匍匐茎末端积累养分，肥大形成球茎，为食用器官，卵形或球形，白色或黄白色。叶互生，戟形，绿色，叶柄长，具浅沟。总状花序，雌雄异花，白色，在广州一般不结籽。

类型：地方品种，有沙姑和白肉慈姑两个品种。

栽培环境与方法：宜栽培于浅水条件下，采用球茎繁殖，2月育苗，7—9月定植，12月至翌年3月收获。

营养及功效：富含淀粉、蛋白质、糖类、无机盐、硒、维生素E、维生素B、维生素C及胰蛋白酶等多种营养成分。味甘、涩，性微温，具清热止血、解毒消肿、散结等功效。

Types: Local cultivars, Shagu Chinese arrowhead and White flesh Chinese arrowhead.

Cultivation environment and methods: Chinese arrowhead perfers shallow water. It is propagated by corm. Corm seedling: February, seedling transplanting: July to September. The harvesting dates are from December to March the following year.

Nutrition and efficacy: Chinese arrowhead is rich in starches, proteins, carbohydrates, minerals, selenium, vitamin E, vitamin B, vitamin C, trypsin, etc.

## 沙姑

Shagu Chinese arrowhead

**品种来源**

地方品种。

**分布地区** | 南沙、番禺、白云、增城等区。

特　　征 | 株高 70~80 厘米，开展度 60~70 厘米。叶长 30 厘米，宽 8 厘米，绿色，叶柄长 65 厘米。球茎卵形，高 5 厘米，横径 3~4 厘米，皮黄白色，肉白色，单个重 40~60 克。

特　　性 | 生长期 110~120 天。含淀粉多，略有苦味。每公顷产量 15 吨左右。

栽培要点 | 2 月用球茎育苗，7—8 月定植，株行距 25 厘米 ×45 厘米。施足基肥，适当追肥，前期灌浅水，后期适当加深水位。10 月下旬至 11 月上旬进行圈根，即割断地下匍匐茎，促进球茎肥大。注意防治蚜虫，及时去除老叶和侧芽。收获期翌年 1—2 月。

## 白肉慈姑

White flesh Chinese arrowhead

**品种来源**

地方品种。

**分布地区** | 南沙、番禺、白云、增城等区。

特　　征 | 株高 70~75 厘米，开展度 80 厘米。叶长 35 厘米，宽 14 厘米，绿色，叶柄微弯曲，长 65 厘米，直径 3.5 厘米。球茎扁圆形，高 4 厘米，横径 5 厘米，皮和肉均白色，单个重 50~70 克。

特　　性 | 生长期 110~120 天。耐贮运。肉质较坚实，稍有苦味，外形美观，质优。每公顷产量 18 吨左右。

栽培要点 | 株行距 40 厘米 ×40 厘米。收获期 12 月至翌年 3 月。其他栽培管理参照沙姑。

# 菱 Water caltrop

　　菱科菱属一年水生草本植物，学名 *Trapa bispinosa* Roxb.，别名菱角，染色体数 $2n=2x=36$。

　　广州 300 多年前已广泛栽培，现南沙、番禺、增城等区有少量种植。

　　**植物学性状**：根系有吸收根和叶状根 2 种。茎蔓生。叶分水中叶和出水叶。水中叶无柄，互生，又称菊状叶；出水叶具长柄，中部有菱形海绵质气囊，轮生，形成叶盆。花腋生于叶盆，乳白色或淡红色。果实为菱角，内含种子，为食用器官，有二角菱、四角菱和圆角菱，广州的菱为二角菱。

　　**类型**：广州的菱角品种有红菱和大头菱，以红菱为主，为地方品种。

　　**栽培环境与方法**：需栽培于浅水条件下，用果实繁殖，11 月育苗，翌年 2—3 月定植，6—9 月采收。

　　**营养及功效**：菱果肉富含淀粉、葡萄糖、蛋白质、维生素 B、维生素 C、氨基酸及微量元素锌、铁、钙等。味甘，性平，具清暑解热、益气健胃、止消渴、解酒毒、利尿通乳等功效。

　　*Trapa bispinosa* Roxb., an annual aquatic herb. Family: Trapaceae. Synonym: Jesuit's Nut. The chromosome number: $2n=2x=36$.

　　Water caltrop has been cultivated widely more than 300 years in Guangzhou. At present, only a few cultivated area of water caltrop is found in Nansha, Panyu and Zengcheng districts.

　　Botanical characters: It has two types of roots: absorption roots and leaf-shaped roots. Stem is a trailing vine with many lateral branches. Two types of leaves: (1) the Chrysanthemun-shaped leaves submerged in water, slender, alternate and without petioles; (2) aerial leaves, rhombic with long petioles and with a spongy gas bag at the middle part, whorl to form leaf-basin. Axillary flowers borne at the leaf-basin are milk-white or light red. Fruit is 2~4 angled or round, with seed inside. Water caltrop in Guangzhou is a 2 angled fruit.

　　Types: Two cultivars are grown in Guangzhou: Red water caltrop (popularly grown cultivar) and Big-headed water caltrop.

　　Cultivation environment and methods: Water caltrop growing in shallow water. It is propagated by fruit. Seedling dates: November. Planting dates: February to March, harvesting dates: June to September.

　　Nutrition and efficacy: Water caltrop is rich in starches, glucose, proteins, vitamin B, vitamin C, amino acids, zinic, iron, calcium, etc.

## 红菱 Red water caltrop

**品种来源**｜农家品种。

**分布地区**｜南沙、番禺、增城等区。

**特　　征**｜茎蔓生。水面株高 10~15 厘米，开展度 30~35 厘米。叶深绿色，叶长 4.5 厘米，宽 6 厘米，海绵质叶柄长 7~8 厘米。果实为二角菱，纵径 2.5 厘米，横径 6.5 厘米，具果柄，嫩果紫红色，老熟时黑色，果肉白色。

**特　　性**｜定植至初收 130 天。耐热。果肉脆嫩，风味佳，品质优。每公顷产量约 15 吨。

**栽培要点**｜播种期为 11 月。将种用菱角浸泡于 5~6 厘米水中让日晒，每 4~5 天换水 1 次，出芽发根后移到田间繁殖，待茎叶长满田后再分苗繁殖。2—3 月大田定植，截取约 30 厘米长茎段作种苗，株行距 1 米 ×2 米，每公顷植 5 000 株左右。注意施足基肥，适当追肥，前期灌浅水，后期适当加深水位。注意防治螟虫。收获期 6—9 月。后期可选留老熟果实作为翌年用种子。

# 荸荠 Chinese water chestnut

沙草科荸荠属多年生水生草本植物，作一年生栽培，学名 Eleocharis tuberosa (Roxb.) Roem. et Schult.，别名马蹄，染色体数 $2n=2x=22$。

广州栽培历史悠久，200多年前已有记述，除蔬食和作水果外，在白云区有企业制作马蹄粉、罐头，为名牌产品，出口到国外。

**植物学性状**：叶退化，以地上叶状茎进行光合作用。叶状茎绿色，丛生直立，管状，中空，内有隔膜。地下分生多数匍匐茎，先端积累养分膨大形成球茎。球茎扁圆形，皮红褐色或黑色，肉白色，为食用器官。穗状花序，花小。果实近圆形，灰褐色，在广州不能开花结果。

**栽培品种**：主要有水马蹄和桂林马蹄两个品种。

**栽培环境与方法**：宜栽培于浅水条件下，采用球茎繁殖，6—7月育苗，7—8月定植，12月至翌年1月收获。

**营养及功效**：富含蛋白质、脂肪、粗纤维、胡萝卜素、维生素B、维生素C、铁、钙、磷和碳水化合物。味甘，性寒，具清热泻火、利湿化痰、凉血解毒、利尿通便、消食除胀等功效。

*Eleocharis tuberosa* (Roxb.) Roem. et Schult., a perennial aquatic vegetable, grown as an annual. Family: Cyperaceae. Synonym: Matai, Waternut. The chromosome number: $2n=2x=22$.

Chinese water chestnut has long been cultivated in Guangzhou. It was mentioned in ancient literatures 200 years ago. It is used as fresh vegetable and fruit. And Chinese water chestnut powder and canned food are produced in Baiyun for export.

**Botanical characters**: The leaves are degenerate. The clustered leaf-shaped stems fulfilled the function of photosynthesis. The above-ground leaf-shaped stem is septum, erect, round, hollow and green. The numerous stolons due to nutrient accumulation at the terminal part become thick to form corms (the edible part), which are flat-round, skin brown-red or black with flesh white. Inflorescence is terminal spikes with small flowers. Fruits are nearly round and brown grey. Neither flowers nor fruits exist in Guangzhou.

**Cultivars**: There are 2 cultivars: the Chinese water chestnut and the Guilin Chinese water chestnut.

**Cultivation environment and methods**: Chinese water chestnut grows in shallow water. It is propagated by corms. Protection should be taken against summer heat in June and July for the care of seedlings, which is to be planted during July to August. Harvesting dates are from December to January the following year.

**Nutrition and efficacy**: Chinese water chestnut is rich in proteins, fats, coarse fibers, carotenoid, vitamin B, vitamin C, iron, calcium, phosphorus and carbohydrates, etc.

## 水马蹄

Chinese water chestnut

> 品种来源

农家品种。

📍 分布地区 | 南沙、番禺、白云、增城等区。

特　　征 | 株高70~90厘米，开展度15~20厘米。叶状茎长而细，长90厘米，横径0.3厘米，绿色。球茎扁圆形，高2.0厘米，横径2.5~3.0厘米，皮黑褐色，肉白色，顶芽较尖长，单个重10克。

特　　性 | 生长期130~140天，生长势旺盛，抗逆性强，较耐贫瘠，耐淹。不耐贮藏，淀粉含量高，多作制淀粉用，出粉率高达16%。每公顷产量23吨。

栽培要点 | 6—7月于阴棚或阴凉处用球茎育苗，苗期25天，定植株行距（70~100）厘米×（70~100）厘米。栽培须灌浅水，注意施足基肥，追肥2次，10月下旬前施完。注意防治鼠害。收获期11—12月。

## 桂林马蹄

Guilin Chinese water chestnut

> 品种来源

地方品种。

📍 分布地区 | 各区均有栽培。

特　　征 | 株高100~120厘米，开展度20厘米。叶状茎长而稍大，长110厘米，横径0.5厘米，绿色。球茎扁圆形，高2.4厘米，横径3~4厘米，皮黑褐色，肉白色，顶芽较粗壮，单个重30克。

特　　性 | 生长期130~140天，生长势壮旺，抗逆性强。耐贮藏，糖分较高，肉质爽脆，品质优，可供鲜食、蔬食或加工用。产量高，每公顷产量25~30吨。

栽培要点 | 定植株行距（80~110）厘米×（80~110）厘米。其余参照水马蹄。

# 豆瓣菜 Water cress

十字花科豆瓣菜属二年水生草本植物，作一年生栽培，学名 *Nasturtium officinale* R. Br.，别名西洋菜，染色体数 $2n=2x=32$，34，36，48，60。

广州已有近百年的栽培历史，现广州各区均有种植，多利用城镇附近低洼田或水田种植，是秋植冬春收获的主要叶菜之一。

**植物学性状**：根系浅，易生不定根。茎近圆形，节较短，分枝多，绿色，匍匐生长，节上生不定根。奇数羽状复叶，顶生小叶较大，小叶 1~4 对，卵圆形或近圆形，深绿色。总状花序，花小，花冠白色。种子扁圆形，细小，褐色。

**类型**：分大叶西洋菜和小叶西洋菜两类。

**栽培环境与方法**：广州豆瓣菜品种不开花结籽，用老茎越夏留种，嫩茎繁殖，或买商品种子育苗。宜栽培于浅水条件下，8—9月种植，9月至翌年4月收获。

**营养及功效**：豆瓣菜的营养物质比较全面，其中超氧化物歧化酶（SOD）的含量很高。味甘、微苦，性寒，具清燥润肺、化痰止咳、利尿等功效。

*Nasturtium officinale* R. Br., a biennial aquatic herb, and grown as an annual. Family: Cruciferae. The chromosome number: $2n=2x=32$, 34, 36, 48 or 60.

Water cress has been cultivated nearly 100 years in Guangzhou and is grown in each district now. It is usually cultivated in the low lands in suburbs of towns and cities. It is one of the principal leaf vegetables which are sown in autumn and harvested in spring and winter.

Botanical characters: Root system is shallow. It produces adventitious roots easily. Stem green, prostrate, nearly-round, with shortened nodes and branching profusely. Terminal odd-pinnately compound leaves, with 1~4 pairs of leaflets, large terminal leaflet ovate or nearly-round and dark green in color. Inflorescence, raceme, small and white flowers. Seeds, flat round, small and brown in color. Young stems and leaves are used as edible parts.

Types: Water cress is divided into two types: large leaf and small leaf water cress.

Cultivation environment and methods: Water cress cultivars in Guangzhou usually do not blossom. The old plants are reserved in summer, then the cuttings of new young stem are used for propagation. Or seedling by seed. Water cress grows in shallow water. Transplanting dates: August to September. Harvesting dates: September to April the following year.

Nutrition and efficacy: Water cress is a comprehensive nutritional vegetable.

# 大叶西洋菜
Large leaf water cress

**品种来源**

地方品种。

**分布地区** | 全市各区。

**特　　征** | 植株匍匐生长，高 40~50 厘米，植株粗大，茎粗 0.7 厘米，易生侧茎。小叶卵形，长 1.7 厘米，宽 1.5 厘米，绿色。叶片大，小叶圆形，全缘，绿色，辛香味较淡，纤维稍多，产量高。

**特　　性** | 定植至初收 20~30 天。适应性强，易生不定根，分枝多，缺肥或遇低温生长缓慢，茎叶变紫红色。在广州地区不开花结果。

**栽培要点** | 8—9 月用嫩茎繁殖。水田整地施基肥，行距 13 厘米，将苗排放在水田中，灌浅水。定植后 30 余天收割。收割后随即翻田、耙平、施基肥再行定植，以后 20 天左右收割一次，每次每公顷可收割 15 吨。每次收割时，留栽植面积 1/4 的植株作繁殖用。收获期 9 月至翌年 4 月，每公顷产量 60~75 吨。

# 小叶西洋菜
Small leaf water cress

**品种来源**

地方品种。

**分布地区** | 全市各区。

**特　　征** | 植株匍匐生长，高 30~40 厘米，茎粗 0.7 厘米，易生侧茎。小叶卵形，长 1 厘米，宽 0.8 厘米，绿色。

**特　　性** | 定植至初收 20~30 天。适应性强，易生不定根，分枝多，缺肥或遇低温生长缓慢，茎叶变紫红色。在广州地区不开花结果。

**栽培要点** | 参照大叶西洋菜。

# 水芹 Water dropwort

伞形科水芹属多年水生宿根草本植物，学名 *Oenanthe javanica* (Bl.) DC.，别名野芹菜、楚葵，染色体数 $2n=2x=22$。

广州已有70多年的栽培历史，现白云、从化等区有零星种植。

**植物学性状**：株高40~50厘米。地上茎和匍匐茎的节上发生须根。茎中空，有小棱沟，浅绿色，老茎变浅紫色，茎上侧芽萌发形成新株。叶互生，奇数二回羽状复叶，小叶椭圆形，边缘锯齿状，深绿色，长3~4厘米，宽2厘米。叶柄细长，长7~8厘米，基部鞘状包住茎节。复伞形花序，开白色小花。

**栽培环境与方法**：宜栽培于浅水条件下，9—12月采用分枝繁殖，当种苗高度达10厘米时，即可移植大田，株行距10厘米×15厘米。前期灌浅水，后期适当加深水位。注意防治蚜虫。10月至翌年4月采收。

**营养及功效**：富含多种维生素、矿物质、蛋白质、脂肪、碳水化合物、粗纤维等。味甘、辛，性凉，具清热解毒、养精益气、宣肺利湿等功效。

*Oenanthe javanica* (Bl.) DC., a perennial aquatic herb. Family: Umbelliferae. The chromosome number: $2n=2x=22$.

Water dropwort has been cultivated more than 70 years in Guangzhou. It is grown sparely in Baiyun and Conghua districts.

Botanical characters: The plant height is about 40~50 cm. The fibrous roots were borne on the nodes of stem and stolon. Stem, ribbed, hollow, light green in color, the elder stem is lavender. A lateral bud on the stem can develop into a new plant. Leaf alternate, odd-bipinnate, with leaflet ovate or oval in shape, margin serrate, dark green, 3~4 cm long and 2 cm wide. Petiole, slender, 7~8 cm long, the basal enclosed the stem with its sheath-shaped involving the basal part. Inflorescence, compound umbelliferous, with small and white flowers. Young leaf and stem are used as cooked vegetable.

Cultivation environment and methods: Water dropwort grows in shallow water. It is propagated by division from September to December. The seedlings are transplanted when its height reaches 10 cm, with the spacing 10 cm×15 cm. Shallow water is kept at early growth stages and a litter deep water is kept at later stages. The harvesting dates: October to April the following year.

Nutrition and efficacy: Water dropwort is rich in vitamins, minerals, proteins, fats, carbohydrates, and crude fibers.

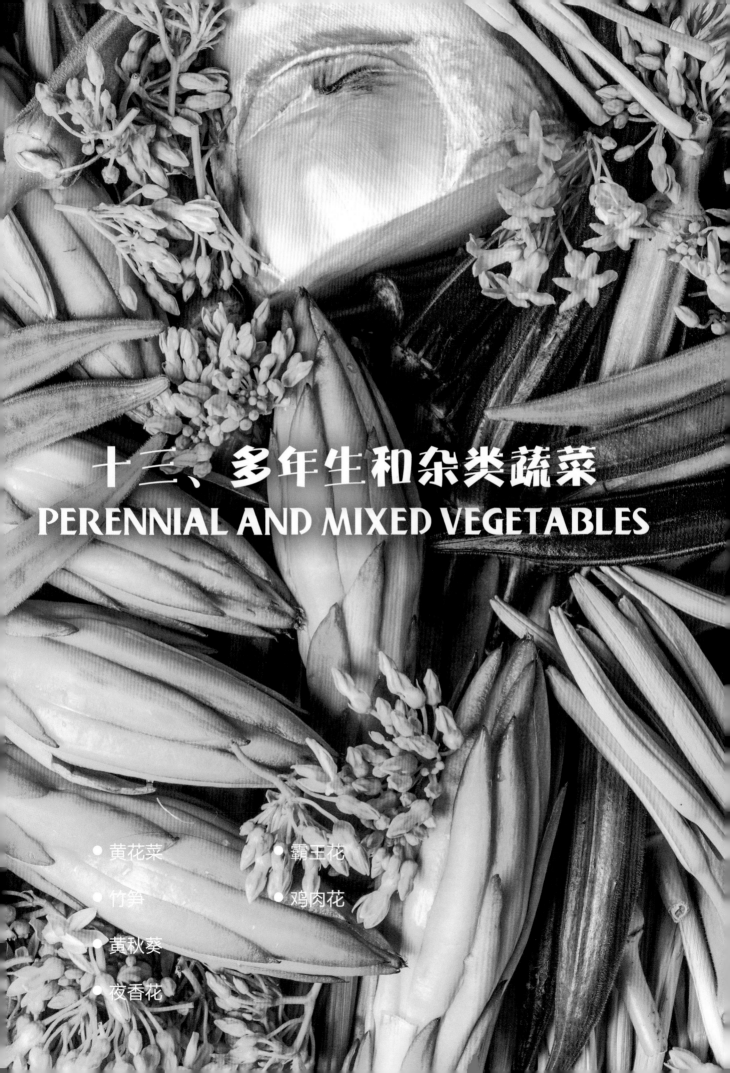

# 十三、多年生和杂类蔬菜
# PERENNIAL AND MIXED VEGETABLES

- 黄花菜
- 竹笋
- 黄秋葵
- 夜香花
- 霸王花
- 鸡肉花

# 黄花菜 Day lily

百合科萱草属多年生宿根草本植物，学名 *Hemerocallis citrina* Baroni，别名金针菜、萱草，染色体数 $2n=2x=22$。

广州栽培历史悠久，多为零星栽培。

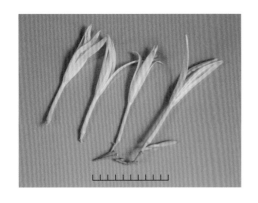

**植物学性状**：短缩茎上发生肉质根和纤细根。叶线形，深绿色，叶鞘抱合成扁阔的假茎，每一假茎及其叶丛成一分蘖。伞形花序，有花 10~30 朵，黄色或红黄色。果为蒴果，成熟后呈暗褐色，每果有种子数粒至 20 多粒，种子成熟后黑色、坚硬。

**栽培环境与方法**：地上部分不耐寒，遇霜即枯死。短缩茎和根在严寒地区的土中能安全过冬。叶丛生长适温为 14~20℃，花梗抽出和开花期间适温 20~25℃。可采用分株、组织培养及种子等方法繁殖。广州多采用分株繁殖，多在春、夏、秋季进行，栽植深度为 10~13 厘米。

**收获**：采收期为 8—11 月，一般采收新鲜花蕾食用，也可蒸制、干燥。干燥后的花蕾称"金针菜"。

**病虫害**：病害主要有叶斑病、锈病、叶枯病和白绢病，虫害主要有蚜虫、红蜘蛛等。

**营养及功效**：富含多种维生素、胡萝卜素等。味甘，性凉，具除湿、宽胸、利尿、止血、通络、下乳等功效。鲜花蕾含有毒物质秋水仙碱，作鲜菜用前应焯水后才能食用。

*Hemerocallis citrina* Baroni, a perennial herb. Family: Liliaceae. Synonym: Tawny day-lily. The chromosome number: $2n=2x=22$.

Day lily has long been cultivated in Guangzhou. It is usually grown sparely.

**Botanical characters**: Stem dwarf. Leaves opposite, linear, dark green. The leaf sheaths braced to form flat broad pesudostem, which with its rosette of leaves become a tiller. Inflorescence: umbelliferous, 10~30 flowers, yellow or red yellow. The mature capsule dark brown, which contains sevwrval to 20 seeds. The mature seeds is black and hard.

**Cultivation environment and methods**: The above ground part of day lily is not tolerant to cold, which will die in frost. The dwarf stem and root in the soil is tolerant to very low temperature. The optimum temperature: 14~20℃ for leaf growth, 20~25℃ for blossoming. The day lily can be propagated by division, tissue culture and seed. Division is mainly used in Guangzhou in spring, summer and autumn. The divisions are planted 10~13 cm in depth.

**Harvest**: The flower buds are edible. The harvest date is from August to November. Flower buds can be steamed or dried.

**Nutrition and efficacy**: Day lily is rich in vitamins, carotene, etc.

# 竹笋

禾本科竹亚科多年生常绿植物，原产中国，广州栽培历史悠久，公元1世纪已有著录。

**植物学性状**：竹秆（地上茎）和竹鞭（地下茎）有4种生长类型：合轴丛生、合轴散生、单轴散生和复轴混生。竹秆由秆柄、秆基和秆茎组成。秆基和节易生须根，节上长出竹枝。叶分竹秆上的秆箨与竹枝上的普通叶。箨叶退化，无明显中脉；普通叶通常条形或矩圆状披针形，具短柄，不常开花。竹笋是植株地下部茎短缩的芽。竹笋的纵切面中部有许多紧密重叠的横隔，相当于竹秆的节隔，两隔之间就是竹秆的节间；包裹在横隔周围的是肥厚的笋肉，相当于竹秆的竿壁；包裹在笋肉外周的笋箨是一种变态叶。笋肉、横隔及笋箨的柔嫩部分可供食用。

**类型**：广州地区笋用竹主要有毛竹和麻竹。

1. 毛竹（*Phyllostachys pubescens* Mazel）又叫楠竹、茅竹、江南竹，属散生型，竹秆高10~15米，横径10~15厘米，竹壁厚，质地坚硬，用途广。

2. 麻竹（*Dendrocalamus latiflorus* Munro）又叫甜竹、大叶乌竹，属丛生型，竹秆高20~25米，横径10~20厘米。

**栽培环境与方法**：喜温暖湿润的气候，在土层深厚、土壤呈微酸性或中性的溪河岸畔、河滩以及在海拔500米以下的丘陵山脚均可生长。一般要求年平均温度≥17.5℃，极端低温＞-6℃，年降水量在1 400毫米以上。在冬季霜冻少、低温时间短的条件下，方可越冬。

**收获**：芽未露土或刚露土时即采收。

**病虫害**：病害主要有竹锈病、枯梢病、水枯病，虫害主要有竹螟、竹蝗、竹斑蛾、笋夜蛾、笋蝇、竹象虫。

**营养及功效**：含丰富的蛋白质、氨基酸、脂肪、糖类、钙、磷、铁、胡萝卜素、维生素$B_1$、维生素$B_2$、维生素C等。味甘，性微寒，具清热消痰、利膈爽胃、消渴益气等功效。

# Bamboo shoot

## 毛竹

Moso bamboo

**品种来源**

地方品种。

📍 分布地区 | 全市各区。

Family: Bambuaoideae. Bamboo shoot, an evergreen perennial plant. It is native to China. Bamboo shoot has long been cultivated in Guangzhou, mentioned in books during the first century.

Botanical characters: The bamboo clum (aerial stem) and rhizome (subterranean stem) can be classified into 4 types of growth habits: sympodial growth in clusters, separate sympodial growth, separate monopodial growth and mixed growth habit. Bamboo culm is composed of 3 parts: the stalk, the base and the stem. Fibrous root easily bome from the clum base and the node, branchs develop at the node. Leaf include the culm-sheath on the clum and ordinary leaf on the branch. The degenerated blade (clum-sheath) is without any obvious midrib. While the leaf is usually linear or rectangular-round lanceolate in shape, with short petiole. A bud emerges from the rhizome, which is called bamboo shoot and is the embryonic form of the clum. There are many intense tabular at the cross section of bamboo shoot, which will develop into inter-node of culm. The tabulars are encompassed by succulent flesh, which will develop into clum wall. Tender flesh, tabular and culm-sheath are edible.

Types: Moso bamboo and dendrocalamus in Guangzhou.

1. Moso bamboo (*Phyllostachys pubescens* Mazel): Rhizome monopodial. It is 10~15 m in height and 10~15 cm in diameter. The bamboo pole is thick and hard.

2. Dendrocalamus (*Dendrocalamus latiflorus* Munro): Clum grow in clusters. It is 20~25 m in height and 10~20 cm in diameter.

Cultivation environment and methods: Bamboo prefers warm and humid climate, it can grow in deep acidic soil at bank and beach of river and the foot of hill below 500 m at altitude. The annual average temperature is required above 17.5 ℃, and extreme low temperature is above -6 ℃. The annual rainfall is required more than 1400 mm.

Harvest: Bamboo shoot should be harvested as it emerge from the mound.

Nutrition and efficacy: The bamboo shoot is rich in proteins, amino acids, fats, carbohydrates, calcium, phosphorus, iron, carotene, vitamin $B_1$, vitamin $B_2$, vitamin C, etc.

**特　　征** | 散生。幼竹被茸毛，节隆起，被毡毛。竹秆端直，鞘部微弯曲，高10~20米，横径8~16厘米，基部节间短，长5厘米，中上部节间长25~40厘米。叶片披针形，长5~10厘米，宽1~2厘米，无毛。笋锥形，长20~25厘米，微弯。冬笋笋壳淡黄色，被浅棕色毛，笋体略呈纺锤形，单个重250~750克；春笋（毛笋）圆锥形，笋壳底色淡黄，有淡紫褐色小斑点，密被淡棕色毛，单个重1~2千克。

**特　　性** | 出笋早，采笋年限长。耐旱。冬笋一般指从农历冬至到翌年春分前埋在地下的笋。冬笋质嫩，脆甜，品质优。春笋在3—4月挖取，春笋有苦味，刚露尖时采收的品质最好。管理精细的竹林，每公顷年产春笋11~15吨。大小年竹林，大年每公顷产春笋约15吨，小年产2~3.8吨。

**栽培要点** | 从冬季到早春，除严寒天气外都可移植。采用整株定植的，每667米²种植20~25株，种后3~4年可采笋。找冬笋，应先掌握壮龄竹鞭的位置和它的伸展规律，新竹绝大多数是从壮龄竹鞭发生的，距新竹竹秆基部弯曲的凹面15~20厘米，其土下就是竹鞭所在处。新竹最下一盘枝叶伸展的方向与竹鞭的走向大体一致。还可根据立竹的形态、色泽来判断冬笋位置，凡枝叶茂盛、叶色浓绿的竹子，光合功能强，与它直接相连的竹鞭养分充分，形成笋芽多，从而冬笋比较多。根据上述规律，找到土表泥块松动或裂缝处挖掘，就可得到冬笋。

# 麻竹

Broad flower dendrocalamus

> 品种来源

地方品种。

📍 分布地区｜全市各区。

特　　征｜地下茎合轴型，丛生。竹秆稍尖，柔软下垂，高 15~20 米，横径 8~18 厘米。节间长 25~40 厘米，新秆绿色，无毛，微被白粉，基部 4~6 节有明显气根或根芽。叶长椭圆状宽披针形，长 15~35 厘米，宽 4~8 厘米，无毛。笋壳黄绿色，被暗紫色毛。笋体圆锥形，长约 25 厘米，基部直径约 12 厘米，单个重 1 千克左右，大的重 3~4 千克。

特　　性｜出笋早，采笋年限长，产量高。出笋期 5—10 月，以 7—8 月最盛。笋肉质较粗，品质一般，主要制笋干和罐头。每公顷年产笋 30~40 吨，高者可达 105 吨。

栽培要点｜清明前后种植。采用整株定植的，每 667 米² 种植 25~30 株，定植第一年和第二年每丛共留健壮新竹 7~8 株，第三年只割笋不留竹，从第四年起每丛每年选留 3~4 株新竹，其余竹笋都可采收，于当年冬季砍除 3~4 株老竹，使每个竹丛经常保持母竹 7~8 株。每年需要对竹丛进行抚育和施肥，每丛施肥量以有机复合肥（N、$K_2O$、$P_2O_5$ ≥ 45%）10~12 千克为宜。笋体由着生在母竹秆基部的大芽发展而成，一株母竹可抽生 5~6 条竹笋，一般仅早抽生的 1~2 条笋能发展成竹。对萌发的笋芽结合施肥进行培土，有利于提高产量和保持笋味。

# 吊丝单

Vario-striata dendrocalamopsis

> 品种来源

地方品种。

📍 分布地区｜全市各区稀有种植。

特　　征｜丛生。竹秆高 5~12 米，顶端弯垂；节间圆柱形或微有膨大，长 30~35 厘米，横径 4~6 厘米。枝簇生，主枝粗大。叶披针形，长 13~25 厘米，宽 1.6~3.0 厘米；笋箨青黄色，被茸毛，箨片直立，三角状披针形，箨耳发达，近相等，长椭圆形。鲜笋单个重 0.5~1.5 千克。

特　　性｜种植后 2~3 年可采笋，采笋年限长，耐旱性较强，产量稳定。笋期 6—10 月。肉质嫩滑，品质优。每公顷产量 15~30 吨。

栽培要点｜清明前后分株繁殖，选 1~2 年生嫩竹连根挖起，截取 1.5 米栽植。竹株行距约 5 米 ×5 米。或 6—9 月选 1~2 年生粗壮、芽饱满、有根点的枝进行育苗，翌年清明前后移苗定植。春天培土施肥，冬季砍伐 3 年以上的老竹和扒土露晒竹头。注意防治竹锈病、枯梢病、水枯病、竹螟、竹蝗、竹斑蛾、笋夜蛾、笋蝇、竹象虫等病虫害。

# 黄秋葵

锦葵科秋葵属一年生草本植物，学名 *Abelmoschus esculentus* (L.) Moench，别名秋葵、羊角豆、补肾果，染色体数 $2n=4x=29$，36。

黄秋葵原产非洲，约 2 000 年前在埃及栽培，李时珍著的《本草纲目》中有记载。目前广州黄秋葵年栽培面积约 200 公顷。

**植物学性状**：直根系，主根发达。茎直立，木质化，圆柱形，高 1~2 米，有分枝。绿色或赤红色，被粗毛。叶互生，有茸毛，叶柄细长，掌状 3~5 裂，裂片披针形至三角形，基部心形。两性花，花萼、花瓣各 5 枚，花瓣上部黄色，基部暗紫色或紫红色，被倒硬毛，单生于叶腋间；雄蕊柱长约 2.5 厘米，平滑无毛；花柱分枝 5，柱头盘状，花常在上午 8：00—9：00 开放，下午凋萎。果为蒴果，圆锥形或羊角形，表面有茸毛，果嫩时绿色或红色，成熟后黄褐色或红色，自然开裂，有种子 50~60 粒或更多。种子近圆球形，皮灰绿色，发芽期限 3~5 年，千粒重 55 克左右。

**类型**：按果皮的颜色分，有绿色和红色两种类型；按果实形状分，有圆锥形果及棱角果两种类型。绿色品种主要有粤海黄秋葵、五福黄秋葵等；红色品种主要有红箭秋葵等。

**栽培环境与方法**：喜温暖，耐热，不耐寒；耐旱，但不耐涝；对光照条件尤为敏感，喜强光，要求光照充足。8℃以下停止生长。种子发芽和生育期适温为 25~30℃。月均温低于 17℃，则影响开花结果；夜温低于 14℃，则生长缓慢，开花少，落花多。直播和育苗移栽均可。广州地区可在 3—8 月种植。生长期间要保持充足的肥水供应，尤其是结果期不能缺水。注意防治蚜虫为害。

**收获**：一般在开花后 4~6 天、嫩果长 7~12 厘米时即可采收。

**病虫害**：病害主要有病毒病、疫病、黑斑病等，虫害主要有蚜虫、斑潜蝇、甜菜夜蛾、红蜘蛛等。

**营养及功效**：嫩荚含有丰富的果胶、膳食纤维、维生素 A、维生素 C、维生素 E、胡萝卜素、黄酮等。味甘，性寒，可助消化，治疗胃炎、胃溃疡，并可保护肝脏及增强人体耐力。

# Okra

*Abelmoschus esculentus* (L.) Moench, an annual herb. Family: Malvaceae. Synonym: Gumbo, Lady's finger, Quiabo. The chromosome number: $2n=4x=29, 36$.

Okra is native to Africa and was cultivated in Egypt 2000 years ago. At present, the cultivation area is about 200 $hm^2$ per annum in Guangzhou.

Botanical characters: Vigorous taproot system. Stem woody, cylinder in shape, 1~2 m in height, branching, green or purple-red in color, with thick hairs. Leaves alternate, palmate 3~5 splits, with hair and long thin petiole. Lobes, lanceolate to triangular with heart-shaped base. Bisexual flowers, 5 sepals and 5 petals. Yellow petals with a large purple area covering the base, covered with obverse bristles. Stamen is 2.5 cm in height, smooth and glabrous, with 5 discoid stigmas. The flowers blossom at 8—9 o'clock. Fruit, capsule, conical or horn in shape. The young fruit is light green fruit and mature fruit is brown or red in color, which contains 50~60 or more seeds. The seeds are round and grayish-green. The weight per 1000 seeds is about 55 g.

Types: According to okra fruit color, there are green and red types. According to fruit shape, there are conic fruit and corniform fruit types. The green fruit cultivars include Yuehai okra, Wufu okra, etc. The red fruit cultivars include Hongjian okra, etc.

Cultivation environment and methods: Okra prefers warm and sunny weather, and is tolerant to hot, drought, while not tolerant to cold and water logging. The growth stagnates below 8℃. Appropriate temperature for seed germination and growth is 25~30℃. Blossoming and fruiting is affected when the monthly mean temperature is below 17℃. Okra grows slowly and blossom slightly when nighttime temperature is below 14℃. Direct seedling or seedling transplanting. Planting dates: March to August.

Harvest: The 7~12 cm young fruits are harvested 4~6 days after anthesis.

Nutrition and efficacy: Okra pods are rich in pectins, dietary fibers, vitamin A, vitamin C, vitamin E, carotenoid, flavonoid, etc.

## 粤海

Yuehai okra

**品种来源**

广州市农业科学研究院育成，2012 年通过广东省农作物品种审定。

**分布地区** | 部分区有栽培。

特　　征 | 株型中等，株高 1.4~1.6 米。主茎绿色，略带紫红色。叶形为深裂缺刻掌状 5 裂，叶缘有锯齿，叶长 35.1 厘米，叶宽 40.2 厘米。花黄色，瓣基褐红色，花萼、花瓣各 5 枚，始花节位 6.0 节。主茎结果为主。果形羊角形，具 5 棱，果皮绿色，果柄略带紫红色，果面茸毛少，果长 11.0~12.0 厘米，横径 2.0 厘米左右，单果重 22.0 克左右。

特　　性 | 生长势强。播种至初收春季 60 天、秋季 56 天，延续采收春季 90 天、秋季 70 天。开花后 5~6 天为最佳采收期，品质优。田间表现较耐病毒病，耐热，耐寒性中等。每 667 米$^2$产量 1 500~2 000 千克。

栽培要点 | 适宜春、夏、秋季种植。播种期为 2—9 月，最适播种期为 3—4 月和 7—8 月。培育壮苗，一般按 1.7 米包沟起高畦，双行植，株距 0.4~0.5 米，行距 0.75 米，每 667 米$^2$种植 2 000 株左右。合理施肥，结果期增施磷、钾肥，不要偏施氮肥。

## 五福

Wufu okra

**品种来源**

引自台湾。

📍 分布地区｜部分区有少量栽培。

特　　征｜株型中等，株高 1.4~1.6 米。主茎绿色。叶形为深裂缺刻掌状 5 裂，叶缘有锯齿，叶片绿色，叶长 36.2 厘米左右，叶宽 43.2 厘米左右。花黄色，瓣基褐红色，花萼、花瓣各 5 枚，始花节位 7.5 节。主枝结果为主。果形羊角形，具 5~8 棱，果皮绿色，果柄略带紫红色，果面茸毛多，果长 11.0~15.0 厘米，横径 1.9 厘米左右，单果重 21.0 克左右。

特　　性｜生长势强。播种至初收春季 63 天、秋季 60 天，延续采收春季 91 天、秋季 64 天。开花后 5~6 天为最佳采收期，品质优。田间表现较耐病毒病，耐热，耐寒性中等。每 667 米$^2$产量 1 500~1 800 千克。

栽培要点｜参照粤海黄秋葵。

## 红箭

Hongjian okra

**品种来源**

引自武汉。

📍 分布地区｜部分区有少量栽培。

特　　征｜株型中等，株高 1.8~2.0 米。主茎绿色。叶形为深裂缺刻掌状 5 裂，叶缘有锯齿，叶片绿紫色，叶脉紫色，叶长 37.9 厘米左右，叶宽 44.8 厘米左右。花黄色，瓣基褐红色，花萼、花瓣各 5 枚，始花节位 8.9 节。主侧枝均可结果。果形羊角形，具 5~6 棱，或圆形，果皮红色，果柄紫红色，果面茸毛多，果长 10.0~13.0 厘米，横径 2.1 厘米左右，单果重 20.0 克左右。

特　　性｜生长势强。播种至初收春季 65 天、秋季 60 天，延续采收春季 90 天、秋季 60 天。开花后 4~6 天为最佳采收期，品质优。田间表现较耐病毒病，耐热，耐寒性中等。每 667 米$^2$产量 1 500~2 000 千克。

栽培要点｜参照粤海黄秋葵。

# 夜香花　Chinese violet

萝摩科夜来香属多年生藤状缠绕草本植物，学名 *Telosma cordata* (Burm. f.) Merr.，别名夜来香、夜兰香、夜香藤等。

广州从 20 世纪 80 年代中期开始，在增城作为商品蔬菜进行生产。

**植物学性状**：根系较强大，侧根多。茎蔓生，圆形，嫩茎浅绿色，被柔毛，老茎灰褐色，分枝力强，左旋性缠绕生长，无卷须。单叶对生，长卵圆形至宽卵形，叶柄长 1.5~5 厘米。聚伞花序，腋生，总花梗短，被毛，花浅黄绿色。花萼 5 裂，雄蕊 5 枚，生于花冠基部，雌蕊由 2 个离生心皮组成，柱头头状或圆锥状。果实为蓇葖果，披针状圆柱形，顶部渐尖。种子宽卵形，顶端具白色绢质种毛。

**栽培环境与方法**：喜温暖、阳光充足、通风良好、土地疏松肥沃的环境，耐旱，耐瘠，不耐涝，不耐寒。可采用扦插、压条、分株或种子育苗等方法繁殖。春夏季栽培，株行距（50~70）厘米 ×（100~150）厘米。幼苗长至 1 米时搭架引蔓。引蔓时每棵留主蔓 2~3 条，离地 50 厘米以上的分枝可留作侧蔓。每年 3 月上旬清园后，重施一次促梢肥，每 667 米$^2$ 用 25 千克进口复合肥混合农家肥，施后培土，以后每隔 20~30 天每 667 米$^2$ 淋施复合肥 15 千克或腐熟的农家肥 500 千克，切勿偏施氮肥。

**收获**：主要采食鲜花和花蕾。

**病虫害**：病害主要有枯萎病，虫害主要有螨类和介壳类。

**营养及功效**：以新鲜的花和花蕾供食用的一种半野生蔬菜。夜香花气味芳香，具清肝明目保健功效。

*Telosma cordata* (Burm. f.) Merr., a perennial lianoid herb. Family: Asclepiadaceae. Synonym: Tonkin jasmine, Pakalana vine, Tonkinese creeper.

Chinese violet was native to China, and was cultivated as vegetable in Zengcheng district from the mid 1980s.

Botanical characters: The vigorous root system with many branches. Stems vine, round, many branched. Young vine with hair, light green while elder vine grayish brown in color. Leaves opposite, long oval to broadly ovate. The petioles are 1.5~5 cm in length. Inflorescence: cyme, axillary, yellow green color. 5 lobed calyxes, 5 stamens. Pistil is composed of 2 separate carpels. Stigma is head-shaped or conical. Individual blooms emerge successively over a period of weeks emitting a rich, heavy fragrance during the day and night. Fruit follicle, lanceolate. Seeds broadly ovate with white silky hairs on apex.

Cultivation environment and methods: Chinese violet prefers to warm, sunny weather and well ventilated, loose and fertile soil. It is tolerant to drought while not tolerant to water logging and cold. Propagated by cutting, mound layer, plant division and seeds. Planting in spring and summer. The planting spacing is (50~70) cm × (100~150) cm. Vine support is required when plants are about 1 m in length. 375 kg per ha compound fertilizer mixed with manure are applied in March. 225 kg compound fertilizer or 7500 kg manure per ha are applied in every 20~30 days.

Harvest: Fresh flowers and flower bud are harvested as vegetable.

Nutrition and efficacy: The flowers are very fragrant and yield perfumed oil. They are tasty, rich in carbohydrate, protein, vitamin A, etc.

# 霸王花

仙人掌科量天尺属多年生肉质攀援草本植物，学名 *Hylocereus undatus* (Haw.) Britt. et Rose，别名剑花、霸王鞭、量天尺、风雨花、假昙花等。

霸王花原产墨西哥、巴西一带，我国主要分布在广东、广西，以广州、肇庆、佛山等为主产区，其中广州的黄埔区长洲镇深井村、白云区人和镇方石村、花都区炭步镇骆村等地盛产。

**植物学性状**：植株肉质，直根系。茎部呈三角形，深绿色，棱边呈波纹状，有小尖刺，具气生根。花大，漏斗形，长约30厘米，直径达11厘米；花萼管状，黄绿色或淡红紫色，花瓣白色，多轮排列；雄蕊多数，柱头分枝状，酷似昙花，略有香气。花期5—11月，一年可开7~8期花，一般晚间开放，早上闭合。

**栽培环境与方法**：喜温暖、阳光充足的气候，耐热，耐旱，但不耐寒。生长适温为25~35℃，低于10℃生长受阻，低于5℃就会受冻害。对土质适应性强，即使在石灰岩隙也能扎根生长。一般在春季选择生长一年以上的肉质茎，切成20~30厘米长的小段，斜插入土5~10厘米进行扦插繁殖。株行距10厘米×80厘米。进入第三年生长盛期，应将茎尖下面长出的芽全部摘除，修剪成"T"形，使其顶芽充分发育、伸长。同时，也要将开花很少的老枝剪掉，以利于通风透光，减少养分消耗。

**收获**：一般采收当天晚上要开放的花苞，加工时纵剖2~4刀，但不使之分离，晒至半干时再放入蒸笼蒸十多分钟，然后取出晒干或直接烘干。每年可采收7~8次。每667米$^2$年产干花200~300千克。

**病虫害**：病害主要有炭疽病、根腐病等，虫害主要有蜗牛等。

**营养及功效**：富含苏氨酸、亮氨酸、异亮氨酸、苯丙酸和赖氨酸等多种氨基酸及钙等元素。其味清香，具清心润肺、清暑解热、除痰止咳等功效。鲜花可蒸食，干花多用于煲汤。

# Nightblooming cereus

*Hylocereus undatus* (Haw.) Britt. et Rose, a perennial succulent climbing herbaceous plant. Family: Cactaceae. It is native to Mexico and Brazil. In China, nightblooming cereus is cultivated in Guangdong, Guangxi province. The main planting area is Guangzhou, Zhaoqing, Foshan, etc.

Botanical characters: Dark green triangular stems. Tiny spines on the undulate margin orifices. There are aerial roots on stem. Large flowers are funnel-shaped, 30 cm in height, 11 cm in diameter. Tubular calyxes are yellow green or reddish purple. White petals are wheel arrangement. Numerous stamens with branched stigmas. Blossoming dates: May to November. It can blossom annually 7~8 times. Flowers open in the evening and close in the morning.

Cultivation environment and methods: Nightblooming cereus prefers warm and sunny weather, and is tolerant to heat, drought and scorching sun while not tolerant to cold. Optimum temperature for growth is 25~35 ℃. The growth stagnates below 10 ℃. Cold injury will occur below 5 ℃. Nightblooming cereus has strong adaptability to soil. Cutting propagation. In spring, more than one year-old succulent stems are cutted into 20~30 cm long sections, then the sections are inserted into soil 5~10 cm in planting spacing 10 cm×80 cm. For nightblooming cereus cultivation, high ridge bricks of 1.5 m in height and 0.3 m in width are built. There is apical dominance in nightblooming cereus. 3 year-old plant are pruned into "T" shape, in order to stimulate the development and elongation of apical bud. The old branch with few blossoms should be removed in order to facilitate ventilation and light transmission, reduce consumption of nutrients.

Harvest: Flower which is going to blossom at night should be harvested. Flowers are cut 2~4 times on the longitudinal section, but don't make them split, steaming and drying or drying directly. It can be harvested annually for 7~8 times. The annual yield of dry flower is 3000~4500 kg per ha a year.

Nutrition and efficacy: Nightblooming cereus is rich in threonine, leucine, isoleucine, phenylalanine, lysine, calcium, etc.

# 鸡肉花 Rose of sharon

锦葵科木槿属多年生灌木或小乔木植物，学名 Hibiscus syriacus L.，原称木槿花，别名佛叠花、鸡腿蕾等。

鸡肉花原产中国中部和印度，在中国各地均有分布，华南地区栽培较为常见，是广州地区特菜之一。

**植物学性状**：根系发达。植株高 3~5 米。茎直立，多分枝。叶绿色，三角形或菱状卵形。花大，单生叶腋，单瓣、重瓣或半重瓣，有紫红色、淡紫色、白色、米黄色等，花多色艳。

**类型**：按花瓣来分有单瓣、重瓣和半重瓣等类型，按花朵颜色分有多种类型，广州主要有重瓣白色花和重瓣紫红色花 2 个种类。

**栽培环境与方法**：喜温暖湿润的气候，耐寒，生长适温为 25~30℃。喜阳光，耐旱忌涝，耐瘠。多采用压条、扦插等无性繁殖。在广州地区，四季可育苗，最适育苗期为冬春季枝条开始落叶至枝条萌芽前。选取成熟、健壮枝条，剪成长 15 厘米左右的小段，插于沙床中或压于花盆或种植槽中，淋水保湿，约 30 天后发根出芽再进行定植。

**收获**：早晨采摘新鲜半开放的花朵，品质最佳。加工产品应在晴天采摘，晒干后置于干燥处贮存。鲜食用花朵宜除去花萼。采收期为 5—11 月，每 667 米$^2$ 年产量约 500 千克。

**病虫害**：病害较少，虫害主要有蚜虫、粉虱等。

**营养及功效**：可观赏、食用兼顾。其花微香，味甘，口感滑，具有清热、凉血等功效。

*Hibiscus syriacus* L., a perennial shrub or small tree. Family: Malvaceae. Synonym: Rose mallow, St Joseph's rod.

Rose of Sharon is native to middle China and India. It distributes all over China. The cultivation in south China is more common. It is one of the special vegetable in Guangzhou.

**Botanical characters**: Strong root system. The plant is upright to 3~5 m in height. Stem erect with many branches. Leaves are green, triangular or ovate in shape. Large flowers axillary. The petals are single, polyphyll or semi-polyphyll, purple, lavender, white and beige in color.

**Types**: There are many varieties of rose of Sharon. According to the petal type, they are classified into single, polyphyll and semi-polyphyll type. The main cultivars in Guangzhou include polyphyll type with white flowers and polyphyll type with purple red flowers.

**Cultivation environment and methods**: Rose of Sharon prefers warm, humid and sunny climate. It is tolerant to cold, drought while not to water logging. The suitable temperature for growth is 25~30℃. Layering and cutting are used for propagation. The suitable seedling dates is winter to spring. The robust shoots are cut into 15 cm sections, and inserted into seedbed. Keep moisture, after 30 days, the roots well developed, the seedlings can be transplanted.

**Harvest**: Rose of Sharon blossoms in summer and autumn. It doesn't bear fruit and fall leaf completely in Guangzhou. The flowers are the edible organ. Harvesting dates is May to November. The flowers should be harvested in sunny days for processing, then be dried and stored. The calyx should be removed for fresh edible flowers.

**Nutrition and efficacy**: The flower with light aroma is sweet and smooth. It can be used for making soup, etc.

# 十四、食用菌类
## EDIBLE FUNGI

- 草菇
- 香菇
- 金针菇
- 蘑菇
- 平菇
- 茶树菇
- 鸡腿菇
- 杏鲍菇
- 木耳
- 猴头菇
- 鸡枞菌
- 香魏蘑

# 草菇 Straw mushroom

光柄菇科小苞脚菇属，学名 *Volvariella volvacea* (Bull. ex Fr.) Sing，别名兰花菇、中国蘑菇、稻草菇、美味包脚菇等。

子实体展开前钟形，平展后伞形，初为白色，后转为淡红色至灰褐色，群生或丛生。

原产热带、亚热带，喜湿润炎热气候，菌丝生长适温为 28~39℃，子实体生长适温为 28~32℃，属典型高温食用菌。

广州地区栽培历史超百年，目前栽培品种有黑菇 V23、白菇 V844 等。栽培季节：露地 4—9 月，室内 4—10 月。

草菇富含蛋白质、氨基酸、维生素及矿物质。味甘，性凉，具补脾益气、清热消暑、化痈肿等功效。

## Straw mushroom

*Volvariella volvacea* (Bull. ex Fr.) Sing. Family: Pluteaceae. Synonym: Chinese mushroom, Paddy straw mushroom.

Sporophore, bell-shaped and umbrella-shaped after opening, grows in clusters. Lamella white at the beginning, and change to light red or red brown afterward.

Straw mushroom is a species of edible mushroom cultivated throughout east and southeast Asia, and is most successfully grown in subtropical climates with high annual rainfall. The suitable temperature for mycelium growth is 28~39℃, and for sporophore growth is 28~32℃.

Straw mushroom has been cultivated more than 100 years in Guangzhou. The strains include Heigu V23 and Baigu V844 at present. The open field cultivation date is from April to September. Indoor cultivation date is from April to October.

Straw mushroom is rich in protein, amino acid, vitamin and mineral.

# 香菇 Shiitake

口蘑科香菇属，学名 *Lentinus edodes* (Berk.) Sing，别名冬菇、香蕈、花菇、香菌、香菰等。

菌盖初期半扁球形，后渐平扁，直径4~15厘米，黄褐色至黑褐色；菌肉白色，肥厚，致密，味美；菌柄近圆柱形或稍扁，长3~6厘米，粗0.5~1厘米，常弯曲。秋冬春季，可寄生于栲栎、柿等200多种阔叶树木上，单生至群生。

广州地区20世纪60年代以前，在山区以段木栽培为主；70年代起，多用人工制作木屑为主的营养袋（菇棒）规模生产。目前，广州栽培品种有粤北地方品种及福建、浙江引进的品种，分低温、中温及高温3种类型。一般在9—10月接种，11月至翌年5月采收。

香菇营养丰富，富含18种氨基酸、维生素D。味甘，性平，具治风破血、益胃助食、理小便失禁等功效。

## Shiitake

*Lentinus edodes* (Berk.) Sing. Family: Tricholomataceae. Synonym: Sawtooth oak mushroom, Black forest mushroom, Black mushroom, Golden oak mushroom, Oakwood mushroom.

Pileus, flat semi-ball shaped and becoming umbrella-shaped afterwards, 4~15 cm in diameter, yellow to deep brown in color. Lamella, white and succulent. Stripe conical or flat, 3~6 cm in length, 0.5~1 cm in diameter.

Shiitake grow in groups on the decaying wood of deciduous trees, particularly shii, chestnut, oak, maple, beech, sweetgum, poplar, hornbeam, ironwood, mulberry, and chinquapin. It is natural distribution includes warm and moist climates in southeast Asia.

Shiitake was mainly produced on the logs in the mountainous area before the 1960s, while after the 1970s, vicarious material such as wood scraps was used to produce shiitake. At present, the strains of shiitake in Guangzhou are introduced from northern Guangdong, Fujian and Zhejiang. These can be divided into low, medium and high temperature types due to different temperature requirement for the development of sporophore. Inoculation dates: September to October. Harvesting dates: November to May the following year.

Shiitake is rich in 18 kinds of amino acid, vitamin D, etc.

# 金针菇 Enokitake

口蘑科类火菇属，学名 *Flammulina velutipes* (Curt. ex Fr.) Sing，别名毛柄金钱菇、冬菇、朴菇等。

菌盖半球形至扁平，直径 1.5~3 厘米，白色、黄白至黄褐色，菌柄圆柱形，柄长 2~10 厘米，黄白色、黄色或黄褐色，呈辐射状排列，丛生。耐寒，喜低温。

广州自 20 世纪 70 年代开始引种栽培，从瓶栽发展为袋栽，10 月至翌年 2 月接种，11 月至翌年 5 月采收。

金针菇菌肉细嫩，软滑可口，富含微量元素，营养价值高。

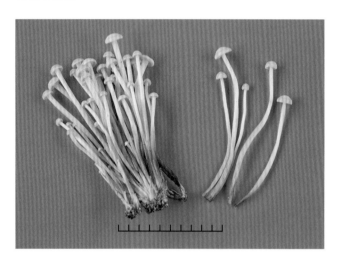

## Enokitake

*Flammulina velutipes* (Curt. ex Fr.) Sing. Family: Tricholomataceae. Synonym: Winter mushrooms, Winter Fungus, Golden needle mushroom, Lily mushroom.

Sporophore grows in cluster. Pileus semi-ball and conical, 1.5~3 cm in diameter, white, yellowish white to yellowish brown in color. Stipe cylindrical, yellowish white, yellow, yellowish brown in color.

Enokitake was introduced to Guangzhou during the 1970s, and cultivated in container and bag. Inoculation dates: October to February the following year. Harvesting dates: November to May the following year.

Enokitake contains antioxidants, like ergothioneine, and is rich in mineral.

# 蘑菇 White mushroom

蘑菇科蘑菇属，学名 *Agaricus bisporus* (Lange) Sing，别名白蘑菇、洋蘑菇、双孢蘑菇等。

菌盖初期扁半球形，后平展，直径 4~15 厘米，白色至淡黄色，不黏，光滑，与菌柄共生，菌柄粗 0.8~3 厘米，长 5~11 厘米，与盖同色。属中低温食用菌。

广州地区于 20 世纪 60 年代开始引种栽培，初期用腐熟堆肥加塘泥进行室内床架栽培，目前发展成营养袋栽培。10—11 月接种生产，12 月至翌年 4 月收获。

蘑菇菌肉肥厚、细嫩，富含多种维生素和氨基酸，有"植物肉"之美誉。

## White mushroom

*Agaricus bisporus* (Lange) Sing. Family: Agaricaceae. Synonym: Common mushroom, Button mushroom, Cultivated mushroom, Table mushroom, and Champignon mushroom.

Pileus flat semi-ball in shape at beginning, and turning to umbrella-shape afterward, 4~15 cm in diameter, white to light yellow in color. Stipe approximately cylindrical, 0.8~3 cm in diameter, 5~11 cm in length, with the same color of the pileus.

White mushroom is cultivated in more than seventy countries, and is one of the most commonly and widely consumed mushrooms in the world. It was introduced to Guangzhou during the 1960s. It is cultivated in bag. Inoculation dates: October to November. Harvesting dates: December to April the following year.

White mushroom is rich in vitamins and amino acids, etc.

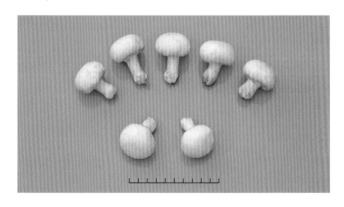

# 平菇 Oyster mushroom

侧耳科侧耳属，学名 *Pleurotus ostreatus* (Jacq. ex Fr.) Quel.，别名北风菌、冻菌、糙皮侧耳等。

子实体大型，扁半球形至平展，直径 4~21 厘米，菌盖白色至灰色，呈覆瓦状丛生。

广州地区以袋栽为主，生产季节为 10 月至翌年 3 月，生长适温为 13~28℃。栽培品种以生产者自主选育为主，如白云区有观兰、黑平等品种。

平菇菌肉白嫩，营养丰富，具鲍鱼风味，是广泛栽培和食用的食用菌。

## Oyster mushroom

*Pleurotus ostreatus* (Jacq. ex Fr.) Quel. Family: Pleurotaceae. Synonym: Oyster shelf, Tree oyster.

Sporophore broad, fan or oyster in shape 4~21 cm in diameter. Pileus white to gray in color, grows in clusters.

Oyster mushroom is widespread in many temperate and subtropical forests throughout the world. Oyster mushroom is mainly cultivated in bag in Guangzhou. Cultivation dates: October to March the following year. The suitable growth temperature is 13~28℃. The strains include Guanlan, Heiping.

The oyster mushroom is best when picked young. As the mushroom ages, the flesh becomes tough and the flavor becomes acrid and unpleasant.

# 茶树菇 Poplar mushroom

粪锈伞科田头菇属，学名 *Agrocybe cylindracea* (DC. ex Fr.) R. Maire，别名柱状田头菇、杨树菇、茶薪菇、油茶菇等。

子实体单生、双生或丛生，菌盖直径 5~10 厘米，光滑，暗红色至黄褐色，菌柄中实，长 4~12 厘米，白色至淡黄色。属中温型食用菌，生长适温为 18~24℃。

广州引种栽培始于 20 世纪 80 年代，可采取覆土栽培或温室床架袋栽，目前多用后者，每年除夏季高温月份外均可生产。

茶树菇既可鲜食，又可制成干品，菇香味浓郁。

## Poplar mushroom

*Agrocybe cylindracea* (DC. ex Fr.) R. Maire. Family: Bolbitiaceae. Synonym: Chestnut mushroom, Velvet pioppino.

Sporophore, single, dual or in cluster. Pileus open and convex, 5~10 cm in diameter, dark red to yellowish brown in color. Stipe, 4~12 cm in length, white to light yellow in color. The suitable temperature for sporophores growth is 18~24℃.

Poplar mushroom was introduced to Guangzhou in the 1980s, and mainly cultivated in bag. The cultivation seasons: spring, autumn and winter.

Poplar mushroom is rich in bioactive secondary metabolites, such as indole derivatives, cylindan and agrocybenine.

## 鸡腿菇 Shaggy mane

鬼伞科鬼伞属，学名 *Coprinus comatus* (Müll. ex Fr.) S. F. Gray，别名毛头鬼伞、鸡腿蘑等。

菌盖圆柱形、桶形或腰鼓形，后成钟形，高6~11厘米，初期洁白，顶部淡红色，后期色渐深，表皮早期光滑，后裂开成平伏的鳞片，与菌柄共生，菌柄基部膨大，向上渐细，长5~20厘米，粗1~2.5厘米。群生或单生。生长适温为20℃以下。

广州地区引种栽培始于20世纪90年代，采用营养袋生产，春、秋、冬季可栽培。

鸡腿菇食味鲜美可口，含17种氨基酸。味甘、滑，性平，具清心安神、益脾胃、助消化等功效。

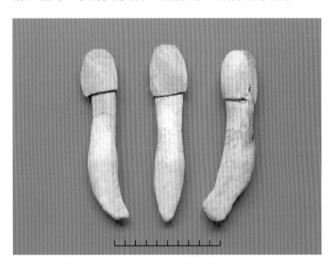

### Shaggy mane

*Coprinus comatus* (Müll. ex Fr.) S. F. Gray. Family: Coprinaceae. Synonym: Shaggy ink cap, Lawyer's wig.

Sporophore, single or in cluster. Pileus, cylinders or drum-shaped, white, light red in color in the beginning, and bell-shaped, covered with scales afterward, 6~11 cm in length. Stipe, 5~20 cm in length, 1~2.5 cm in diameter. The suitable temperature for sporophore growth is below 20℃.

Shaggy mane was introduced to Guangzhou in the 1990s, and mainly cultivated in bag. Cultivation seasons: spring, autumn and winter.

## 杏鲍菇 King oyster mushroom

侧耳科侧耳属，学名 *Pleurotus eryngii* (DC. Fr.) Quél.，别名刺芹侧耳、雪耳、干贝菇等。

子实体单生或群生，菌盖宽2~12厘米，初呈半球形，后渐平展，表面有丝状光泽，平滑，细纤维状，菌肉白色，菌柄长2~8厘米，粗0.5~3厘米，偏心生或侧生。属中低温食用菌，生长适温为12~22℃。

广州地区自20世纪90年代引种，采取室内菌袋层架栽培。生产季节为11月至翌年4月，目前已利用温控设备进行周年生产。

杏鲍菇是集食用、药用、食疗于一体的新品种，肉质肥厚，口感鲜嫩，带杏仁清香味。

### King oyster mushroom

*Pleurotus eryngii* (DC. Fr.) Quél. Family: Pleurotaceae. Synonym: King trumpet mushroom, French horn mushroom, King brown mushroom.

Sporophore, single or in cluster. Pileus, semi-ball in shape in the beginning, and flat afterward, 2~12 cm in diameter. Smooth surface. Lamella, white in color. Stipe, 2~8 cm in length, 0.5~3 cm in diameter. The suitable temperature for sporophore growth is 12~22℃.

King oyster mushroom was introduced to Guangzhou in the 1990s, and mainly cultivated in bag. Cultivation seasons: November to April the following year.

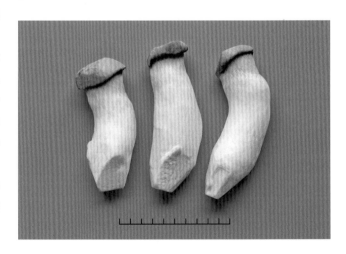

# 木耳 Jew's ear

木耳科木耳属，学名 *Auricularia auricula* (L. ex. Hook.) Underw.，别名黑木耳、云耳等。

子实体耳状或叶片状，胶质，半透明，耳片薄，有弹性，直径 3~12 厘米，表面光滑或有脉状皱纹，初呈红褐色，干后收缩变深褐色至黑色。喜阴凉，属低温型食用菌。

广州地区栽培木耳历史超百年，早期是采集寄生在段木上的木耳，目前采取菌袋栽培。生产季节为 11 月至翌年 3 月。

木耳脆嫩润滑，风味独特。味甘，性平，具益气、滋阴益胃、止痛止血、通便等功效。

## Jew's ear

*Auricularia auricula* (L. ex. Hook.) Underw. Family: Auriculariaceae. Synonym: Wood ear, Jelly ear.

Sporophore, floppy ear in shape, translucent, gelatinous, elastic in texture, 3~12 cm in diameter. The outer surface smooth or undulating with folds and wrinkles, bright reddish-tan-brown with a purplish hint in color, and becomes darker with age. The inner surface grey brown in color and smooth. It is sometimes wrinkled, again with folds and wrinkles, and may have "veins", making it appear even more ear-like.

Jew's ear has been grown more than 100 years in Guangzhou. In the past, Jew's ear was mainly produced on the logs, and cultivated in bag at present. Cultivation dates: November to March the following year.

# 猴头菇 Lion's mane mushroom

猴头菌科猴头菌属，学名 *Hericium erinaceus* (Bull.) Pers.，别名刺猬菇、花菜菇等。

子实体肉质，外形头状或倒卵状，似猴子头，故名猴头菇。鲜时全白色，干后变淡黄色或淡褐色，直径 3.5~20 厘米，无柄，除基部外均密布肉质的刺。属中低温型食用菌，子实体生长最适温为 18~20℃。

广州地区自 20 世纪 60 年代引种栽培，初期采用阔叶枯立或腐木段栽培，近年多采取菌袋生产。适宜栽培季节为 11 月至翌年 4 月。

猴头菇干品每百克蛋白质含量是香菇的 2 倍，属典型的高蛋白质、低脂肪山珍。

## Lion's mane mushroom

*Hericium erinaceus* (Bull.) Pers. Family: Hericiaceae. Synonym: Satyr's beard, Bearded hedgehog mushroom.

Sporophore, succulent, head-shaped or obovate, covered with dense prickle, 3.5~20 cm in diameter. Fresh sporophore is white, and dry sporophore is light yellow to light brown in color. The suitable temperature for sporophore growth is 18~20℃.

Lion's mane mushroom was introduced to Guangzhou in the 1960s. It was mainly produced on the logs, while cultivated in bag at present. Cultivation dates: November to April the following year.

Lion's mane mushroom contains a number of polysaccharides, such as $\beta$-glucan, heteroglucans, heteroxylans.

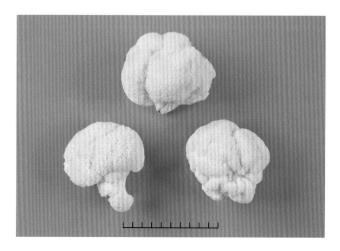

# 鸡枞菌 Termite mushroom

鹅膏科鸡枞菌属，学名 *Termitomyces albuminosus* (Berk.) Heim，别名蚁枞、伞把菇等，广东人称夏至菌，野生的叫荔枝菌。

菌盖初期圆锥形、斗笠状，后伸展，中央有显著的乳头状突起，直径 3~25 厘米，褐色至淡黄色，菌柄组织与菌盖相连，常扭曲，具很长假根，柄长 3~15 厘米。常见于针阔叶林中地上，荔枝树下最多，散生至群生，其生长依赖白蚁巢物质及其渗出液。生长温度要求略高，子实体在 18~26℃才能形成。

广州地区大量收获时期在 5—6 月，夏至前后最多。可人工栽培，4—9 月为生产季节。其产品在食味上比野生的逊色许多。

鸡枞菌清香四溢，鲜甜可口，细嫩爽脆，是食用菌界的奇珍异品，含氨基酸种类及数量均极丰富。

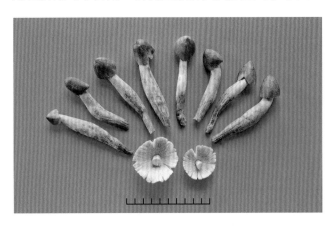

## Termite mushroom

*Termitomyces albuminosus* (Berk.) Heim. Family: Amanitaceae.

Pileus, cone in shape at the beginning, and flat with obvious centrical papilla afterward, brown to light yellow, 3~25 cm in diameter. Stipe, usually contorted, 3~15 cm in length. Sporophore, single or in cluster. The suitable temperature for sporophore growth is 18~26℃. They are found growing from termitaria in grassy fields, hills, or forest borders in China.

The harvest peak season of wild termite mushroom is May to June in Guangzhou. Termite mushroom can be cultivated from April to September.

# 香魏蘑 *Lentinula edodes* × *Pleurotus nebrodensi*

侧耳科侧耳属，别名香魏菇，是香菇与阿魏蘑杂交，20 世纪 90 年代育成的新品种。

子实体大，肉厚，柄短白色，菌盖灰色，产量高。适应温度广，10~26℃可生长，最适温为 15~23℃。

广州地区引种于 20 世纪 90 年代，采用木屑、玉米秆、棉籽壳等培养料进行袋栽，生产季节为 10 月至翌年 5 月。

香魏蘑集香菇、阿魏蘑两者优点，具两者营养成分，有香菇香味和阿魏蘑风味，口感嫩脆。

## *Lentinula edodes* × *Pleurotus nebrodensi*

The hybrid of *Lentinula edodes* × *Pleurotus nebrodensi*. Family: Pleurotaceae.

Sporophore, large and succulent, with short white stipe. Pileus, grey in color. The suitable temperature for sporophore growth is 15~23℃.

*Lentinula edodes* × *Pleurotus nebrodensi* was introduced to Guangzhou in the 1990s, and usually cultivated in bag. Cultivation dates: October to May the following year.

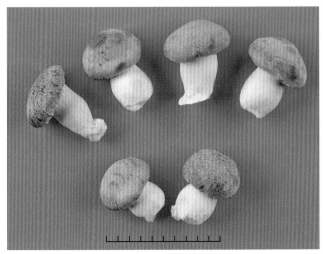

# 十五、菜用玉米
## VEGETABLE CORN

禾本科玉蜀黍属一年生草本植物，学名 Zea mays L.，别名苞谷、粟米等，染色体数 $2n=2x=20$。

菜用玉米是玉米的特殊类型，包括甜玉米和糯玉米，主要以采摘乳熟末期的鲜嫩果穗用于鲜食和加工。广东是我国菜用玉米种植大省，其中甜玉米面积约占全国的50%。广州为广东省菜用玉米的主要产区之一，目前年种植面积约16 700公顷。

**植物学性状**：根发达，须根系，茎基部多个节位易生不定根。茎秆直立，由许多茎节构成，每节着生一片叶。叶多互生，剑形，平行脉，有叶鞘。每个叶鞘内均有腋芽分化，但通常只有上部第6~7个节上的1个或2个腋芽能分化成雌穗。雌穗为肉穗花序，着生成行排列的小花，每朵小花经过双受精发育成一个具有胚和胚乳的完整籽粒。雄花序着生于茎秆顶端，由主轴和若干个分枝组成，其上具有成对排列的小穗。雌雄同株异花，异花授粉。果穗多为筒形或锥形，乳熟期果穗籽粒可呈现不同的颜色，鲜果穗千粒重300~350克。甜玉米种子千粒重120~180克，糯玉米种子千粒重220~280克。

**类型**：可分为甜玉米、糯玉米以及甜加糯玉米3种类型。

1. 甜玉米　甜质型籽粒，携带与碳水化合物代谢有关的隐性突变基因（例如：$sh_2$、$su$ 等）。籽粒以黄色和白色居多，含糖量高，口感甜。常作鲜食、冷冻及罐头产品。

2. 糯玉米　糯质型籽粒，携带隐性突变基因 $wx$ 基因，籽粒中支链淀粉含量接近100%。籽粒有白色、黄色、红色、紫色、黑色等多种颜色。常作鲜食或淀粉加工。

3. 甜加糯玉米　携带 $wx$ 糯质基因和 $sh_2$、$su$ 等甜质基因，果穗中同时有甜籽粒和糯籽粒，口感兼具甜和糯的特点。常作鲜食、冷冻产品。

**栽培环境与方法**：喜温，生长适温20~28℃，低于10℃停止生长；喜光，广州地区一般于2—4月或7—9月播种。直播和育苗移栽均可，每667米$^2$用种量：甜玉米1 000~1 500克，糯玉米1 500~2 000克。田间要求品种间相互隔离。

**收获**：授粉后23天左右，玉米处于乳熟期，花丝完全变褐，可采收鲜苞。

**病虫害**：病害主要有锈病、大小斑病、纹枯病、丝黑穗病、茎腐病、矮花叶病等，虫害主要有玉米螟、黏虫、玉米蚜、小地老虎等。

**营养及功效**：富含亚油酸、维生素E、谷氨酸、谷胱甘肽。味甘，性平，具调中开胃、益肺宁心、清湿热、利肝胆等功效。

*Zea mays* L., an annual herb. Family: Gramineae. The chromosome number: $2n=2x=20$.

Vegetable corn include sweet corn and waxy corn. The immature ears (milk stage) are used for fresh or processing. There is the largest cultivation area of vegetable corn in Guangdong, which is more than 50 percent in China. In Guangdong, the major vegetable corn cultivation region is in Guangzhou, the annual acreage is about 16,700 $hm^2$.

Botanical characters: Fibrous root system, adventitious roots easily born at the lower nodes. Stem, erect and with a lots of nodes, one leaf born on each node. Leaves, opposite and ensiform, with paralled veins and leaf sheaths. Axillary buds differentiate in leaf sheath, one or two axillary buds in the sixth or seventh node on the top will develop into pistillate inflorescence, istillate inflorescence, a spadix with florets borne in line. Staminate inflorescence, apical and consists of main axis and branches, spikelets borne in line. It is monoecious. The ear is usually tubular or tapered. The kernel has different color at the milk stage, fresh weight per 1000 grain is about 300~350 g. The weight per 1000 seed is 120~180 g for sweet corn, and 220~280 g for waxy corn.

Types: There are three types of vegetable corn: sweet corn, waxy corn and sweet-waxy corn.

1. Sweet corn: Sweet kernels, with recessive mutant genes, such as $sh_2$ and *su*, which relate to metabolism of carbohydrates. The kernels usually yellow or white in color, rich in sugar and taste sweet. It is used for fresh, frozen and can food.

2. Waxy corn: Waxy kernels, with recessive mutant gene *wx*, and the ratio of amylopectin is up to 100 percent. The color of kernel is white, yellow, red, purple, black or others. It is used for fresh or starch processing.

3. Sweet-waxy corn: Sweet and waxy kernels born on the same ear, with genes *wx*, $sh_2$, *su*. It is used for fresh and frozen.

Cultivation environment and methods: The suitable temperature for vegetable corn growth is 20~28℃, and growth stunt below 10℃. Sowing dates: February to April and July to September in Guangzhou. Seeding or seedling transplanting. Different cultivars should be separated for planting. The seed per ha is 15,000~22,500 g for sweet corn and 22,500~30,000 g for waxy corn.

Harvest: The ears are harvested about 23 days after pollination, when the silk turn into brown at the milk stage.

Nutrition and efficacy: Vegetable corn is rich in linoleic acid, vitamin E, glutamic acid, glutathione.

# 甜玉米 Sweet corn

## 粤甜 9 号
Yuetian No. 9 sweet corn

### 品种来源
广东省农业科学院作物研究所育成的杂种一代，2004年通过广东省农作物品种审定。

📍 分布地区 ｜ 增城、从化、白云等区。

特　征 ｜ 株型半紧凑，株高 205~224 厘米，穗位高 84~96 厘米。叶片宽大。果穗筒形，穗长 17.3~19.0 厘米，穗粗约 5.0 厘米。单苞重 273~327 克，千粒重约 320 克。籽粒黄色，果皮较薄。

特　性 ｜ 三交种，中熟，春、秋植播后约 85 天、75 天可采收。生长势较强，早生快发，后期保绿度较好。抗纹枯病和大、小斑病，抗倒性较强。甜度较高，清甜爽脆，适口性好，品质优。每公顷鲜苞产量约 14 吨。

栽培要点 ｜ 春播 2 月底至 3 月初，秋播 7—8 月，每 667 米² 种植约 3 200 株，可直播或育苗移栽。施足基肥，轻施苗肥，适施拔节肥，重施攻苞肥。注意防旱排涝，重点抓好苗期地下害虫和中后期玉米螟的防治工作。

## 粤甜 13 号
Yuetian No. 13 sweet corn

### 品种来源
广东省农业科学院作物研究所育成的杂种一代，2006年和2010年先后通过广东省和国家农作物品种审定。

📍 分布地区 ｜ 增城、花都、白云等区。

特　征 ｜ 株型平展，株高 170~181 厘米，穗位高 48~52 厘米。叶色浓绿，叶距大。果穗筒形，穗长 20.7~21.2 厘米，穗粗 4.6~4.8 厘米。单苞重 302~320 克，千粒重 354~386 克。籽粒黄白相间，基本无秃顶。

特　性 ｜ 早熟，播后 68~71 天可收。抗旱性、耐涝性较差，中抗纹枯病和小斑病，高抗茎腐病，抗大斑病。甜度较高，适口性较好，品质优。每公顷鲜苞产量约 13 吨。

栽培要点 ｜ 春播 3 月上中旬，秋播 8 月中旬，每 667 米² 种植可达 4 000 株。选择高肥力地块种植，加强苗期管理，及时查缺补苗，尤其注意排除田间积水。

# 粤甜 16 号
Yuetian No.16 sweet corn

**品种来源**

广东省农业科学院作物研究所育成的杂种一代，2008年和2010年先后通过广东省和国家农作物品种审定。

📍 **分布地区** | 增城、从化、白云、南沙等区。

**特　征** | 株型半紧凑，株高 218~230 厘米，穗位高约 82 厘米。果穗筒形，穗长约 20 厘米，穗粗约 5.4 厘米，秃顶长 0.7~1.5 厘米。单苞重 356~366 克，千粒重 388~393 克。籽粒黄色。

**特　性** | 中熟，春、秋植播后约 85 天、75 天可收。植株生长整齐，前、中期生长势强，后期保绿度好。中抗纹枯病和小斑病。甜度较高，适口性较好，品质较优。丰产性好，每公顷鲜苞产量 15 吨以上。

**栽培要点** | 每 667 米² 种植约 3 300 株，矮花叶病发生区慎用。其余参照粤甜 9 号。

# 正甜 68
Zhengtian 68 sweet corn

**品种来源**

广东省农科集团良种苗木中心、广东省农业科学院作物研究所育成的杂种一代，2009 年通过广东省农作物品种审定。

📍 **分布地区** | 花都、增城、从化、白云、番禺、南沙等区。

**特　征** | 株型平展，株高约 215 厘米，穗位高约 74 厘米。果穗筒形，穗长约 20.5 厘米，穗粗 5.2 厘米。单苞重 339~376 克，千粒重 355~385 克。籽粒黄色，果皮较薄。

**特　性** | 中熟，春、秋植播后约 85 天、75 天可收。前、中期生长势强，后期保绿度好。高抗纹枯病、茎腐病和大、小斑病。甜度高，适口性较好，品质较优。丰产性较好，每公顷鲜苞产量约 15 吨。

**栽培要点** | 春播 3 月上中旬，秋播 7—8 月，每 667 米² 种植约 3 300 株。其余参照粤甜 9 号。

## 农甜 88

Nongtian 88 sweet corn

**品种来源**

华南农业大学农学院育成的杂种一代，2009 年通过广东省农作物品种审定。

📍 **分布地区** | 增城、花都、南沙等区。

特　　征 | 株型平展，株高 210~218 厘米，穗位高约 70 厘米。果穗筒形，穗长约 21 厘米，穗粗 5.0~5.3 厘米，秃顶长 1.6 厘米左右。单苞重 308~345 克，千粒重 386~394 克。籽粒黄白相间，粒大皮薄。

特　　性 | 早熟，播后 68~76 天可收。中抗纹枯病和小斑病，抗纹枯病、茎腐病和大斑病。甜度高，适口性好，品质优。每公顷鲜苞产量约 14 吨。

栽培要点 | 春播 2 月中旬至 4 月，秋播 8—9 月，每 667 米² 种植约 3 300 株。田间只留植株最上部果穗，吐丝期及时去掉下部果穗。授粉后 18~22 天采收为佳。

## 华宝甜 8 号

Huabaotian No.8 sweet corn

**品种来源**

华南农业大学生命科学学院育成的杂种一代，2007 年通过广东省农作物品种审定。

📍 **分布地区** | 增城、花都、南沙、白云等区。

特　　征 | 株型半紧凑，株高约 200 厘米，穗位高约 62 厘米。果穗筒形，穗长约 20 厘米，穗粗 4.7 厘米，秃尖小。单苞重约 300 克，千粒重 354~375 克。籽粒黄白相间，饱满，色泽鲜亮。

特　　性 | 较耐湿热，可夏播，早熟，春、夏、秋植播后约 80 天、60 天、70 天可收。高抗茎腐病，适应性广，易种。皮薄清甜，适口性好，品质优。每公顷鲜苞产量约 14 吨。

栽培要点 | 2 月中下旬至 9 月可播种，每 667 米² 种植 3 000~3 300 株，夏种密度不宜超过 3 000 株。其余参照粤甜 9 号。

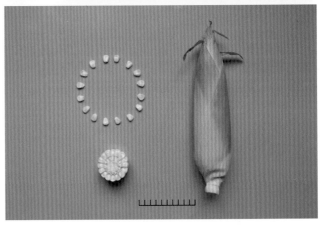

## 新美夏珍

Xinmeixiazhen sweet corn

**品种来源** 广东省珠海市鲜美种苗发展有限公司育成的杂种一代，2005 年通过广东省农作物品种审定。

**分布地区** | 花都、增城、从化、白云、番禺、南沙等区。

**特　　征** | 株型半紧凑，株高 223~239 厘米，穗位高 72~91 厘米。果穗筒形，穗长约 19.4 厘米，穗粗 5.0 厘米，秃顶长 0.8 厘米。单苞重约 312 克，千粒重 391~393 克。籽粒淡黄色，皮较薄，排列整齐。

**特　　性** | 中熟，春、秋植播后约 85 天、75 天可收。前中期生长势强，后期保绿度好，整齐度好。高抗纹枯病和茎腐病，中抗小斑病。抗倒性较强，适应性广。甜度较高，适口性较好，品质较优。每公顷鲜苞产量约 15 吨。

**栽培要点** | 春播 2 月底至 3 月初，秋播 7—8 月，每 667 米² 种植 3 000 株为宜，行距 75 厘米，株距 30 厘米。施足基肥，每 667 米² 施过磷酸钙 50 千克、复合肥 15 千克。苗期注意防治斜纹夜蛾、甜菜夜蛾和玉米螟。

## 穗甜 1 号

Suitian No.1 sweet corn

**品种来源** 广州市农业科学研究院育成的杂种一代，1998 年和 2003 年先后通过广东省和国家农作物品种审定。

**分布地区** | 花都、增城、从化、白云、番禺、南沙等区。

**特　　征** | 株型平展，株高 190~200 厘米，穗位高约 65 厘米。果穗筒形，小苞（旗）叶较长。穗长 21~22 厘米，穗粗 4.5~5.0 厘米，穗行数 14~16 行。单苞重 300~400 克，千粒重约 360 克。籽粒黄色，皮薄，穗轴白色。

**特　　性** | 中早熟，春、秋植播后约 80 天、70 天可收。高抗大、小斑病和矮花叶病，抗茎腐病。较耐热，抗逆性强。清甜爽口，品质优。每公顷鲜苞产量约 13.5 吨。

**栽培要点** | 春播 2 月中旬至 4 月上旬，秋播 7 月中旬至 9 月上旬，每 667 米² 种植约 3 500 株。重施基肥和攻苞肥，适施苗肥和壮秆肥，全生育期内保持土壤湿润。及早防治地下害虫，大喇叭口期至心叶末期防治玉米螟 2~3 次。

## 广甜 2 号

Guangtian No.2 sweet corn

**品种来源**

广州市农业科学研究院育成的杂种一代,2003 年和 2004 年先后通过广东省和国家农作物品种审定。

**分布地区** | 花都、增城、从化、白云、番禺、南沙等区。

**特　　征** | 株型半紧凑,株高 190~200 厘米,穗位高约 70 厘米。果穗筒形,无秃顶,穗长约 20 厘米,穗粗 5.0 厘米,穗行数 16~18 行。单苞重 300~400 克,千粒重约 320 克。籽粒黄白相间,皮薄,有光泽。

**特　　性** | 中早熟,春、秋植播后约 80 天、70 天可收。抗倒伏,较耐热。抗大斑病,中抗小斑病。甜度高,柔嫩性好,风味佳,品质较优。每公顷鲜苞产量 15 吨以上。

**栽培要点** | 春播 2 月底至 3 月,秋播 7—8 月,每 667 米$^2$种植约 3 500 株。重施有机肥和攻苞肥。苗期注意防治地下害虫,大喇叭口期及时防治玉米螟,茎腐病及矮花叶病发生区慎用。

## 广甜 3 号

Guangtian No.3 sweet corn

**品种来源**

广州市农业科学研究院育成的杂种一代,2005 年通过广东省农作物品种审定。

**分布地区** | 花都、增城、从化、白云、番禺、南沙等区。

**特　　征** | 株型半紧凑,株高 202~209 厘米,穗位高约 65 厘米。果穗筒形,无秃顶。穗长 21.4 厘米,穗粗 5.1~5.3 厘米,穗行数 14~16 行。单苞重 337~380 克,千粒重 394~413 克。籽粒黄色,皮薄。

**特　　性** | 中熟,春、秋植播后约 85 天、75 天可收。前、中期生长势强,后期保绿度较好。高抗茎腐病,中抗纹枯病和小斑病。抗倒性较强,适应性广。渣少,口感好,品质较优。每公顷鲜苞产量 15 吨左右。

**栽培要点** | 春播 2 月底至 3 月,秋播 7—8 月,低肥力田块每 667 米$^2$种植约 3 000 株,高肥力田块可达 3 500 株。其余参照穗甜 1 号。

## 华美甜 8 号

Huameitian No.8 sweet corn

## 华美甜 168

Huameitian 168 sweet corn

### 品种来源

华南农业大学农学院育成的杂种一代，2010 年通过广东省农作物品种审定。

📍 **分布地区** | 花都、增城、白云、南沙等区。

**特　征** | 株型紧凑，株高 196~199 厘米，穗位高 73~82 厘米。果穗筒形，穗长约 18.5 厘米，穗粗约 4.8 厘米，秃顶长 0.9~1.5 厘米。单苞重 298~305 克，千粒重 346~358 克。籽粒黄白相间，均匀度好，果皮薄。

**特　性** | 中早熟，春、秋植播后约 80 天、70 天可收。植株整齐、健壮，前、中期生长势强，后期保绿度好。中抗纹枯病和小斑病。甜度高，适口性好，品质较优。每公顷鲜苞产量约 15 吨。

**栽培要点** | 春播 2 月底至 3 月，秋播 7—8 月，每 667 米² 种植约 3 500 株。抽穗前后 10 天为水分敏感期，保持土壤湿润，注意防旱防排涝。一般最佳采收期在授粉后 20 天左右。

### 品种来源

华南农业大学科技实业发展总公司育成的杂种一代，2008 年通过广东省农作物品种审定。

📍 **分布地区** | 花都、增城、从化、白云、番禺、南沙等区。

**特　征** | 株型紧凑，株高 217 厘米，穗位高约 82 厘米。果穗筒形，穗长约 19.7 厘米，穗粗约 5.1 厘米，秃顶长 0.9~1.5 厘米。单苞重 306~343 克，千粒重 396 克。籽粒淡黄色，粒大饱满。

**特　性** | 中熟，春、秋植播后约 85 天、75 天可收。中抗纹枯病和小斑病。甜度高，适口性好，品质较优。每公顷鲜苞产量约 15 吨。

**栽培要点** | 春播 2 月底至 3 月，秋播 7—8 月，每 667 米² 种植约 3 000 株。施足基肥，轻施苗肥，适施拔节肥，重施攻苞肥，尤其应重视施用磷钾肥，拔节期和抽雄开花前结合施肥培土 2~3 次。

## 金银粟 2 号

Jinyinsu No.2 sweet corn

**品种来源**

广东省农业科学院蔬菜研究所育成的杂种一代，2005年通过广东省农作物品种审定。

**分布地区** | 花都、增城、白云等区。

**特　征** | 株型披散，株高 200 厘米以上，穗位高约 58 厘米。果穗筒形，穗长 20.3 厘米，穗粗 5.0 厘米，单苞重 319~338 克，千粒重 367~380 克。籽粒黄白相间，皮薄。

**特　性** | 三交种。中早熟，春、秋植播后约 80 天、70 天可收。前、中期生长势较强，后期保绿度好。高抗茎腐病，抗纹枯病和小斑病。抗倒性较强，适应性广。甜度高，适口性好，品质较优。每公顷鲜苞产量 14 吨左右。

**栽培要点** | 每 667 米² 种植 2 800~3 000 株，吐丝期遇不良天气须进行人工辅助授粉。其余参照广甜 3 号。

## 先甜 5 号

Xiantian No.5 sweet corn

**品种来源**

先正达（泰国）公司育成的杂种一代，2006年通过广东省农作物品种审定。

**分布地区** | 花都、增城、从化、白云、番禺、南沙等区。

**特　征** | 株型半紧凑，株高 220 厘米以上，穗位高 57~76 厘米。叶片宽大，浓绿。果穗筒形，穗长 18.7~20.2 厘米，穗粗 4.7~5.4 厘米。单苞重 296~438 克，千粒重 339~415 克。籽粒黄色，果皮较薄。

**特　性** | 迟熟，春、秋植播后 90 天、80 天可收。植株整齐、健壮，前、中期生长势强。抗病性和抗倒性强，适应性广。甜度较高，适口性较好，品质较优。每公顷鲜苞产量可达 16 吨。

**栽培要点** | 春播 3 月至 5 月初，秋播 7 月至 9 月初，每 667 米² 种植不宜超过 3 000 株。生长前期控制氮肥用量，5 叶期、拔节期和大喇叭口期每 667 米² 分别追施复合肥 8 千克、10 千克和 15 千克。授粉后 18~20 天为最佳采收期。

## 仲鲜甜 3 号

Zhongxiantian No.3 sweet corn

> 品种来源

仲恺农业工程学院作物研究所 2013 年育成的杂种一代。

📍 **分布地区** | 花都、增城等区。

**特　征** | 株型半紧凑，株高 219~231 厘米，穗位高 58~64 厘米。果穗筒形，穗长 19.9~20.9 厘米，穗粗 4.9~5.2 厘米。单苞重 379~383 克，千粒重 297~318 克。籽粒淡黄色，果皮较薄。

**特　性** | 迟熟，春、秋植播后约 90 天、80 天可收。中抗纹枯病和小斑病。适口性较好，品质较优。丰产性和稳产性好，每公顷鲜苞产量 16 吨以上。

**栽培要点** | 每 667 米$^2$ 种植约 3 500 株。其余参照广甜 3 号。

## 珍珠

Zhenzhu sweet corn

> 品种来源

广州市番禺区绿色科技发展有限公司从美国引进。

📍 **分布地区** | 番禺、南沙、从化等区。

**特　征** | 株型半紧凑，株高 170 厘米左右，穗位高约 55 厘米。果穗筒形，穗长约 19 厘米，穗粗约 5.0 厘米。籽粒纯白色，果皮薄。

**特　性** | 早熟，春、秋植播后约 75 天可收。前期生长势较弱。抗逆性较差，耐热性较差，抗倒性较强。甜度高，爽脆，渣少，适口性佳，品质优，为水果型玉米。每公顷鲜苞产量 13~14 吨。

**栽培要点** | 春播 2 月底至 3 月，秋播 8 月中旬至 9 月，每 667 米$^2$ 种植约 3 200 株。育苗移栽，田间管理要求高。施足基肥，注意排涝，重点抓好中后期玉米螟的防治工作。

# 糯玉米 Waxy corn

## 广糯 1 号
Guangnuo No.1 waxy corn

**品种来源**

广州市农业科学研究院育成的杂种一代，先后于 2002 年、2004 年通过广东省和国家农作物品种审定，是目前唯一获广东省名牌产品的糯玉米品种。

📍 **分布地区** | 花都、增城、从化、白云、番禺、南沙等区。

**特　　征** | 株型半紧凑，株高约 200 厘米，穗位高 70 厘米。叶片浓绿，叶大质硬。果穗筒形，穗长约 20 厘米，穗粗约 5 厘米，无秃尖，穗行数 12~14 行。单苞重约 350 克，千粒重 368 克。籽粒白色，皮薄。

**特　　性** | 中早熟，春、秋植播后约 80 天、70 天可收。前期早生快发，后期保绿度高。抗纹枯病和大、小斑病，较耐湿热，适应性广。黏软，糯香，品质优。丰产稳产性好，每公顷鲜苞产量 15 吨左右。

**栽培要点** | 春播 2 月下旬至 3 月中旬，秋播 8—9 月，每 667 米² 种植约 3 500 株，秋季种植或肥力高田块可密些，重施有机肥和攻苞肥。苗期注意防治地下害虫，大喇叭口期注意防治玉米螟。

## 广糯 2 号
Guangnuo No.2 waxy corn

**品种来源**

广州市农业科学研究院育成的杂种一代，2005 年通过广东省农作物品种审定。

📍 **分布地区** | 花都、增城、从化、白云、番禺、南沙等区。

**特　　征** | 株型半紧凑，株高约 190 厘米，穗位高约 65 厘米。果穗筒形，穗长约 20 厘米，穗粗 4.5 厘米，穗行数 14~16 行。单苞重 300~400 克，千粒重 309 克。籽粒紫黑色，皮较薄，硬粒型。

**特　　性** | 中熟，播后 75~85 天可收。植株整齐度较好，后期保绿度高。中抗小斑病，抗纹枯病。抗倒性较强，适应性广。糯性强，口感好，品质较优。每公顷鲜苞产量 15 吨左右。

**栽培要点** | 每 667 米² 种植约 3 500 株。注意及时收获，籽粒由白转紫时可开始采收。其余参照广糯 1 号。

## 仲糯 1 号

Zhongnuo No.1 waxy corn

**品种来源**

仲恺农业工程学院作物研究所育成的杂种一代，2005年通过广东省农作物品种审定。

📍 **分布地区** | 增城、花都等区。

特　征 | 株型半紧凑，株高约 220 厘米，穗位高 78~81 厘米。叶宽厚浓绿。果穗筒形，穗长 20.3 厘米，穗粗约 4.6 厘米。单苞重 282~352 克，千粒重 311 克。籽粒白色，有秃尖，半马齿型。

特　性 | 中早熟，春植播后约 80 天可收。植株整齐、健壮，生长势强。抗小斑病和纹枯病，抗倒性强，适应性好。品质较优。每公顷鲜苞产量 13~14 吨。

栽培要点 | 春播 2 月下旬至 3 月中旬，秋播 8—9 月，中上肥力地块每 667 米$^2$ 种植 3 000~3 200 株，高肥力地块可种植 3 500 株。早施苗肥，重施拔节肥（占总量的 35% 左右），重施穗肥（10~11 叶，占总量的 45%~55%）。

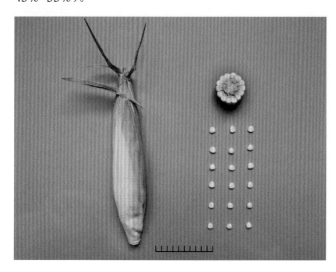

## 粤彩糯 2 号

Yuecainuo No.2 waxy corn

**品种来源**

广东省农业科学院作物研究所育成的杂种一代，2012年通过广东省农作物品种审定。

📍 **分布地区** | 花都、增城、白云等区。

特　征 | 株型紧凑，株高 175~202 厘米，穗位高 57~74 厘米。果穗锥形，穗长约 18.5 厘米，穗粗 4.5 厘米，秃顶长 0.3~0.8 厘米。单苞重 278~300 克，千粒重 305~347 克。籽粒紫白相间。

特　性 | 中熟，春、秋植播后约 85 天、75 天可收。植株整齐、健壮，后期保绿度高。高抗纹枯病、茎腐病和大、小斑病。糯性好，品质优。每公顷鲜苞产量 13~14 吨。

栽培要点 | 春播 2 月下旬至 3 月中旬，秋播 8—9 月，每 667 米$^2$ 种植约 3 500 株。重点防治地下害虫和玉米螟，注意防治锈病。

## 粤白糯 6 号
Yuebainuo No.6 waxy corn

**品种来源**

广东省农业科学院作物研究所、广东金作农业科技有限公司 2013 年育成的杂种一代。

📍 **分布地区** | 增城、花都、白云等区。

**特　征** | 株型半紧凑，株高 195~222 厘米，穗位高 69~86 厘米。果穗筒形，穗长 18.7 厘米左右，穗粗 4.8~4.9 厘米，秃顶长 0.5~0.8 厘米。单苞重 288~316 克，千粒重 345~354 克。籽粒白色，皮薄饱满，排列整齐。

**特　性** | 中熟，春植播后约 80 天可收。植株整齐度好，前、中期生长势强。高抗茎腐病，中抗纹枯病和大、小斑病。糯性强，适口性好，品质较优。每公顷鲜苞产量约 14 吨。

**栽培要点** | 春播 2 月下旬至 3 月中旬，秋播 8—9 月，每 667 米$^2$ 种植 3 000~3 300 株。

## 美玉糯 7 号
Meiyunuo No.7 sweet waxy corn

**品种来源**

海南绿川种苗有限公司育成的杂种一代，2007 年通过广东省农作物品种审定。

📍 **分布地区** | 花都、增城、南沙等区。

**特　征** | 株型半紧凑，株高约 190 厘米，穗位高 66~73 厘米。果穗筒形，穗长 19.9~20.8 厘米，穗粗 4.7 厘米，秃顶长约 0.5 厘米，单苞重 249~290 克，千粒重约 300 克。籽粒白色，糯粒、甜粒比为 3∶1，皮薄无渣。

**特　性** | 早熟，春、秋植播后约 80 天、70 天可收。抗茎腐病，抗锈病能力较差。抗倒性较强。糯性好，有香甜味，品质优。每公顷鲜苞产量 13 吨左右。

**栽培要点** | 春播 2 月下旬至 3 月中旬，秋播 8—9 月，每 667 米$^2$ 种植 3 000~3 500 株，增施有机肥和钾肥可提高品质。精细管理，避免干旱，防止积水。田间注意防治纹枯病、小斑病及锈病。严格掌握收获期，采收期以糯粒成熟度为准。

# 附录 APPENDICES

## 附录 1
## 广州蔬菜产销概况

Appendix 1 The General Survey of Vegetable Production and Market Consumption in Guangzhou

广州地处珠江三角洲，属南亚热带季风气候，四季温和，日照充足，雨量充沛，一年四季均可露地栽培蔬菜。其种菜历史悠久，品种繁多，不少本地特产、品质优良、风味别致、大宗生产的蔬菜，如菜心、芥蓝、芥菜、节瓜、丝瓜、苦瓜、莴苣、芋头、莲藕等，既为本市人民喜爱，又可出口创汇。广州蔬菜栽培技术精湛，有传统适合本地自然地理气候的独具特色的栽培方式，如水坑、旱畦和水塘栽培，也有现代化的栽培技术，现在水坑栽培已逐渐被现代化的喷滴灌所代替。在轮间套种方面，也因地制宜采用水旱轮作、高矮结合、精粗结合、长短结合等，使广州的蔬菜生产具有品种多、产量高、质量佳等优点，进而使广州成为我国南方重要的蔬菜资源和生产基地之一。

2013年广州蔬菜常年种植面积3.38万多公顷，播种面积15.05万公顷，年产量347万吨，总产值83亿元，除满足全市1 000多万人口的消费外，当年还有16万吨左右蔬菜北运和出口，但有些本地不宜生产和淡季时蔬菜的供应则由外地的蔬菜进行补充。广州每年冬末春初常出现低温湿冷、光照不足，夏、秋季有热带气旋风暴和高温多雨，病虫害较严重，蔬菜生长困难，常形成3—4月的春淡和8—10月的秋淡。

中华人民共和国成立以前和成立初期，广州的蔬菜购销形式是自产自销，多渠道流通，价格随行就市，蔬菜供应比较充足。1956—1984年，采用统购包销的形式，实行计划生产、计划供应，造成品种单调、细菜少、粗菜多、质量差、淡季明显，不能满足市场的需求。

1978年8月，广州实行的蔬菜购销体制逐步改革开放，至1984年11月购销体制全部放开。市政府对蔬菜的宏观调控实行"三管三放"，即管蔬菜面积和购销网点、管基地菜农与蔬菜公司的优惠政策、管对菜区的"四联系"，放上市任务、放价格、放流通渠道，同时每年投入大量资金加强菜区农田水利、市场流通和科技建设，广州蔬菜生产走上了高产、优质、高效益的发展道路。近年来，广州蔬菜种植面积稳定，标准化建设的菜田面积逐年增加，耕作条件逐步改善，初步形成了近郊以叶菜类为主、中郊以叶菜类与瓜豆类并举、远郊以度淡菜和干口菜为主的多层次相结合的商品菜基地，蔬菜科技和产销服务网络逐步建立和健全。围绕市场需求，培育、引进和推广了许多适应性广、抗性强、品质优的品种，大面积推广应用了冬春季薄膜小拱棚覆盖、夏秋季遮阳网覆盖及各类大棚、喷滴灌、无土栽培、工厂化育苗、配方施肥、水肥一体化等设施及高产优质综合配套栽培技术措施，使全市的科学种菜水平得到较大的提高。积极抓好蔬菜专业村和产业化生产基地建设，狠抓产品质量安全。

蔬菜产销体制的改革和各项措施的配合，调动了产、供、销各方面的积极性，同时积极向外地发展蔬菜生产基地，使蔬菜种植面积稳定，上市量和均衡供应得到保证，品种增加，质量提高，产品安全，流通渠道多，买卖方便。蔬菜价格稳定，旺淡差别不明显，实现了全市人民多年来期盼的优质、高产、多品种、均衡上市的愿望。

（有关照片见282~285页）

塑料温室栽培

连栋温室

现代蔬菜栽培基地（南沙）

菜心栽培

甘蓝栽培

芥蓝栽培

冬瓜栽培

番茄栽培

葱水坑栽培

苦瓜棚架栽培

温室黄瓜无土栽培

漂浮育苗

离地高设管道水耕栽培

生菜温室栽培

附录 1　广州蔬菜产销概况

品种展示

采后加工

蔬菜批发市场

蔬菜种子市场

# 附录 2
# 广州气候概况

Appendix 2  The Climatic Conditions of Guangzhou

广州包括中心六区及番禺、增城、花都、从化等区，位于东经 112°57′~114°03′、北纬 22°35′~23°35′的区域内，属南亚热带季风气候区。广州市气候与农业气象中心根据 5 个国家气象站 20 年（1994—2013 年）的资料统计，整理广州地区的气候特点如下：

广州地处低纬度区域，地表接受太阳辐射量较多，同时受季风的影响，夏季在海洋暖湿气流的影响下形成高温、高湿、多雨的气候，冬季在北方大陆冷风的影响下形成低温、干燥、少雨的气候。

## 一、气温

广州年平均气温为 21.7~23.0℃，北部为 21.7℃，中部为 22.5℃，南部为 23.0℃。最热的 7—8 月平均气温为 28.2~29.2℃，极端最高气温为 39.3℃；最冷月为 1 月（个别年份为 2 月），平均气温为 12.7~14.3℃，极端最低气温为 -2.9℃。每年 1—7 月平均气温逐渐上升，11 月下旬至翌年 2 月中旬可能出现霜冻。

## 二、降水量

广州年平均降水量为 1 680.2~1 993.8 毫米，北部多于南部，每年自 1 月起雨量逐增，4 月激增，5—6 月雨量最多。4—9 月是广州的汛期，4—6 月为前汛期，多为锋面雨，7—9 月为后汛期，多为热带气旋雨，其次为对流雨（热雷雨），10 月至翌年 3 月则是少雨季节。

## 三、光照

1. 太阳辐射  广州年平均太阳辐射值为 3 295~5 008 兆焦/米$^2$，年内辐射以 2 月最低，7 月最高。

2. 日照时数  广州年平均日照时数为 1 461~1 803 小时，年日照百分率为 34%~41%，中部花都区、增城区多，南北的从化区、番禺区少。季节上以秋季最多，夏季次之，冬季再次，春季最少。

## 四、风

因受季风的影响，广州年内冬季（1 月）受冷高压控制，多偏北风和东北风；春季（4 月）风向较零乱，而以东南风较多；夏季（7 月）受副热带高压和南海低压影响，以偏南风为主；秋季（10 月）由夏季风转为冬季风，以偏北风为主。

在平均风速方面，广州以冬、春季风速比较大，夏季风速较小，但夏、秋季常有热带气旋侵袭，风速可急剧增大为 8 级以上的大风。

### 广州气候因素统计

| 项目 | | 月份 | | | | | | | | | | | | 全年 |
|---|---|---|---|---|---|---|---|---|---|---|---|---|---|---|
| | | 1 | 2 | 3 | 4 | 5 | 6 | 7 | 8 | 9 | 10 | 11 | 12 | |
| 气温/℃ | 平均值 | 13.6 | 15.6 | 18.3 | 22.6 | 25.9 | 27.8 | 28.8 | 28.6 | 27.3 | 24.6 | 20.1 | 15.3 | 22.4 |
| | 极端最高温 | 28.6 | 30.3 | 33.1 | 34.3 | 35.8 | 38.9 | 39.3 | 38.5 | 38.0 | 36.7 | 34.1 | 29.7 | 39.3 |
| | 极端最低温 | -2.9 | 0.0 | 1.9 | 8.3 | 13.8 | 18.1 | 22.1 | 21.7 | 15.7 | 8.4 | 1.5 | -1.6 | -2.9 |
| 降水量/毫米 | | 37.8 | 51.1 | 89.0 | 204.6 | 289.6 | 373.1 | 238.9 | 259.6 | 171.3 | 61.0 | 36.8 | 40.4 | 1 853.2 |
| 辐射值/兆焦·米$^{-2}$ | | 270 | 233 | 257 | 292 | 386 | 383 | 466 | 444 | 425 | 425 | 347 | 312 | 4 240 |
| 日照 | 时数/小时 | 113 | 80 | 72 | 74 | 115 | 128 | 184 | 173 | 170 | 186 | 170 | 157 | 1 622 |
| | 百分率/% | 33 | 25 | 19 | 19 | 28 | 32 | 45 | 44 | 47 | 52 | 52 | 48 | 37 |

注：本资料由广州市气候与农业气象中心艾卉提供。

# 附录 3
## 广州主要名特产蔬菜营养成分

Appendix 3  The Nutrient Composition of Guangzhou Main Vegetables

广州主要名特产蔬菜营养分析

| 蔬菜品种 | 还原糖 | 可溶性固形物 | 蛋白质 | 维生素C/（毫克/100克） | 纤维素 |
|---|---|---|---|---|---|
| 四九-19号菜心 | 0.34% | 2.6% | 1.9% | 42.4 | 0.78% |
| 油绿702菜心 | 1.33% | 4.3% | 0.94% | 138.7 | 0.66% |
| 增城迟菜心 | 1.37% | 3.1% | 1.8% | 49.1 | 0.68% |
| 矮脚黑叶小白菜 | 0.26% | 2.5% | 1.7% | 27.1 | 0.7% |
| 顺宝芥蓝 | 0.49% | 3.5% | 2.7% | 67.6 | 0.66% |
| 竹芥（吕田大芥菜） | 1.22% | 3.2% | 2.1% | 77.5 | 0.97% |
| 白梗蕹菜 | 0.36% | 3.1% | 1.7% | 39.8 | 1.24% |
| 西洋菜 | 0.54% | 3.5% | 2.4% | 44.4 | 0.61% |
| 长丰白瓜 | 1.98% | 4.1% | 0.87% | 12.2 | 0.3% |
| 冠星2号节瓜 | — | 3.94% | 0.66% | 60.67 | — |
| 黑皮冬冬瓜 | 2.04% | 2.4% | 0.41% | 15.4 | 0.48% |
| 大顶苦瓜 | 0.68% | 2.5% | 0.78% | 114.0 | 1.14% |
| 丰绿苦瓜 | 0.36% | 2.1% | 0.7% | 102.9 | 0.98% |
| 雅绿六号丝瓜 | — | 3.4% | 0.77% | 9.6 | 0.63% |
| 美味高朋丝瓜 | — | 4.1% | 1.08% | 19.4 | 0.42% |
| 穗青豆角 | 2.8% | 5.9% | 2.5% | 40.4 | 1.32% |
| 穗丰8号豆角 | 2.8% | 5.1% | 1.9% | 30.4 | 1.22% |
| 耙齿萝卜 | 2.8% | 3.8% | 0.66% | 23.4 | 0.44% |
| 海南洲（新垦莲藕） | 0.19% | 8.2% | 2.5% | 51.3 | 0.92% |
| 霸王花（干） | — | — | 13.5% | — | 7.44% |
| 槟榔芋（炭步芋头） | 0.46% | 16.3% | 1.6% | 4.4 | 0.7% |
| 细叶粉葛 | 0.32% | 8.4% | 2.95% | 27.8 | 0.94% |

# 附录 4
# 广州蔬菜主要病虫害与防治简介

Appendix 4  Synopsis of Main Vegetable Diseases and Insects of Guangzhou and Their Control Methods

## 主要病害与防治

| 名称 | 为害对象 | 发病条件 | 主要症状 | 防治方法 |
|---|---|---|---|---|
| 病毒病 | 茄果类、瓜类、豆类、叶菜类等 | 干旱及蚜虫、蓟马、粉虱、叶蝉多的季节易发病。初侵染源是带毒的植物或种子，借虫媒和人为传播 | 有5种类型。花叶型：叶片黄绿相间，皱缩；蕨叶型：叶片变小，叶尖变成线状；条斑型：病叶、茎、果实出现各种云纹斑状；卷叶型：叶脉间黄化，叶缘上卷；丛枝型：病株矮化，分枝极多，呈丛枝状，叶小，色淡 | 选种抗病或耐病品种；避免种子、种苗带毒；发现病株及时拔除；施足基肥，增强植株抗病能力；及时喷杀蚜虫、蓟马、粉虱、叶蝉；苗期用药防病，药剂有病毒通、病毒A、植病灵、菌毒清等，交替使用 |
| 猝倒病 | 瓜类、茄果类、豆类等 | 气温15~20℃、高湿、苗床低洼、光照不足、通风不良易发病，特别是2—5月多雨潮湿天气发病严重。病菌借流水和风雨传播 | 为害幼苗嫩茎，茎基部或中部初呈水渍状，后变黄褐色，逐渐缢缩成线状，叶片尚未凋萎，幼苗已猝倒伏地，有时幼苗出土胚轴和子叶腐烂变褐枯死。湿度大时病部及周围土壤长出白色棉絮状菌丝 | 选用耐寒或耐风雨的品种；苗床床土消毒；加强苗床管理；发病初期及时喷药，药剂有金雷多米尔、代森锰锌、普力克、克露等，交替使用 |
| 霜霉病 | 黄瓜、丝瓜、白菜、菜心、芥蓝、菠菜等 | 气温20℃以上，多雨雾重，湿度大，偏施氮肥，种植感病品种易发病，特别是1—4月阴雨天多发病严重。病菌借风雨及流水传播 | 为害叶片，初生褪绿小点或黄斑，后扩大变淡褐色或黄褐色，病斑受叶脉限制成多角形。叶菜类和瓜类等病斑背面有暗灰色霉，白菜等叶菜类有白色霉 | 种植抗病品种；加强田间管理，施足基肥；降低地下水位。药剂防治可用瑞毒霉锰锌、普力克、杀毒矾等，交替使用 |
| （绵）疫病 | 辣椒、黄瓜、节瓜、冬瓜、苦瓜、南瓜、番茄、茄子等 | 气温28~30℃，高湿、地势低洼、排水不良、浇水过勤的黏土地易发病，特别是4—6月大雨后发病严重。病菌借风雨及流水传播 | 幼苗多从叶尖、茎基部发病，呈暗绿色水渍状。成株期主要在嫩茎和节部，水渍状，变软，缢缩，患部以上叶片萎蔫。果实上病斑初为暗绿色水渍状，后变褐软腐，湿度大时有白霉 | 种植抗病品种；轮作；加强田间管理，清除病叶、病茎和病果；初花期开始喷药预防，药剂有甲霜灵、可杀得、克露、普力克等，交替使用 |
| 晚疫病 | 马铃薯、番茄、茄子等 | 低温、阴雨、湿度大、雾重、气温10~24℃、相对湿度75%~100%时往往大发生。病菌借风雨传播 | 为害茎、叶和果实，以叶和青果为害最甚。病斑水渍状，不规则，暗绿色，后变褐色，潮湿时边缘有白霉 | 选用抗病品种；水旱轮作；发病初期喷药防治，药剂有金雷多米尔、杀毒矾M8、普力克等，交替使用 |
| 枯萎病 | 瓜类、茄果类、豆类等 | 气温20~30℃时盛发；多雨或冷暖交替反复出现，种植感病品种或不注意轮作易引起流行。种子、肥料、土壤为初侵染源，以土壤为主 | 苗期发病，子叶变黄干枯，茎、叶、叶柄萎蔫或根茎基部变褐色，缢缩或猝倒。成株期发病，茎基部纵向褐斑可延及数节，凹陷或裂开，基部叶片变黄，向上扩展，根和茎部维管束变褐色，全株萎蔫下垂 | 水旱轮作；种植抗病品种；加强田间管理，施足基肥；及时拔除病株；淋药防治；药剂有多菌灵、恶霉灵、敌克松、木霉菌等，交替使用，防止蔓延 |

(续表)

| 名称 | 为害对象 | 发病条件 | 主要症状 | 防治方法 |
|---|---|---|---|---|
| 炭疽病 | 节瓜、冬瓜、黄瓜、丝瓜、辣椒、茄子、白菜、菜心等 | 4—10月，高温、多雨、多露、多雾、连作地、地势低洼、排水不良、通风不好等易发病。病菌借风雨及流水传播 | 为害叶、茎、果实。叶上初生黄白色圆形斑点，后变褐色有同心圆纹，干枯时易破裂。果实上的病斑下陷，中央龟裂。茎和叶柄的病斑为菱形，下陷 | 选用抗病品种；种子消毒；选择排灌良好的土地高畦种植；清除病残体；及时喷药防治，可用甲基托布津、百菌清、易斑净、施保功等，交替使用 |
| 黑斑病 | 十字花科蔬菜及葱类等 | 11月至翌年4月均有发生，连续阴雨天气发病严重，一般在植株生长中后期为重。土壤及种子传播病害 | 叶、茎、花梗及荚均可发病，老叶发生最多，初生近圆形褪绿斑，扩大成灰褐色或黑褐色病斑，有明显的同心轮纹，严重时半叶或整叶枯死；茎或叶柄的病斑长梭形，下陷；果实病斑稍凹陷，有明显的边缘，湿度大时病斑上生暗褐色霉层 | 轮作；种子消毒；加强田间管理，清除病残体；发病初期喷药防治，可用多菌灵、百菌清、代森锰锌、易斑净等，交替使用 |
| 早疫病 | 番茄、茄子、马铃薯等 | 多发生在5月前后和9—10月，多雨高温为诱发此病的重要因素。病菌借风雨传播 | 主要为害叶片，有时也为害叶柄、果梗和茎。初呈暗褐色水渍状小斑，后扩大稍有凹陷，圆形，有同心轮纹，病斑上有黑色绒毛状霉 | 选用早熟或抗病品种；轮作；合理密植；注意肥水管理，使植株健壮；发病初期喷药防治，药剂有丙森锌、甲基托布津、代森锰锌等，交替使用 |
| 褐纹病 | 茄子 | 3—10月均可发生，高温多雨水季节或连续阴雨条件下为害严重。病菌借风雨及昆虫传播 | 为害叶、茎、果实。叶片病斑褐色、圆形；茎部病斑灰褐色、不规则，病斑上有轮纹，表面密生黑色粒点，严重时茎枝病斑成圈枯死；果实上的病斑凹陷，轮纹明显，密生小黑粒 | 合理轮作；种子消毒；植前撒施石灰或用改土肥料改良土壤；初发病时喷施百菌清、多菌灵、易斑净、扑海因等，交替使用 |
| 白粉病 | 瓜类、茄果类、豆类、叶菜类等 | 气温15~30℃、相对湿度80%以上、通风条件不良发病严重。病菌主要借气流传播，其次是雨水 | 为害叶片、叶柄、茎和果实，初呈大小不一的白色粉斑（叶片正反面均有），后扩大至全叶，变灰白色 | 选用抗病品种；避免偏施氮肥；发病初期及时喷药防治，可用胶体硫、苯醚甲环唑、嘧菌酯、粉锈宁等，交替使用 |
| 锈病 | 菜豆、长豇豆 | 温暖高湿、有夜露或叶面上有水滴、通风条件不良易发病。病菌借风雨传播 | 主要为害叶片，严重时茎、蔓、叶柄及荚均可受害，初为黄白色斑点，后病斑隆起呈深黄色小脓疱，表皮破裂后，散出红褐色粉末，后出现紫黑色疱斑 | 选用抗病品种；清洁田园，适当密植，及时摘除病叶，加强田间管理；发病初期及时喷药防治，可用粉锈宁、烯唑醇等，交替使用 |
| 菌核病 | 生菜、白菜、甘蓝、菜心、黄瓜等 | 气温20℃左右，相对湿度85%以上条件下易发病，连作、排水不良、偏施氮肥、田间通透性差易发病。病菌主要借流水及风雨传播 | 主要为害茎蔓和果实。茎蔓染病多在近地面的茎部或主侧枝分杈处，果实染病多在残花部，病斑初呈浅色水渍状、腐烂，并长出白色菌丝，后菌丝结成黑色菌核。病茎髓部腐烂中空，内生黑色鼠粪状菌核，纵裂干枯 | 播种前用10%盐水浸种；轮作；深耕及土壤处理；覆盖地膜；发病初期喷药防治，可用咪鲜胺、异菌脲、农利灵、嘧霉胺等，交替使用 |
| 软腐病 | 叶菜类、茄果类等 | 气温25~30℃、多雨天气易发病，特别是1—3月回南天发病严重。病菌多从伤口入侵 | 叶菜类为害初期叶柄基部水渍状，后软化，全株腐烂，有强烈的腐臭味。茄果类主要为害果实，初生水渍状暗绿色斑，后内部变褐腐烂，具恶臭，病果后期脱落或变白干枯，挂在枝杈或茎上 | 避免连作；清除病株；用石灰消毒病穴，喷药预防，可用农用链霉素、新植霉素、络氨铜、氧氯化铜等，交替使用 |

（续表）

| 名称 | 为害对象 | 发病条件 | 主要症状 | 防治方法 |
|---|---|---|---|---|
| 青枯病 | 茄科蔬菜 | 气温30~37℃、微酸性土壤、植株生长不良、久雨或大雨后转晴易发病，6—10月高温发病重。病菌从根部或茎基部伤口入侵，借流水及土壤传播 | 初感病植株在白天烈日时萎垂，傍晚恢复；在晴天高温时2~3天便全株萎垂，垂叶仍是绿色 | 水旱轮作；种植抗病品种；加强田间管理，防止积水；及时拔除病株，并用药剂淋病穴，可用农用链霉素、络氨铜、新植霉素等，交替使用 |
| 黑腐病 | 十字花科蔬菜，以甘蓝类为害重 | 连作，气温25~30℃，高温多雨天气易流行。由种子及土壤病残体带菌，通过灌水、农事操作及害虫造成的伤口入侵 | 病菌从叶缘向里发展，形成网状黄脉，叶脉变黑，形成"V"形黄色枯斑；空气干燥时病部干脆，潮湿时则腐烂，发出浓烈的腐臭味 | 水旱轮作；种子消毒；加强栽培管理，减少伤口，收获后及时清园；发病初期喷药防治，可用农用链霉素、络氨铜、加瑞农、可杀得等，交替使用 |

## 主要虫害与防治

| 名称 | 为害对象 | 发生季节及为害情况 | 防治方法 |
|---|---|---|---|
| 小菜蛾 | 十字花科蔬菜 | 一年发生19~20代，3—5月及9—12月发生较多；春、秋干旱时发生严重；雨水过多或暴雨后发生量减少。成虫产卵于叶背，初孵幼虫多集中在心叶及嫩叶潜食叶肉，3~4龄幼虫将叶食成孔洞，严重时叶片成网状，并钻入种荚为害 | 轮作；彻底清园；幼虫盛孵前及时喷药防治，药剂有高效Bt、甘蓝夜蛾核型多角体病毒、阿维菌素、多杀霉素等，交替使用 |
| 菜粉蝶 | 十字花科蔬菜 | 一年发生10~14代，2—4月、10—11月为高峰期，以幼虫咬食叶片 | 收获后清园，降低虫口密度；低龄期喷药防治，可用高效Bt、多杀霉素、菊酯类农药等，交替使用 |
| 黄曲条跳甲 | 十字花科蔬菜 | 一年发生7~8代，四季均有发生，以3—5月、10—11月为害最重，湿度高的菜田重于湿度低的菜田。成虫、幼虫均可为害。成虫食叶成小孔洞、缺刻，严重时只留叶脉；幼虫为害根部，严重时全株凋萎 | 与非十字花科蔬菜轮作；播前深耕晒土；及时清洁田间；播种前喷淋杀虫双、辛硫磷等防治幼虫；防治成虫用杀虫双、辛硫磷、敌敌畏、茵蒿素等，交替使用 |
| 斜纹夜蛾 | 十字花科、葫芦科、豆科、旋花科、天南星科、藜科、睡莲科等蔬菜 | 杂食性，一年发生9代，盛发期6—8月和10—11月，卵多产于叶背，初孵幼虫群集为害，3龄后分散，4龄后为暴食期，白天躲在暗处或土隙，傍晚爬出为害。严重时，可吃光叶肉，仅留叶脉，甚至剥食茎秆皮层 | 诱杀成虫；摘除卵块；捕捉低龄幼虫；在3龄前喷药防治，可用菊酯类农药、抑太保、敌敌畏、茵蒿素、斜纹夜蛾核型多角体病毒等，交替使用 |
| 甜菜夜蛾 | 百合科、苋科、十字花科、葫芦科、豆科、伞形科、旋花科等蔬菜 | 多食性，一年发生6~8代，7—8月发生多，高温、干旱年份更多，卵多产于叶背，初孵幼虫群集为害，3龄后分散，4龄后为暴食期，白天躲在暗处或土隙，傍晚爬出为害。严重时，可吃光叶肉，仅留叶脉，甚至剥食茎秆皮层。常和斜纹夜蛾混发 | 诱杀成虫；摘除卵块；捕捉低龄幼虫；在3龄前喷药防治，可用菊酯类农药、灭幼脲、米螨、茵蒿素、甜菜夜蛾核型多角体病毒等，交替使用 |

（续表）

| 名称 | 为害对象 | 发生季节及为害情况 | 防治方法 |
|---|---|---|---|
| 蓟马 | 主要为害节瓜、冬瓜和茄子，其次是白瓜、南瓜、丝瓜、黄瓜、辣椒、枸杞等 | 蓟马种类多，主要有瓜蓟马、葱蓟马。瓜蓟马一年发生17~18代，葱蓟马8~10代，3月底至4月初始见，5月下旬至9月为高峰期，夏秋季受害严重。成、若虫集中锉吸嫩叶、花、幼果汁液，使嫩叶萎缩丛生，受害果茸毛变黑，表皮粗糙，呈锈褐色伤疤状，造成落果，严重影响产量和质量 | 避免连作；进行地膜覆盖栽培；以防为主，始见期喷药防治，药剂有啶虫脒、阿维菌素、吡虫啉等，交替使用 |
| 烟粉虱 | 茄科、葫芦科、豆科、十字花科等蔬菜 | 一年发生11~15代，四季均有发生，秋季为发生高峰期。成虫和若虫吸食植物汁液，被害叶片褪绿、变黄、萎蔫，甚至全株枯死 | 清洁田园；用黄色黏虫板诱杀成虫；发生初期喷药防治，药剂有吡虫啉、阿维菌素、啶虫脒、噻嗪酮等，交替使用 |
| 瓜实蝇 | 葫芦科蔬菜，以苦瓜、丝瓜受害重 | 一年发生7~8代，主要发生在5—11月瓜类生长中后期，夏秋植比春植严重。成虫在果实上产卵，幼虫蛀食果肉，引起腐烂和落果 | 及时清除受害烂瓜；利用毒饵或性诱剂、黄色黏虫板诱杀成虫；套袋护瓜；喷药防治可用敌敌畏、菊酯类农药等，交替使用 |
| 守瓜 | 丝瓜、节瓜、冬瓜、白瓜、黄瓜等 | 一年发生1~2代，越冬成虫于3月中旬飞到幼苗上为害，产卵于近植株基部表土。幼虫食根，严重时致植株枯死。成虫食叶，有时嫩芽、幼果也被害 | 冬季清洁田园；基质育苗；基部地面撒施草木灰或石灰，覆盖薄膜防虫；喷药防治可用敌百虫、菊酯类农药等，交替使用 |
| 蚜虫 | 豆科、茄科、葫芦科、十字花科等蔬菜 | 蚜虫种类多，主要有甘蓝蚜、萝卜蚜、桃蚜、瓜蚜、豆蚜。一年发生20代以上，以孤雌繁殖，生育快。以成、若虫刺吸植物汁液，使植株卷叶萎缩，还能传播病毒病 | 用银灰色地膜驱蚜或用黄色黏虫板诱杀；药剂防治有乐果、克蚜星、吡虫啉、阿维菌素等，交替使用 |
| 美洲斑潜蝇 | 以葫芦科、茄科、豆科蔬菜为害最重 | 杂食性，一年发生14~17代，主要发生在5—11月温暖的季节，30℃左右世代周期最短，发生数量较多。卵产于叶肉。幼虫潜食叶肉，形成白色隧道 | 合理布局，轮作间种；适当疏枝和疏叶，改善田间通透性；用黄色黏虫板诱杀成虫；发生初期喷药防治，药剂有阿维菌素、灭蝇胺、吡虫啉等，交替使用 |
| 螨类 | 茄果类、豆类、瓜类、莴苣等 | 有红蜘蛛及茶黄螨两类，以茶黄螨为主。一年发生20多代，5月中旬至6月、8~9月为盛发期；温暖高湿环境有利于茶黄螨的发生，高温干旱环境有利于红蜘蛛的发生。成、若螨吸食嫩叶和幼果汁液，叶片被害后皱缩，易落叶，对产量影响很大 | 注意检查虫情，及时喷药防治，药剂有哒螨酮、阿维菌素、三唑锡、克螨特、抗螨23等，交替使用 |
| 豆荚螟 | 长豇豆、菜豆 | 一年发生7代以上，4月下旬至10月均可发生。幼虫蛀花、荚，使花蕾和嫩荚脱落，老熟幼虫钻进荚内为害种子 | 轮作；及时清除落花、落荚，摘除被害的卷叶和果荚；上午开花时及时喷药防治，药剂有阿维菌素、高效Bt、农地乐、菊酯类农药等，交替使用 |

## 参考文献
References

[ 1 ] 中国农业科学院蔬菜花卉研究所. 中国蔬菜栽培学 [ M ]. 2版. 北京：中国农业出版社，2010.
[ 2 ] 方智远，张武男. 中国蔬菜作物图鉴 [ M ]. 南京：江苏科学技术出版社，2011.
[ 3 ] 关佩聪，李碧香，陈俊权. 广州蔬菜品种志（1993）[ M ]. 广州：广东科技出版社，1994.
[ 4 ] 关佩聪，刘厚诚，罗冠英. 中国野生蔬菜资源 [ M ]. 广州：广东科技出版社，2013.
[ 5 ] 广东省农业科学院作物研究所，广东省科学技术厅农村科技处. 广东甜糯玉米新品种新技术 [ M ]. 广州：广东人民出版社，2011.
[ 6 ] 徐鹤林，李景福. 中国番茄 [ M ]. 北京：中国农业出版社，2007.
[ 7 ] 叶华谷，彭少麟. 广东植物多样性编目 [ M ]. 广州：世界图书出版公司，2006.
[ 8 ] 赵培洁，肖建中. 中国野菜资源学 [ M ]. 北京：中国环境科学出版社，2006.
[ 9 ] 张平真. 中国蔬菜名称考释 [ M ]. 北京：北京燕山出版社，2006.
[ 10 ] 孔庆东. 中国水生蔬菜品种资源 [ M ]. 武汉：湖北科学技术出版社，2005.
[ 11 ] 邹学校. 中国辣椒 [ M ]. 北京：中国农业出版社，2002.
[ 12 ] 刘纪麟. 玉米育种学 [ M ]. 北京：中国农业出版社，2001.
[ 13 ] 绕璐璐. 名特优新蔬菜129种 [ M ]. 北京：中国农业出版社，2000.
[ 14 ] 张亚光. 中国常见食用菌图鉴 [ M ]. 云南：云南科技出版社，1998.
[ 15 ] 中国科学院植物研究所. 新编拉汉英植物名称 [ M ]. 北京：航空工业出版社，1996.
[ 16 ] 《中国传统蔬菜图谱》编委会. 中国传统蔬菜图谱 [ M ]. 杭州：浙江科学技术出版社，1996.